TROUBLESHOOTING & REPAIRING

SATELLITE TV SYSTEMS

RICHARD MADDOX

TAB BOOKS Inc.
Blue Ridge Summit, PA 17214

FIRST EDITION

SECOND PRINTING

Printed in the United States of America

Reproduction or publication of the content in any manner, without express
permission of the publisher, is prohibited. No liability is assumed with respect to
the use of the information herein.

Copyright © 1985 by TAB BOOKS Inc.

Library of Congress Cataloging in Publication Data

Maddox, Richard.
Troubleshooting and repairing satellite
TV systems.

Includes index.
1. Earth stations (Satellite telecommunication)
—Maintenance and repair—Amateurs' manuals.
I. Title.
TK9962.M33   1985   621.388'8   85-14744
ISBN 0-8306-0977-6
ISBN 0-8306-1977-1 (pbk.)

Front cover photograph courtesy of Gould/Dexcel Division.

Contents

Preface   v

Acknowledgments   vi

Introduction   vii

1 The World of Home TVRO   1

Satellite Link—FM Broadcasting Via Satellite—Early Systems—Current Outlook

2 Basic System Types   10

Dual-Conversion Systems—Single-Conversion Systems—Blockconversion Systems—DBS Systems

3 Terrestrial Interference   29

TI Profile—Out-Of-Band TI—Other Inband TI—Combating TI

4 Cables and Connectors   34

Coax Cabling—Insulated Wiring—Custom Cabling—Cable/Connector Sealing—Checking Cables—Splicing Cables—Connectors

5 Parabolic Reflectors   46

System Gain—Satellite Spacing—Dish Materials

6 Polarization and Polarity Control   **57**
Polarization Devices—Matching Feed to Dish—Troubleshooting Guide

7 LNAs, LNBs, LNCs, and LNFs   **71**
Low-Noise Amplifiers—Down Converters—Low-Noise Converters—Low-Noise Blockconverters—Low-Noise Feed

8 Power Supplies   **83**
Receivers—Antenna Actuators—IC Regulators—Safety—Preventing Failures—Troubleshooting

9 Intermediate Frequency Circuits   **97**
70 MHz Amplifiers—70 MHz Bandpass Filters—Limiter Circuits—Troubleshooting—70 MHz I-F Circuitry

10 Video Processing   **115**
Demodulator Circuits—Video Circuits—Clamp Circuits—Video Signal—Measuring the Video—Interpreting VITS

11 Audio Processing   **136**
Audio Subcarrier Specifications—Audio Circuits—Stereo Broadcasting Methods—Audio Broadcasting Systems—Troubleshooting Guide

12 RF Modulators   **152**
TV Signal Standards—Types—Auxiliary Components—Troubleshooting

13 Miscellaneous Receiver Circuits   **165**
Indicator Circuits—Tuning Circuits—Remote Controls

14 Antenna Positioning Systems   **179**
Fundamentals—Motor Controllers—Feedback Circuits—Sophisticated Controllers—APS Trouble Areas—Troubleshooting Guide—Troubleshooting Tips for Specific APSs

15 Setting Up a Test Bench   **200**
Fundamentals—Hooking Up—Hooking in LNAs and DCs—Making a DC Test Generator

16 Troubleshooting a System   **210**
Test Bench on Wheels—Test Procedure for Dexcel LNCs—Troubleshooting Guide

17 Specialized Components   **223**
Diodes—Transistors—Field Effect Transistors—Integrated Circuits—Hybrid Components—Surface Acoustic Wave Filters

18 Product Information   **237**
Amplica—Avcom—Channel Master—Conifer—Drake—Dexcel—Luxor—Sat-Tec—STS—Winegard

Appendix A   Active Component Guide   **361**
Appendix B   Equations, Calculations, and Charts   **378**
Appendix C   References and Resources   **384**
Appendix D   Common Abbreviations   **388**
Index   **389**

Preface

The purpose of this book is two-fold: to provide a training and reference source on home satellite TVRO systems, and to provide a troubleshooting manual that can be used on almost any TVRO receiving system. It is assumed you already have an understanding of basic electronics, as well as some knowledge of TV and FM broadcasting.

If necessary, there are many fine reference volumes that discuss such topics as electronics basics, the NTSC TV system, AM and FM signal transmission, and detection and basic troubleshooting skills. A listing of some of these will be found in Appendix C.

Over the past year as Customer Service Manager for Dexcel, I've answered technical questions from literally thousands of TVRO dealers and end-users from all across the country. I've answered questions at trade shows, at technical seminars, and on planes flying across the country. Out of all the questions came the seed for this book.

Acknowledgments

A book of this nature could not have been written without the technical assistance of many of the major TVRO manufacturers. I want to thank everyone who assisted in this effort. I hope the book lives up to your expectations. I'd also like to thank all the various engineers who've put up with my 20 questions concerning areas that I needed to become more familiar with.

I'd like to especially extend my sincere appreciation and gratitude to: Bill Howell, Dexcel's video wiz and cover boy, Bob Pizzi, Amplica's man on the run, Gus Gravemberg, the Mamou guru, Kenn "with 2N's" Hadermann, Luxor's man about town, Mike Gustafson and Chris Schultheiss of STV magazine, Tom Zehr of SRS, Jim Sterling of Horizon, Fred Graham of GCI, Gil Cunningham of Winegard, Van Stevenson of Birdview, Dale Hemmie of Conifer, Wiley Reed and Tay Howard of Chaparral, William Frost of R.L. Drake, Andy Hatfield of AVCOM, David Broberg of ICM, Jay LaBarge of MFC, and last but certainly not least, to the guy that started all this craziness, Bob Cooper, who introduced me and probably at least a million other people, either directly or indirectly, to the joys (and sorrows) of the TVRO biz.

Lastly, since every book needs a dedication, and since every writer needs a wife and kids, I'd better dedicate this work to the wife and kids, who've had to put up with a 24 hour-a-day TVRO maniac in their midst for these past six months.

"When the book's done, we'll go out to dinner and a movie, I promise . . ."

Introduction

In 1980, there were less than 1,000 home earth terminals installed anywhere in the world. Most of these installations used surplus military gear and home-brew equipment, and for the most part were owned by the experimenters and amateur radio buffs that built them.

By early 1985, the composition of the typical home TVRO owner had changed drastically. First, it was estimated that there were between 800,000 and 1,000,000 home TVRO systems installed in the United States alone. Second, almost all of the system owners are typical homeowners who could care less about the technology. All they're interested in is the programming.

Most owners don't have the slightest notion how their home TVRO system works. A lot of them don't even know how to properly operate their system. And unfortunately, a large percentage will experience problems with their equipment. Anything can happen: from component failures and operator errors, to the neighbor's pine tree growing another foot or the family pet chewing on the cable.

As an industry, the Home TVRO business is just now learning to walk. It survived its infancy, when receivers were built on kitchen tables, and the press thought we were all pirates, and HBO was about the only regularly scheduled program to be found. The industry has gone from being Research and Development driven (by engineers and entrepreneurs) to being a Sales and Marketing (corporate) entity.

Today, there's a lobbying group in Washington (SPACE), several industry trade organizations (STTI, NASDA, et al), a whole slew of magazines dedicated to TVRO information (STV, Coop's Digest, Home Satellite Marketing, Satellite TV Opportunities, Private Cable, et al), and certainly enough trade shows to keep everyone busy.

With the October, 1984 passage of the Home TVRO viewing rights bill by the U.S. House and Senate, the days of Pirate Satellite TV are over. We're all legal now, and we are gathering in sufficient numbers to attract both admirers and detractors. To keep the steamroller moving, an industry must keep advancing its numbers, but increasing the numbers and increasing the knowledge are two different things. The days of the one on one question and answer session with people like Taylor Howard and Bob Cooper are over; sheer numbers prevent that. So, we must settle for the next best thing: a technical manual that answers the most often asked questions that newcomers and oldtimers have asked.

The book is written so that anyone with a relatively light technical background should be able to follow along and understand the basics the first time through. I begin by giving a little background on how this whole Home TVRO business (or should I say madness) got started, and a little bit on who started it. Chapter 2 covers a signal flow and component makeup in the three most common types of systems: single or separate down converter systems, dual down conversion systems, and block conversion systems.

An entire chapter is devoted to the ever present threat of Terrestrial Interference, or TI. Chapter 3 goes into how to identify it, catagorize it, and finally, how to combat it.

The next chapter covers the simplest part of a system, or so most people think. Some of the topics in Chapter 4 include how to properly assemble a type-N connector, the do's and don'ts on handling type-N connectors, and how to seal connectors.

Chapter 5 gets into dishes, or parabolic reflectors. How to install them, align them, and tweak them up for optimum performance are all covered. Two-degree spacing and its effects on smaller dishes is covered, as is how to figure the f/D ratio, focal point, G/T and C/N of any dish.

Chapter 6 explains the four types of satellite polarity. All the various types of polarization devices (PDs) are covered, with the Polarotor I and its control circuitry being discussed in detail, since it's probably used in 75% of all installations.

Next, I cover the microwave parts of the system. Although most microwave components aren't field repairable, they can be troubleshot to determine if they are the source of trouble. Some simple test gear that can be cooked up to help in your troubleshooting is also outlined.

The next six chapters cover the TVRO receiver. I start at the power supply and then work my way through the i-f strip, the video detection and processing circuits, the audio detector and processing circuits (including stereo broadcasting and detection), the rf modulator and rf in general, and finally end up with the *everything that couldn't be classified chapter*, or miscellaneous receiver circuits. It covers such items as LED displays, channel tuning methods and remote controls.

Chapter 14 is on what many people in the business consider the most troublesome part of any system, the APS or Antenna Positioning System. The fundamentals of what an APS is and what it does are covered. There is also a test controller you can build for checking out motor drives. The chapter concludes with a troubleshooting guide for several top name APSs.

The next chapter is dedicated to setting up a TVRO test bench. Now that the various parts of a system have been covered, it's only fair to help you set up shop. The chapter not only includes descriptions of "store-bought" test equipment, but also describes the smaller pieces like waveguide adapters and attenuators and all the other bits and pieces that will make or break a TVRO repair shop. A block diagram of how to get off-air test signals into the shop for troubleshooting LNAs, DCs, APSs and receivers is also included, as is a homebrew 4 GHz signal generator.

Chapter 16 contains troubleshooting tips on systems, both for in the field and on the bench troubleshooting. There is also a step-by-step troubleshooting guide for many of the common TVRO problems.

The next chapter is on the specialized components that make up a TVRO receiver. Everything from diodes and transistors to SAW filters and hybrid components is covered.

The last chapter concentrates on several major receiver manufacturers and their products. Included are installation tips, troubleshooting tips, technical bulletins, schematics, and other important service information.

There are appendices that list common equations and calculations that are used every day during installs and repairs, pinouts and block diagrams for many of the ICs used in TVRO receivers, and references and resources covering books, magazines, and mail order parts houses.

Throughout the book I've used abbreviations for many of the common TVRO components. You've already come across several of them in this introduction. Probably the most used abbreviations are DC, for Down Converter, LNA, for Low Noise Amplifier, dc, (note the lower case) for direct current, PD, for Polarization Device (typically meaning the Polarotor I) and APS, for Antenna Positioning System. There is a complete abbreviation listing in Appendix D.

The World of Home TVRO

T O FULLY APPRECIATE WHERE THE HOME TVRO IN-
dustry has come from, and why systems designed
for home use are designed the way they are, let me provide some home TVRO background history.

Up until 1976, the FCC required that every Earth terminal be licensed, and that it be 9 meters (29 and a half
feet) in diameter. This didn't make sense to some people. Especially to two antenna manufacturers, Prodelin
(who later merged with M/A Com) and AFC (who later
became Microdyne/AFC). Both companies demonstrated
4.5 (15 foot) and 5 meter (16 and a half feet) systems that
met FCC video and audio performance specifications. It
was their demonstrations, coupled with the prompting by
the CATA (Community Antenna Television Association)
and by CATJ (Community Antenna TV Journal)
magazine, that predicated the FCC to obsolete the 9 meter dish size rule in December of 1976.

Meanwhile, Bob Cooper, the editor of CATJ
magazine, put in a home system, and he set out to prove
that satellite reception did not even need 4.5 or 5 meter
antennas. He had a 20 foot, a 10 foot and a 6 foot dish
at his house, as well as a full test lab for evaluating the
performance of the dishes.

Cooper, like many other early experimenters and
owners, also wanted he FCC to drop the required licensing of TVRO terminal. But it was mainly because
of networks like Mutual Radio and ABC, who wanted to
install thousands of ARO (Audio Receive Only) terminals,
that the FCC was finally convinced that licensing was
really unnecessary for receive-only terminals. On October
18, 1979, the FCC unanimously decided to deregulate
TVRO and ARO terminals from licensing requirements.
Thus the stage was set for anyone to install a satellite
dish without having to do a frequency coordination check
and go through the licensing procedure.

SATELLITE LINK

It was really AT&T that was the moving force behind
the concept of telecommunication satellites. For them,
it was a natural progression from their microwave links
across the United States to a microwave link across the
Atlantic.

This means that the frequency band assigned to
AT&T for their terrestrial (land lines) microwave links,
is the same as that assigned to their satellite links. At
the time that this was decided, no one at the FCC ever
envisioned Home TVRO, which unfortunately, is the
reason some installations suffer from TI (Terrestrial Interference). TI is a form of crosstalk mainly caused by

1

interference from the telephone company's microwave transmissions. It is covered in more depth in Chapter 3.

The total C-band TVRO bandwidth is 500 MHz (Megahertz or million cycles per second) wide. This band of frequencies is divided into several channels that, for domestic satellites, are typically 40 MHz wide. There can be 12 channels (12 × 40 or 480 MHz) in this spectrum with 20 MHz left over for satellite control signals. But this 500 MHz wide band can be used twice if the satellite broadcasts the signals using dual polarity. Thus 24 television channels, each 40 MHz wide can be obtained from one satellite. Polarization is covered in detail in Chapter 6.

The C-band stretches from 3 to 10 GHz. The bottom of the band is used to downlink (from satellite to Earth), while the uplink (from Earth to satellite) is done in the middle of the band, from 5.9 to 6.4 GHz. The GHz stands for Gigahertz or billion cycles per second (refer to Table 1-1). This frequency range is slightly more than 10 times higher in frequency than the top of the VHF (Very High

Table 1-1. The Electromagnetic Spectrum.

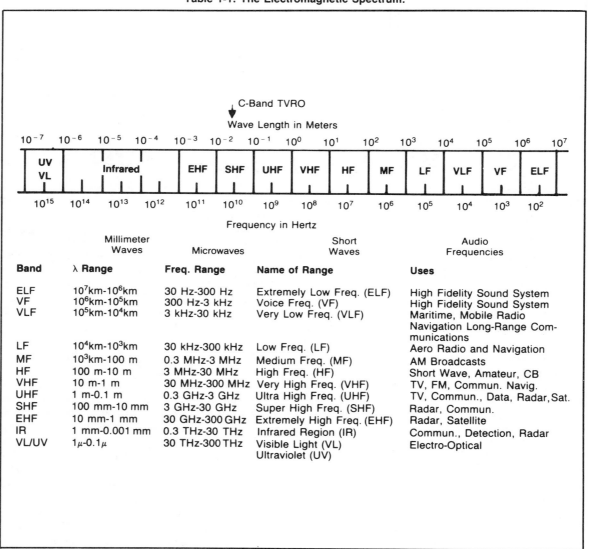

Band	λ Range	Freq. Range	Name of Range	Uses
ELF	10^7km-10^6km	30 Hz-300 Hz	Extremely Low Freq. (ELF)	High Fidelity Sound System
VF	10^6km-10^5km	300 Hz-3 kHz	Voice Freq. (VF)	High Fidelity Sound System
VLF	10^5km-10^4km	3 kHz-30 kHz	Very Low Freq. (VLF)	Maritime, Mobile Radio Navigation Long-Range Communications
LF	10^4km-10^3km	30 kHz-300 kHz	Low Freq. (LF)	Aero Radio and Navigation
MF	10^3km-100 m	0.3 MHz-3 MHz	Medium Freq. (MF)	AM Broadcasts
HF	100 m-10 m	3 MHz-30 MHz	High Freq. (HF)	Short Wave, Amateur, CB
VHF	10 m-1 m	30 MHz-300 MHz	Very High Freq. (VHF)	TV, FM, Commun. Navig.
UHF	1 m-0.1 m	0.3 GHz-3 GHz	Ultra High Freq. (UHF)	TV, Commun., Data, Radar, Sat.
SHF	100 mm-10 mm	3 GHz-30 GHz	Super High Freq. (SHF)	Radar, Commun.
EHF	10 mm-1 mm	30 GHz-300 GHz	Extremely High Freq. (EHF)	Radar, Satellite
IR	1 mm-0.001 mm	0.3 THz-30 THz	Infrared Region (IR)	Commun., Detection, Radar
VL/UV	1μ-0.1μ	30 THz-300 THz	Visible Light (VL) Ultraviolet (UV)	Electro-Optical

Table 1-2. Transponder Conversion Chart.

Channel/ Dial Number	Receiving Frequency MHz	Polarization			
		Normal		Reversed	
		SATCOM 1,2,3R&4	COMSTAR 1,2,3&4	WESTAR 4&5	WESTAR 1,2&3 ANIK 2,3&B
1	3720	1 (V)	1V (V)	1D (H)	1 (H)
2	3740	2 (H)	1H (H)	1X (V)	
3	3760	3 (V)	2V (V)	2D (H)	2 (H)
4	3780	4 (H)	2H (H)	2X (V)	
5	3800	5 (V)	3V (V)	3D (H)	3 (H)
6	3820	6 (H)	3H (H)	3X (V)	
7	3840	7 (V)	4V (V)	4D (H)	4 (H)
8	3860	8 (H)	4H (H)	4X (V)	
9	3880	9 (V)	5V (V)	5D (H)	5 (H)
10	3900	10 (H)	5H (H)	5X (V)	
11	3920	11 (V)	6V (V)	6D (H)	6 (H)
12	3940	12 (H)	6H (H)	6X (V)	
13	3960	13 (V)	7V (V)	7D (H)	7 (H)
14	3980	14 (H)	7H (H)	7X (V)	
15	4000	15 (V)	8V (V)	8D (H)	8 (H)
16	4020	16 (H)	8H (H)	8X (V)	
17	4040	17 (V)	9V (V)	9D (H)	9 (H)
18	4060	18 (H)	9H (H)	9X (V)	
19	4080	19 (V)	10V (V)	10D (H)	10 (H)
20	4100	20 (H)	10H (H)	10X (V)	
21	4120	21 (V)	11V (V)	11D (H)	11 (H)
22	4140	22 (H)	11H (H)	11X (V)	
23	4160	23 (V)	12V (V)	12D (H)	12 (H)
24	4180	24 (H)	12H (H)	12X (V)	

*Polarization for each transponder denoted in parenthesis.

(V) = Vertical

(H) = Horizontal

Frequency) band used to broadcast standard TV. C-band is part of what is known as the SHF or Super High Frequency band.

FM BROADCASTING VIA SATELLITE

As can be seen from the frequency chart in Table 1-2, channel 1 is broadcast on a carrier with an unmodulated frequency of 3.720 GHz. This means that channel 1's frequency bandwidth, which is 40 MHz, stretches from 3.700 GHz to 3.740 GHz. Channel 3 starts at 3.740 GHz and stretches to 3.780 GHz and so on. Refer to Fig. 1-1.

Ideally information could be carried in the entire 40 MHz, but in the real world, there must be some separation between different signals. Thus the actual information carrying part of a 40 MHz wide transponder, or satellite channel, is only 36 MHz wide, with a 2 MHz guard band on either side.

When using Frequency Modulation (FM) to transmit information, most of the critical components of the signal are found within ± 10 MHz of the carrier frequency (see Fig. 1-2). Thus most of the energy being broadcast on transponder 1, actually only stretches from 3.710 GHz

3

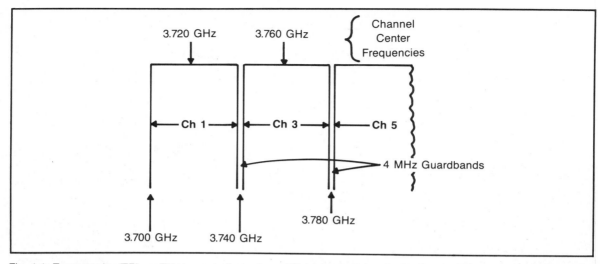

Fig. 1-1. Transponder (TR), or TV channel, dimensions. TR 1 is centered at 3.720 GHz and TR 3 is centered 40 MHz above that, at 3.760 GHz. Each transponder is 40 MHz wide with a 2 MHz guard band at each side. This leaves 36 MHz that may be used for transmitting information.

to 3.730 GHz. There is still important information out to 3.702 GHz and 3.738 GHz, but its energy level is much lower than the ± 10 MHz bandwidth. This is one reason why the satellite's transponders can overlap each other when using dual polarity without causing interference.

Again referring to Fig. 1-2, channel 2 is centered at 3.740 GHz, so its highest energy portion is from 3.730 GHz to 3.750 GHz, which is exactly where channel 1 drops off and channel 3's energy level starts to come up.

In order to keep interference down between the satellite signals and the terrestrial microwave links, a low frequency signal is applied to the video signal before it is sent up to the satellite. This signal is called *dithering*. It is a 30 Hz triangle waveform imposed on the video. It spreads the signal across the transponder bandwidth, to prevent any "hot spots" from developing. This signal is required because Ma Bell, who had the C-band before satellites came along, wanted to ensure that the satellite signals would not interfere with the terrestrial carriers. This waveform is removed by the receiver's video clamping circuitry.

EARLY SYSTEMS

Since the earliest TVRO systems were assembled from surplus communications and military surplus materials, it naturally follows that their parameters would be used as a starting point. Also since commercial systems used certain tried-and-true designs, their way of doing things

became a model as well.

Most commercial systems consisted of a very large dish and a control house with the receiver located right

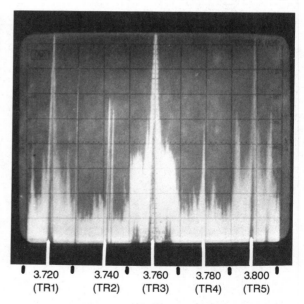

Fig. 1-2. A spectrum analyzer photo showing TR 1 through TR 5. TR 2 and TR 4 are crosspolarized and thus are lower in level than TR 1, 3 and 5. The center line is 3.760 GHz and there is 10 MHz/div. The top line (reference level) is −54 dBm.

at the dish. A parametric amplifier was used to preamplify the reflected satellite signals so that they could be sent into the receiver where the 4 GHz signals would be down converted and demodulated.

This worked fine if the receiver was physically 100 feet or less from the output of the LNA. A longer run was virtually impossible because the signals that were carried were 4 GHz in frequency. Once the signals were inside the receiver they were down converted to a lower frequency to allow demodulation. In the Telco TD-2 the

i-f (intermediate frequency) was 70 MHz.

First Generation

The first generation of home satellite systems consisted of a fairly large (by today's standards) parabolic dish, an LNA, and some kind of low-loss coaxial cable which carried the signals inside the house to the receiver. A dual conversion down converter inside the receiver lowered the signals to a final i-f of 70 MHz where signal

Fig. 1-3. The ADM-11, one of the first mass-produced home TVRO dishes. It consists of 12 stamped panels that are bolted together at the dish site. It is a shallow prime focus dish. Courtesy of ADM.

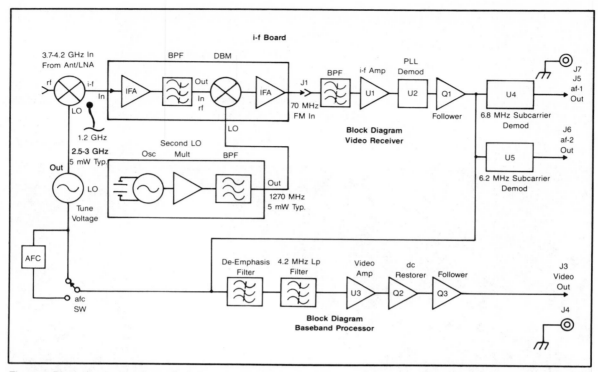

Fig. 1-4. Block diagram of the typical early model dual conversion receiver. It used two oscillators and mixers to down convert the signal for demodulation. The diagram is of a TV-4200 receiver. Courtesy of ICM.

demodulation took place. One of the first dish manufacturers whose product was designed especially for home TVRO use was made by ADM. Figure 1-3 is a photo of the ADM-11, first introduced in 1980.

Figure 1-4 is the block diagram from the ICM TV-4200, an early first generation receiver also introduced in 1980. It was designed to be used with a 50 dB gain LNA and about 75 feet of RG-8 (or similar) cable, which plugged into the receiver. The receiver also provided power for the LNA.

The channels were tuned by changing the output frequency of the 1st LO (Local Oscillator) from 2.5 GHz to 3 GHz. This signal, when mixed with the incoming satellite signals produced a high i-f (Intermediate Frequency) of 1.2 GHz. The 1.2 GHz was amplified (i-f A) and then bandpass filtered (BPF) and sent to a Double Balanced Mixer (DBM) where it was mixed with a 1.270 GHz fixed frequency 2nd LO. To separate out the 70 MHz signal, another bandpass filter was used. The 70 MHz signal was then amplified to drive the demodulator circuitry. Finally a video processing circuit and audio demod circuit completed the package.

The mark of a first generation receiver is the location of the down converter—inside the receiver. Clyde Washburn was one of the first people to change this concept. He designed his receiver (now known as the Earth Terminal receiver) with the down converter in a separate box. This meant that it could be placed separate from the receiver and closer to the dish. This meant that less interference within the receiver from LO signal leakage would occur as well as less expensive cable could now be used to carry the signal into the receiver.

The next step was to design the down converter so that it could be placed at the dish. Thus was born the second generation of satellite receivers.

Second Generation

The predominant difference at this time was the change in location of the down converter. Of secondary nature was the almost complete change to single conversion. Single conversion is the process whereby the tunable LO is running either 70 MHz above or 70 MHz below the desired channel. Thus the LO output would be, for

Fig. 1-5. One of the standards in the industry for several years, the COM-3 series of receivers were available with an internal DC or with an external DC that was mounted at the dish. Courtesy of AVCOM of Virginia.

a high side LO, 3.790 GHz to 4.290 GHz. Figure 1-5 is an AVCOM COM-3R receiver which straddled the two generations by offering the receiver with either an internal or external down converter.

Single conversion was more economical than the dual conversion used in 1st generation products. No longer was hard-line or expensive cable necessary to run 4 GHz signals into the house. Only a short 10 to 20 foot length of expensive cable from the LNA to the down converter, which was mounted behind the dish, would be required. From the down converter, less expensive cable could be run, such as RG-59 or RG-6. Plus the dish could

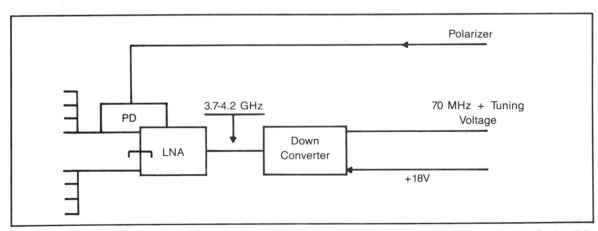

Fig. 1-6. Block diagram of the active outside receiving equipment in a separate DC system. The Polarization Device (PD) is a Polarotor I or II (or equivalent). The LNA power is supplied through the DC via a coax cable or separate wire. The channels are selected in the DC by a tuning voltage that is supplied up the 70 MHz coax cable.

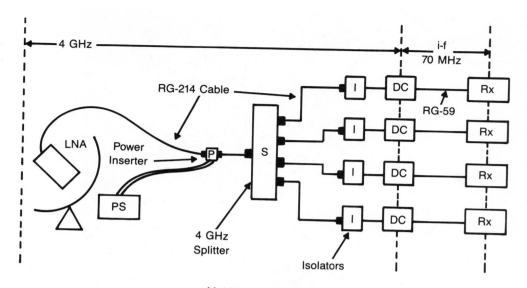

Multiple Receivers Using Separate DCs
"The Old Way"

RX = Receiver
PS = Power Supply

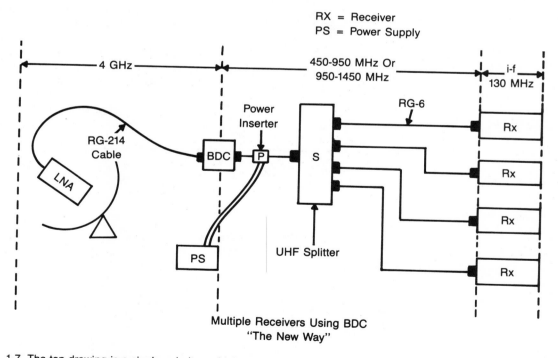

Multiple Receivers Using BDC
"The New Way"

Fig. 1-7. The top drawing is a single polarity multiple receiver system using separate DCs. All connectors between the LNA and the DCs are type-N using RG-214-type cable. The bottom drawing is a BDC system. The only RG-214 cable is between the LNA and the BDC, from there to the receiver, RG-6 or RG-11 is used. There are two dominant block frequencies, 450 to 950 MHz and 950 to 1450 MHz.

8

be located farther from the receiver. In some cases, the down converters could drive cable lengths of up to 1,000 feet without the need for line amps.

Figure 1-6 shows the basic single conversion down converter system that has been the defacto standard for about 4 years now. It consists of an LNA, a Polarization Device (PD), and a down converter (DC). All three are located at the dish. The DC tunes in a single channel, which is centered at 70 MHz, and amplifies it enough to drive a coax cable, which runs to the receiver. The receiver supplies the operating voltage for the DC, the LNA and the PD as well as supplies a tuning voltage and PD control pulse.

There were several advantages to this scheme and some unforeseen disadvantages as well. The biggest plus was that the cable loss inherent in the 1st generation receiver was now cut to a minimum, thus down converters no longer had to have a lot of gain to compensate for this loss. In most cases, this meant cleaner pictures. Another plus was that there was no LO in the receiver proper, meaning less shielding required in the receiver to keep out LO leakage.

The disadvantages were immediately apparent the first time cold weather struck: channel drifting. The LOs had to be temperature compensated since they were now subjected to extremes of temperature. The down converter boxes also had to be weatherproofed, either by their design or by their placement inside a weatherproof box. The savings in hard-line coax cost was often offset by the increased cost of a weatherproof box.

Second generation receivers are the predominant group in the field today, although a third generation is rapidly coming in to take their places.

Third Generation

The third generation receiver is called the BDC (Block Down Conversion) system. BDC is fast becoming the industry buzzword. "Is this receiver available in a block system?" must be the most often asked question confronting manufacturers today.

The third generation receiver has its roots back at the first generation. If we were to reverse the position of the 2nd LO, which is fixed, with the 1st LO, which is tunable, then we would essentially have a block system. In a block system, the whole satellite band from 3.7 GHz to 4.2 GHz is moved to a lower frequency. The most popular frequency ranges are in the upper UHF spectrum, from 450 to 950 MHz and 950 to 1450 MHz.

This band of frequencies is then sent into the receiver, where a second down converter is located. Thus the channel tuning is done in the receiver, and, as long as the LO at the dish in the BDC (Block Down Converter) is stable, there should be no temperature related drift since the actual tuning circuit is located indoors.

Figure 1-7 is a typical BDC system outline which is compared with a standard 4 GHz splitter and separate DC multiple receiver system. At first glance, they look similar, but once the cost of the isolators, type-N connectors and RG-214 jumpers are added up, the cost of the separate DC system is many times more expensive than the block system which only uses two type-N connectors, the remainder being type-F. Also RG-6 or RG-11 is used to run all cabling between the receivers and the BDC.

CURRENT OUTLOOK

Today's Home TVRO systems have evolved a long way from the sheet metal boxes with homebrew electronics and surplus dishes that mark its origins a mere 5 years ago. Even the venerable 70 MHz i-f is on its way out because of technological advances in high frequency components such as SAW filters and PLL chips that operate using i-fs in the 500 MHz and above region.

Channel drift, which plagued some of the earlier receivers is now a thing of the past with the new synthesized tuning receivers. Fully programmable receivers are now in the wings, coming from the factory with the satellite channels preset for correct video and audio tuning and with automatic polarity peaking and even automatic dish peaking.

But regardless of what the future brings, today's reality is that there are still more than 1,000,000 installed systems functioning in the United States today. The repair market for these receivers is enormous as most first and second generation products were designed not for long life but for quick delivery to a rapidly changing marketplace.

Basic System Types

BEFORE A SYSTEM FAILURE CAN BE DIAGNOSED, the technician must be familiar with the individual components that make up the different systems and with the signals to be found within them. This chapter begins the TVRO familiarization process by describing the signal flow in the three most commonly used systems.

The three systems previously referred to as the first, second and third generation receivers have, in many cases, overlapped one another in time. There is definitely not a linear progression in their development, but there was in their marketing. Basically the dual DC receiver came first, then the image-reject mixer was introduced, which allowed single conversion to be feasible, and then block DC was introduced to simplify multiple receiver installations.

Many designers felt that BDC was the way to go originally, especially those that were investigating 12 GHz DBS (Direct Broadcast Satellite) receivers and LNAs. All 12 GHz systems use BDC. At 12 GHz, one doesn't just use RG-213, or even Belden 9913, to run 12 GHz signals down to the DC, since the signals would be attenuated beyond use after a very short run. Economically the BDC took a few years to be fully developed, and now with SAW resonators and DSOs (Dielectrically Stabilized Oscillators)

being available at microwave frequencies, BDC has begun to "explode" in the marketplace.

But you cannot forget what has come before, as this is what will be coming in for repair for many years to come. If you look at the basic stripped down block diagrams for all three kinds of receiving systems, as shown in Fig. 2-1, you will see that most of the blocks look the same. Only a few minor changes in placement are what determines the type of receiver and the signals that will be found at various points in the circuit.

Starting from the right side and working our way back to the dish, all three have demodulation sections to remove the carrier and detect the video and audio information. All three have an i-f amplifier to drive the demodulation circuit. Here is where there could be a difference in frequency between types. Typically both the single and dual conversions use 70 MHz as an i-f frequency. Block receivers seem to be settling on 130 or 140 MHz i-fs, although there are i-fs that are found up to 650 MHz, as well as down around 40 MHz.

In between the first and second i-f amps is the i-f filter. Most i-f filters are similar in what they do, although they do differ in design. For the most part, the i-f is band-pass filtered to somewhere between 22 and 30 MHz in

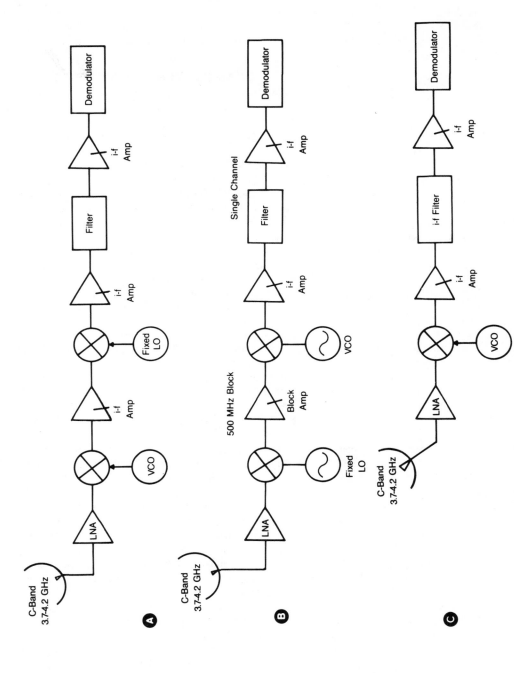

Fig. 2-1. The block diagram in a is a dual conversion receiving system. The amplified output of the LNA is mixed with the output of a tunable VCO, which results in the high i-f. This is amplified and then mixed with a fixed oscillator (LO) to yield the final i-f, which is usually 70 MHz. The i-f signal is filtered and amplified to drive the demodulator once it arrives in the receiver. The block diagram in B is a BDC system. It reverses the fixed LO and VCO position. This results in the satellite signals being converted enmass to a lower frequency. This block is amplified and sent to the receiver where they are mixed with the VCO's output to form the final i-f, which is typically 130 or 140 MHz. The block diagram in C is a single conversion system. The channel is tuned at the dish in a separate DC. The mixer, VCO and i-f amp are located in the DC. In some systems, the LNA, mixer, VCO, and i-f amp are in one housing called an LNC.

11

bandwidth. If the receiver is designed for Intelsat reception, then this bandwidth would be cut down even further, to 15 to 18 MHz for half-transponder reception. A similar situation occurs on Spacenet 1. It has several transponders that are 72 MHz wide, but that are mainly used in a half-transponder format (36 MHz). Thus a regular receiver would not need any modifications, unless someone was to broadcast using the full 72 MHz, which is unlikely at this time.

So far, all the components have been located within the receiver. Occasionally there will be a receiver that has its i-f filter in the DC, but for the most part the second i-f amp and the i-f filter are in the receiver. The next stage back is the last mixer, or final i-f mixer. Here is where the roads diverge. In the block conversion system (Fig. 2-1B), that mixer is driven by a VCO (Voltage Controlled Oscillator) and the channel is actually tuned at that point. In the dual conversion system (Fig. 2-1A), that point is driven by a fixed oscillator, labeled *fixed LO*. Both of these stages are located at or in the receiver. The single conversion system, (Fig. 2-1C) takes the same approach as the block system in that a mixer and a VCO is used, but they are located at the dish; in an LNC or in a separate DC.

At this point, the block converter has another stage where there is another mixer and a fixed LO to knock the entire satellite band down to a lower frequency. The dual conversion system looks similar, but there is a world of difference. There is only one channel (or 40 MHz) being amplified by the i-f amp between the two mixers in a dual conversion system. In the block system, the same amplifier must amplify all 12 channels of one polarity (or 500 MHz).

The LNA is virtually the same in all three systems. In dual conversion systems (both oscillators and mixers are inside the receiver) the LNA has to have 50 dB gain to drive the cable that runs into the receiver. In some single conversion systems, the LNA, VCO, mixer, and one stage of the i-f amp are put into the same box. When this is done it is called an LNC system.

The other new component is the LNB or Low Noise Blockconverter. This grew out of 12 GHz research, and it is basically the LNA, a fixed LO, mixer, and block i-f amplifier in one housing. The output is all 12 channels of one polarity, and is compatible with any receiver that uses the same high i-f frequency band. This is most often from 450 to 950 MHz or from 950 to 1450 MHz.

The last combination is an LNC that uses dual conversion. This has only been used, to my knowledge, by one receiver manufacturer, USS/Maspro, in their SR-1

and SR-2 receiver systems. In both of them, the LNC tunes the channel and outputs it at 1.2 GHz. This is then sent to a DC, usually located behind the dish, which down converts the channel to a final i-f of 70 Mhz using another LO. This signal is then run in to the receiver.

DUAL-CONVERSION SYSTEMS

In the beginning was dual conversion. There was no choice. No one had come up with an image reject mixer yet and single conversion from 4 GHz to 70 Mhz just wasn't feasible without it. So instead two frequencies were used to lower the signals to a more manageable frequency.

The frequencies chosen were designed to not cause interference between receivers. Most often the frequencies were between 800 MHz and 1.5 GHz for the first down conversion and then. 870 to 1.570 GHz for the second, fixed LO. A VCO was used to output a sinewave that was mixed with the 4 GHz band. Almost all of these systems used a low side LO, meaning that the oscillator was running somewhere between .8 to 1.5 GHz below the desired channel. For instance, in the DC-60 dual conversion down converter from ICM, the VCO tuned from 2.86 to 3.36 GHz.

An advantage to dual conversion over most single conversion designs is the image rejection, which was typically 30 to 40 dB in a good dual-conversion DC. Maximum rejection in the single-conversion world is 25 db with the typical image rejection more like 15 to 20 dB. This is cutting it close because at 12 dB rejection, an image will start to be faintly visible in the background of the picture.

By 1980/81, single conversion had made its appearance and was a real threat to dual-conversion systems that still had to have 4 GHz coming into the house. To counteract the single-conversion invasion, the dual DC was separated from the receiver and put into a weather resistant housing so it too could be stuck out at the dish. This helped somewhat as now both single- and dual-conversion designs were equivalent in cabling and hook-up. Plus dual conversion had an advantage; no isolators were needed between the splitter and the DC if multiple receivers were used.

Unfortunately this advantage was really no advantage except in multiple receiver systems. End users didn't care that the design was better and that the image rejection was better and that the drift was less. They only cared about price, and single DC systems were cheaper. So, to cut down the cost, the manufacturers took out a

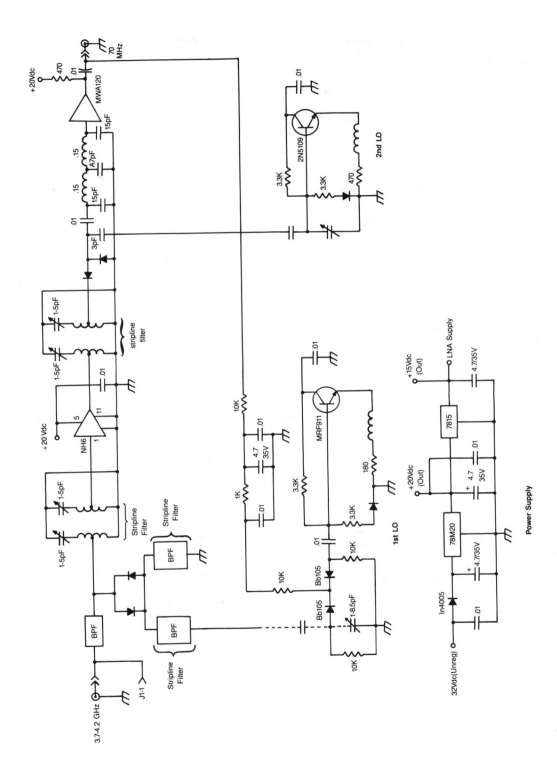

Fig. 2-2. Schematic of a DC-60, a dual-conversion dc from ICM. The first oscillator is mixed with the satellite signals to yield an i-f of 850 MHz. This is amplified and filtered by the NH6 and the two stripline filters. It is mixed with the 2nd LO to yield an output of 70 MHz. Courtesy of ICM.

VCO, the high i-f filter and amp, and then put in a stripline image-reject mixer and lowered the price. Magically they were back in the market again. Thus the dual-conversion system was all but abandoned in favor of the simpler and cheaper single conversion receiver.

ICM DC-60

Figure 2-2 is the schematic of the DC-60 down converter from ICM. It was designed to be used with the TV-4400 receiver, but it was used on many other homebrew and early consumer receivers as well. It is fairly typical of the early 1980's approach to satellite systems.

The DC-60 is enclosed in a 1 pound environmental protective case ($3'' \times 4'' \times 7''$), which is basically a metal box with one open side that the actual DC assembly is slid into. The down converter is held in by 6 sheet metal screws and retaining clips. It is fairly water resistant if drip loops are used and if the connector side is pointed down. Of course by now, any that come in for repair will probably have all the hardware rusted up and the connectors corroded from the weather.

The DC-60 uses a 6-prong, male, Cinch Jones plug for dc interconnection. The pins are in two rows; the left row, from the top down contains pins 1, 3, and 5, while the other side has pins 2, 4, and 6, also from the top down. Refer to Fig. 2-3A for a rear view of the DC-60 showing the connector locations. Pin 1 is tied to the type-N connector, pin 2 is the LNA dc supply voltage output (regulated + 15 Vdc), pin 3 is a dc test point, pin 4 has no connection, pin 5 is the unregulated input voltage and pin 6 is the ground.

To power the LNA, the cable requires a jumper to be placed between pins 1 and 2. To power a side-powered LNA, connect pin 2 to the LNA power connector. Do not connect anything to pin 1.

A type-N connector is used to bring in the 4 GHz signals, and a BNC jack is used to output the 70 MHz signals. Tuning voltage is supplied up the coax cable. There is also an RCA jack for plugging in a 1 milliamp meter movement.

The optimum input level is − 25 to − 40 dBm. The maximum noise figure is 12 dBm. The DC-60 can be powered by any unregulated voltage from + 24 to + 40 Vdc. The TV-4400 receiver supplied + 32 Vdc. The tuning voltage is from + 5 to + 18 Vdc.

Referring to Fig. 2-2, the 1st LO is tuned from 2.860

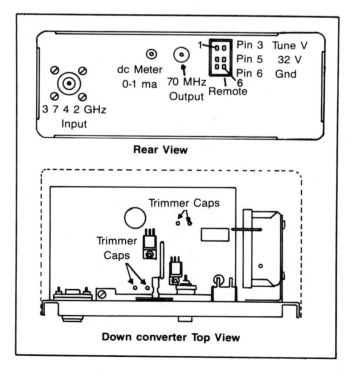

Rear View

Down converter Top View

Fig. 2-3. DC-60 down converter back panel has the 6-pin dc connector, the 70 MHz jack (BNC), the meter jack (RCA) and the LNA input jack (type-N). Looking inside the DC from the top shows the four trimmer caps that adjust the 850 MHz filter.

to 3.360 GHz. It is controlled by the voltage applied to the junction of the two varactor diodes. As their capacitance changes, so does the output frequency of the oscillator. The satellite band (3.7 to 4.2 GHz) passes through a bandpass filter (BPF) which keeps the 1st LO from radiating out and also serves to minimize interference from out of band sources from entering the mixer. The LO goes through a narrow bandpass filter, and is mixed by the two diodes with the incoming satellite signals.

The 840 MHz output signal is passed through the two filters and the NH6, which is a hybrid amplifier made by Newton Electronics (340 Middlefield Rd., Mountain View, CA 94043, 415/967-1473). It is then mixed with the fixed 2nd LO output which is 910 MHz, again by using two diodes. It is low pass filtered so that only those frequencies below about 90 MHz are passed onto the MWA120, which is a three terminal VHF amplifier transistor. This is coupled to the BNC connector through the .01 μF cap, which stops the tuning voltage from entering the MWA120. The 70 MHz in turn is filtered by the RC network on the tuning voltage line so that it does not modulate the 1st LO.

If the signals seem to be weaker than normal or if there is poor picture quality, the DC may need realignment. Before attempting any adjustments, allow the DC to warmup for at least 30 minutes. During the warm-up period check all connections between the LNA and DC, and between the DC and the receiver for corrosion. Clean any suspect connectors. Check to make sure the feedhorn is clean and that there are no insect webs or nests inside it. By observing a signal strength meter hooked to the RCA jack on the DC, try biasing the feed in all directions to see if the signal can be peaked any higher. Try adjusting the dish back and forth, and up and down slightly in azimuth and elevation.

If the above procedures fail to improve reception, then follow these steps to realign the DC:

1. Once it is warmed up, remove the DC from its housing and if not already done, attach a 1 mA meter movement to the RCA jack. Be careful not to short any part of the circuitry to ground.

2. There are four trimmer cap adjustments, per the drawing in Fig. 2-3B, that can be made internally. These control the tuning of the 840 MHz i-f filter. These must be made with a nonmetallic jeweller's screwdriver. Do not adjust the tuning slug on either LO without having the proper test gear to measure the output frequency.

3. With the receiver tuned to a strong channel, ad-

just each control for a peak meter reading. Repeat this step 2 or 3 times to ensure that all four are set correctly.

4. If a TV-4400 receiver is used, remove its cover and turn the agc control fully ccw (looking from the front of the receiver). Tune in the strongest channel (unless it is in a fixed tune installation, in which case leave it on its channel). Fine tune for best picture. Adjust the input control fully cw and then back it down until the front panel meter just begins to drop.

5. Tune to a weak channel and adjust the agc control for best picture. Recheck the strongest channel.

USS/Maspro SR-1 and SR-2

The SR-1 was designed in 1981 and was first introduced in mid-1982. The SR-2 was introduced in late 1983. It is basically a new-improved SR-1. There was also a SR-1A, which cleaned up some of the graininess that the SR-1 had in its picture by changing the i-f strip slightly.

The SR-1 and SR-2 systems are the only dual-conversion LNC (Low Noise Converter) systems on the market. The only differences between the two systems is in the video detection circuit. The SR-2 uses a higher frequency PLL circuit and a wider bandwidth i-f than does the SR-1. This yields improved picture quality.

One unique aspect of the USS/Maspro system is that only one cable is used to connect the receiver with the 2nd DC. On this cable are several carriers, in addition to the 70 MHz carrier that is bringing the signal into the receiver. There is a 10.7 MHz carrier for the afc correction voltage, a 4 MHz carrier for the channel tuning voltage and a 38 KHz carrier for the Polarotor control signal.

Both systems use an LNC with a high i-f output of 1.2 GHz. The LNC's noise temperature is typically less than 100 degrees. A block diagram of the LNC and 2nd DC is shown in Fig. 2-4 and Fig. 2-5. The VCO, which runs at about 1 GHz, but which is multiplied to about 5 GHz before it is mixed with the satellite signals, has a total tuning drift of about 1 channel (maximum ± 15 MHz), but because of the quartz synthesized tuning circuit and afc in the DC, this will not affect performance.

The tuning voltage is modulated on the 4 MHz carrier that is sent from the receiver up the rf cable. This carrier is phase detected or compared with the divided down output from the VCO. A difference signal is obtained and is fed to the tuning input of the VCO. Thus the VCO's output frequency follows any changes in the 4 MHz signal. The VCO's signal is then multiplied by 5 to yield a variable output of 4.9 GHz to 5.4 GHz. This is mixed with the incoming satellite signals to yield the

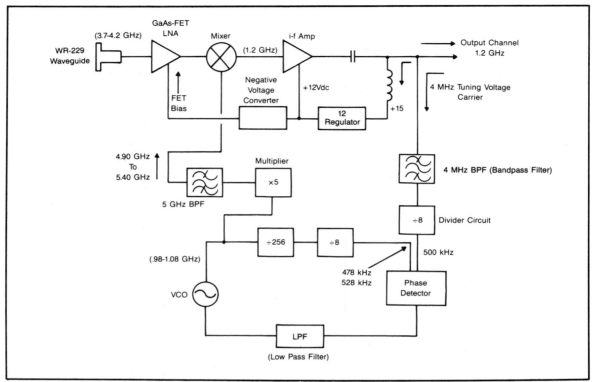

Fig. 2-4. Block diagram of the USS/Maspro LNC. It amplifies and down converts one channel onto an i-f of 1.2 GHz. The tuning voltage enters the LNC up the coax cable on a 4-MHz carrier. The dc is also supplied up the coax cable.

tuned channel at 1.2 GHz.

The separate, 2nd down converter, shown in Fig. 2-5, takes the 1.2 GHz i-f signal and lowers it to an output of 70 MHz. Thus allowing standard 75 ohm cable to be used to send the signal into the receiver. The 2nd DC has about 45 dB gain for a total gain of about 75 dB between LNC and DC (after counting in cable losses). The DC also detects the 38 KHz carrier with the Polarotor control pulses on it and outputs these pulses on the white wire that connects to the Polarotor. The afc voltage on the 10.7 MHz carrier is also detected and used to adjust the frequency of the 1.270 GHz VCO that mixes with the incoming 1.2 GHz signal so that it follows any drift that occurs in the LNC's VCO. The 4 MHz tuning voltage is also passed on through to the LNC. The DC has a test point for checking signal level and for dish peaking that is 10 dB below the output signal level.

The Maspro remote control (shown in the block diagram in Fig. 2-6) is unique in the TVRO world. Instead of using an infrared or a UHF carrier to transmit the remote control information, it uses wires, but not like in

old-fashioned wired remotes. It uses the ac wiring throughout the house to couple the signal back to the receiver. The principle is the same as that used by the remote ac line controllers from BSR and Radio Shack. The remote is plugged into any outlet in the house to remotely control the receiver. It does this by sending a series of pulses on a 38 kHz carrier that it sends out on the ac line. When any of the buttons on the remote are pressed, a distinct series of coded pulses is sent through the house wiring to the receiver.

At the receiver, there is a detector tuned to 38 kHz and a decoder which translates the instructions into actions. Unfortunately, if your next door neighbor uses a Maspro receiver, his pulses would be received by your receiver, unless there is a power line pole transformer between the houses. The BSR and Radio Shack controller allow you to set different code levels to prevent this from happening. At this time, the Maspro doesn't have that capability.

The receiver, which is shown in block diagram form in Fig. 2-7, uses a 26 MHz SAW filter in the i-f and a

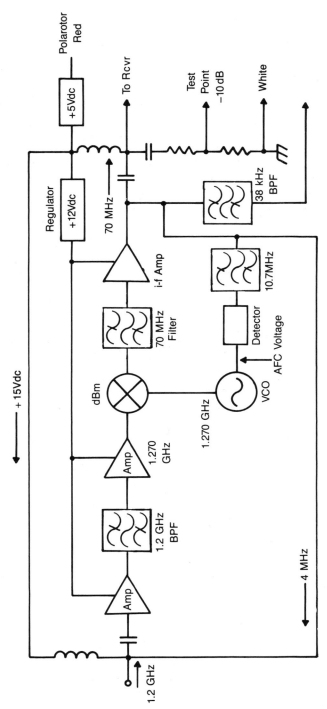

Fig. 2-5. The Maspro 2nd DC block diagram. The 2nd DC takes the 1.2 MHz i-f and mixes it with a 1.270 GHz variable oscillator whose frequency is controlled by the afc voltage. It also creates the PD's control signals and operating voltage.

17

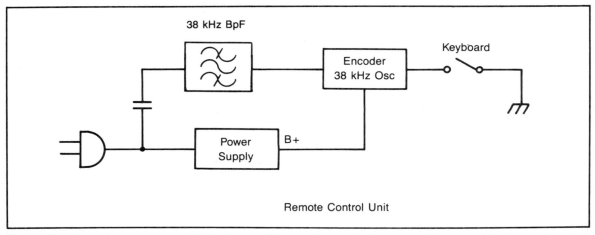

Fig. 2-6. The Maspro's remote control block diagram. The remote uses the house ac wiring to send the instructions back to the receiver. They are sent on a 38 kHz carrier. Unfortunately there is no way to independently control two receivers in the same house, as the remotes all use the same codes.

PLL chip to detect the video. It uses an FM radio chip operating at 10.7 MHz to detect the audio. The video amp is a common LM733 balanced amplifier, whose output is selected by a back panel video polarity switch. The rf Modulator can be ordered to output either channel 3, 4, 5, or 6.

SINGLE-CONVERSION SYSTEMS

A typical single conversion receiver system that is available in both LNC and LNA/separate DC packages is the Gould/Dexcel DXR 1100. It has been around since 1981, first as the DXR 1000, and then after a few circuit changes, the DXR 1100, making it one of the longest running receiver designs in the business. There are between 50,000 and 75,000 of the receivers currently in the field.

The DXR 1100 receiver was sold in a package with either a DC or an LNC, and 125 feet of preassembled cable. Both the LNC and the DC contain built-in isolators. In the DC, the isolator is used to block the LO, which in a single conversion system is in-band, from being radiated out the input port. In the LNC, the isolator protects the first GaAs-FET from static electricity, provides proper matching between the feedhorn and the LNA electronics, and also stops any spurious signals from being broadcast out the waveguide and back into the dish. The block diagrams of the LNC and the DC are shown in Fig. 2-8 and Fig. 2-9.

In a separate DC system, a standard LNA is used. From the LNA to the DC RG-213-type cable is used. This type of cable typically loses about .215 dB/foot at 4 GHz.

So for a normal 15 foot length, about 3.25 dB will be lost. There will also be some loss in each connector so an average loss figure will be under 5 dB. If one of the N connectors is poorly installed, any amount of gain could be lost, up to and including complete signal blockage. The hook-up of a typical separate DC is shown in Fig. 2-10.

For an LNC, the down converter is contained in the same housing as the LNA, so there's no need for any RG-213 cable or N connectors, since the LNC output is at 70 MHz. Gould/Dexcel, however, uses N connectors on the output of the LNC for two other reasons: waterproofing and mechanical strength.

Signal Levels

To graphically show the varying signal levels found in the typical single conversion system, I'll use a system consisting of an 8-foot dish, LNA, separate DC, and receiver. A 55 percent efficient 8-foot dish will show a gain of 37.5 dB. Figure 2-11 is the chart of gain and losses to be found within the system. It starts with the satellite's signal level as it is broadcast. There is about 196 dB of free space loss between the satellite and the dish. This includes loss for spreading of the signals and for attenuation by the atmosphere. The signal is at a level of −120 dBm as it strikes the dish. The dish has about 40 dB of gain and the typical LNA about 50 dB of gain. Thus the output signal from the LNA is typically around −30 dBm.

There is about 5 to 15 dB of cable and connector loss between the LNA and DC. There is gain within the DC

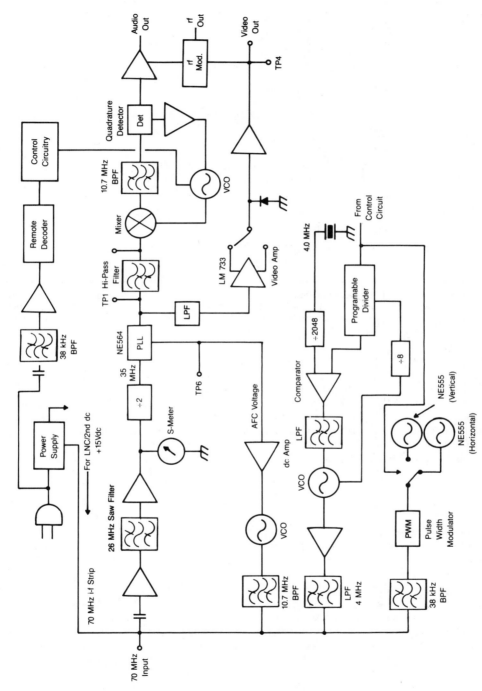

Fig. 2-7. Block diagram of the SR-1 receiver. It uses a 26 MHz SAW filter in the i-f and PLL video detection. Only one cable is used to go to the DC. The tuning voltage, afc, and PD control signals are all sent on carriers on the 70 MHz coax cable. The carrier frequencies are 38 kHz, 4 MHz and 10.7 MHz.

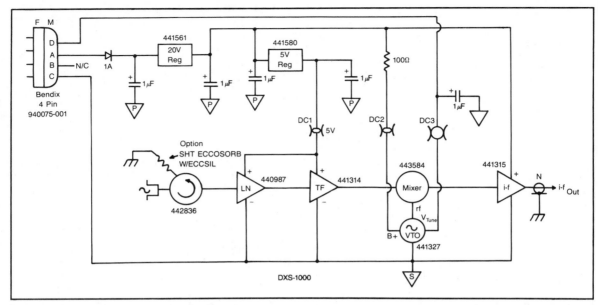

Fig. 2-8. Block diagram of a DXS 1000 LNC from Dexcel. It is an LNA and DC in the same housing. It uses a +24 Vdc power supply voltage and a tuning voltage of +10 to +20 volts. All voltages are regulated to provide lightning protection. The circuitry consists of an isolator (442836), a Low Noise module (LN), a Thick Film gain block (TF), an image reject mixer, a VTO and an i-f amplifier. The i-f is 70 MHz.

because of frequency conversion and i-f amplification of typically 15 to 20 dB. Thus the signal is at a level of −15 to −10 dBm. The cable run into the house shows from 3 to 20 dB of loss according to the length. Typically it is a loss of 15 dB. Thus our signal level entering the receiver is at −30 dBm. Most receivers will work with input levels of from −45 to −15 dBm. The i-f gain in the receiver boosts the level to that required by the video detection circuit.

VTO

The heart of a down converter is the VTO, or Voltage Tuned Oscillator (refer to Fig. 2-8 and Fig. 2-9). This oscillator is controlled by a tuning voltage from the receiver. The oscillator's output is heterodyned with the incoming satellite signals in a mixer. Out of the mixer comes an array of signals, but the only one of interest is the difference signal between the VTO and the satellite signals. This difference signal is centered at 70 Mhz. By using a wideband bandpass amplifier that is tuned to 70 Mhz, you can now separate out one satellite channel from the band of channels that entered the mixer.

As an example, if I want to watch transponder 15, the receiver has to send out the proper tuning voltage to

the DC. The VTO in the DC then has to output a frequency that is 70 MHz above channel 15's center frequency. Since channel 15 is 4.00 GHz, the VTO output is 4.070 GHz. The VTO's signal is then mixed with the incoming satellite signals. Out of the mixer comes channel 15, centered at 70 Mhz, channel 13, centered at 110 MHz, and channel 17, centered around 30 MHz and so on. By only allowing the signals from about 55 MHz to about 85 MHz to be amplified by the i-f amp, the down converter has effectively separated out one channel from the 12 channels that originally entered into it. This same process occurs regardless of whether an LNC or a DC is used.

The 70 MHz signal is then sent down RG-59 cable to the receiver. The typical receiver, in block form is shown in Fig. 2-12. The first section it encounters is the i-f strip. It is here that the 70 MHz signals are filtered to their final bandwidth. Even though the total transponder width is 36 MHz, a bandwidth of only 26 MHz is wide enough to get good video reception and yet is narrow enough to have good noise rejection. Most receivers have at least a 26 MHz bandwidth.

After filtering, the i-f signal is amplified and goes into a limiter. Because the satellite signals are FM, or Frequency Modulated, any AM noise riding on top of the

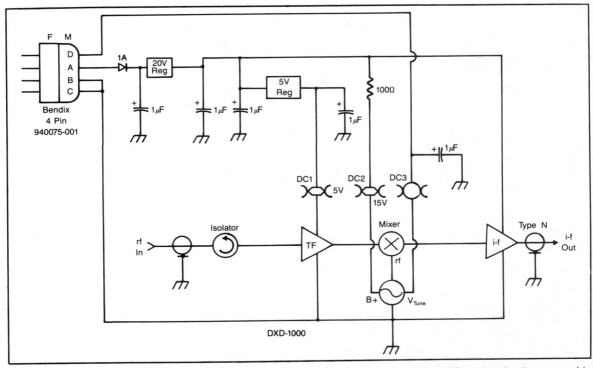

Fig. 2-9. Block diagram of a DXD 1000 DC from Dexcel. It has all the components of the LNC except for the waveguide and the Low Noise Module. The integral isolator is used to stop LO leakage from going into the LNA or power divider. Tuning and power supply voltages are identical to the LNC.

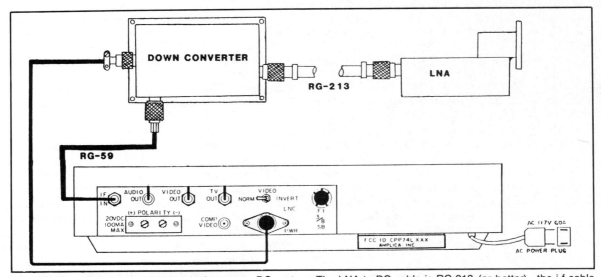

Fig. 2-10. Installation wiring for a typical separate DC system. The LNA to DC cable is RG-213 (or better), the i-f cable is RG-59 (or better), and the dc voltage is sent through shielded insulated wiring. In some receivers, the power is on a terminal strip, on others (like the Amplica R-20 shown) it is done via a 4- or 5-pin jack.

21

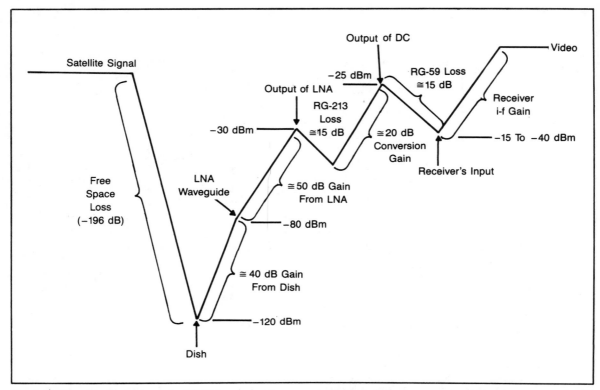

Fig. 2-11. Signal levels in the typical TVRO system. There are four places that gain takes place; the dish, the LNA, the DC and the i-f strip in the receiver. There are many places or reasons that loss occurs in the system. Dish mispointing, feed misalignments, 4 GHz connector and cabling loss, splitter loss, i-f cable loss and i-f filter loss.

signal is stripped off by limiting. At this point, the 70 MHz carrier is stripped off by the detector circuit.

The Gould/Dexcel DXR 1100, the DXR 1200, and the DXR 1300 were all designed around the LNC concept. There are three types of detection circuits used in the Gould/Dexcel satellite receiver line. PLL, or Phase Lock Loop, detection is used in the DXR 1100. The DXR 1200 and DXR 1300 use a delay line discriminator, while the DXR 900 and Econ 1 uses a monolithic quadrature detector.

At the output of the video demodulator or video detector is baseband video. This is a range of frequencies from 30 Hz to about 9 MHz. This includes all the components of the video waveform plus every audio subcarrier that is broadcast along with the video.

The baseband video is then lowpass filtered, to remove the frequencies above 4.2 MHz, clamped, to remove the dispersion signal added on during uplinking, and amplified and buffered to send to a VCR or video monitor. It is also sent to the rf modulator, which

reheterodynes the video back up to between 54 MHz and 72 MHz, or TV channels 2, 3, or 4 (depending on the type of rf modulator).

The audio is separated from the video signals, by a highpass filter which passes only signals above 4.5 MHz. The band of signals from 5.0 Mhz to 8.5 MHz contain the audio subcarrier(s). To separate out one channel requires that this band of frequencies be mixed with a tunable oscillator, in basically the same manner as the video detector or down converter.

The output of the audio detector is then sent to a buffer amplifier. This isolates the detector stage and also boosts the signals which are then sent to the rf modulator. The rf modulator reheterodynes the audio to produce a 4.5 MHz carrier frequency which is mixed with the video carrier and is then sent to the TV, where it is detected and amplified again.

The audio is also routed to the line-out jacks where it can be sent to a Hi-Fi or VCR. In the DXR 1100, 1200, and 1300, there is also a matrix stereo decoder built-in.

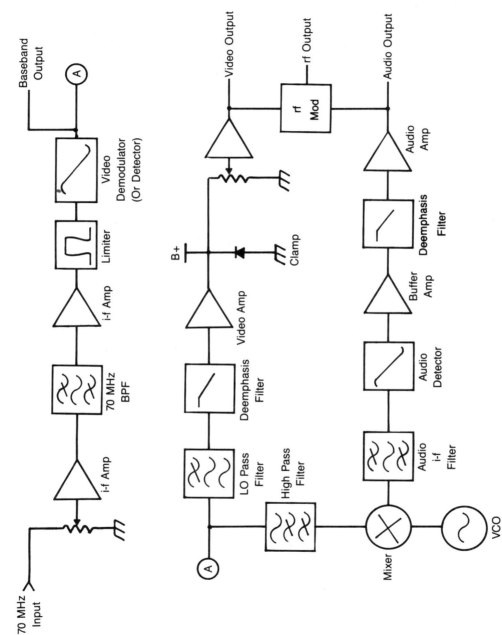

Fig. 2-12. Typical TVRO receiver's block diagram. Most receivers use a 70 MHz i-f which is bandpass filtered (BPF), amplified and limited before it is sent to the video detector. Once the signal is detected it is again filtered to provide a compatible signal for a video monitor. The audio is detected from the video signal. It is also processed to provide the proper signal for a stereo system. Most receivers also have an rf modulator for use with a TV set.

Both the DXR 1200 and 1300 can also be set to receive discrete stereo and narrowband FM radio stations.

Multiple Receivers

To use multiple receivers with single conversion DCs requires a system consisting of an LNA and 4 GHz power splitter and isolators on each DC. This can be done easily in theory, but it is pretty complicated in practice.

A single-polarity system would consist of a dish with a circular to rectangular feedhorn, a 50 dB minimum LNA, a power supply for the LNA, a power inserter, a power splitter (to divide up the signals) and an isolator, dc block, and DC for each receiver. This would be assembled as shown in Fig. 2-13.

For a dual-polarity system, the feedhorn would have to be an orthocoupled model. Another LNA, power inserter and power divider would have to be added. In this set-up, each receiver can only tune 12 channels unless a 4 GHz coaxial switch (either a relay or PIN diode) is used. This allows the receiver to be switched between the two LNAs. It also means two more RG-214 jumpers

are used for each receiver. This type of installation is outlined in Fig. 2-14.

When using any type of multiple receiver system, always put the power dividers, isolators, and down converters in a waterproof locking box. The box is usually located at the dish, although longer 4 GHz cable runs could be used if the accumulative dB drop is accounted for. Remember that for each split of the signal in a divider, the level drops by at least 3 dB, and for every ten feet of RG-213 that is used, there is a 2.5 dB loss. Add to this the .5 dB loss in each N connector and in the 4 GHz relay (if used), to find the total loss before the DC.

To simplify the multiple, separate DC system, several manufacturers have come out with splitters that have most of the separate items built into them. One LNA manufacturer, California Amplifier, has two splitters, the C4010-2A (2-way split) and the C4010-4A (4-way split) which have an amplifier that compensates for the 3 or 6 dB loss in the splitter. They also have built-in isolators on each output and the capability to pass dc to the LNA from any input. This means that dc blocks and power inserters are not necessary. The improved splitter version

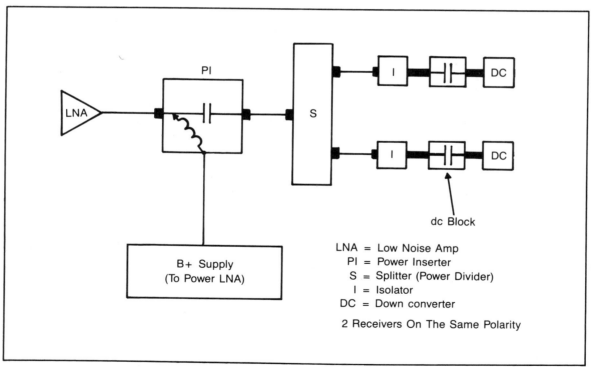

dc Block

LNA = Low Noise Amp
PI = Power Inserter
S = Splitter (Power Divider)
I = Isolator
DC = Down converter

2 Receivers On The Same Polarity

B+ Supply
(To Power LNA)

Fig. 2-13. Diagram of a multiple receiver system using separate DCs. All receivers can independently tune any channel on one polarity at a time. For more receivers, a larger splitter would be used. Splitters typically have 2, 4, 8 and 16 outputs.

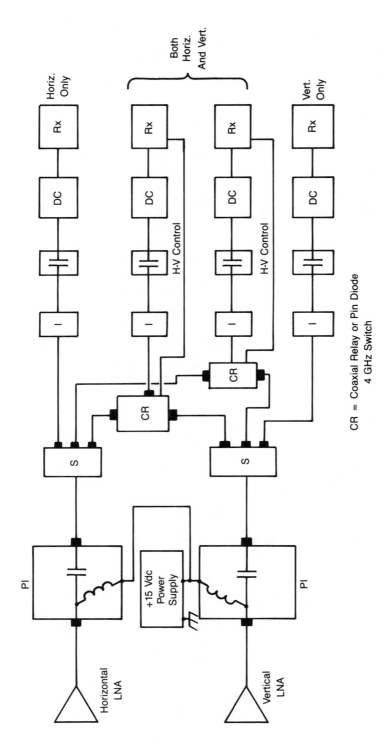

Fig. 2-14. A multiple receiver system with full independent 24 channel selection. The coaxial relays can be controlled by some receivers automatically, others require a separate switch or a modification.

CR = Coaxial Relay or Pin Diode
4 GHz Switch

for four receivers is shown in Fig. 2-15. Note that it still requires the 4 GHz switch, although I'm sure that there will soon be models that will have the switching capability built-in as well.

Multiple LNB or LNC receivers can also be intermixed with separate DC systems if a device called a coax waveguide adapter and some attenuation is used. The coax waveguide adapter is basically the waveguide part of an LNA with an N connector directly attached to the waveguide probe. It is used to connect an LNA, LNB, or LNC to a 4 GHz coax cable. This type of set-up is especially useful in showroom applications where several brands of receivers are to be shown. It is also useful in repair shops. All that is required to troubleshoot any LNB, LNC or separate DC system is a 4 GHz line coming into the shop from the dish, a variable 4 GHz attenuator, and a waveguide adapter with which to hook in the LNA, LNB or LNC.

An LNA can be used as a line amplifier or an LNC or LNB could be inserted after a 4 GHz splitter to allow multiple receivers from different design families to coexist, all by using the coaxial waveguide adapter. A multi-ple receiving system using coaxial waveguide adapters is shown in Fig. 2-16. The system would be the same as for any multiple receiver system, except that where the separate down converter would be placed, there would be a waveguide adapter and an LNB or LNC. This same type of system would be well suited for locations where there is a long cable run, as the LNB or LNC could then be used as a line amplifier.

One thing to remember when using this type of multiple receiver installation is that the LNB or LNC could be overdriven by a high gain LNA. Some attenuation must be placed in the line between the divider and the LNB or LNC. This can be done by using some RG 8 cable (which drops about 3 dB/10 feet) as an attenuator or by using a standard in line 4 GHz attenuator.

BLOCK CONVERSION SYSTEMS

As stated previously, block systems are really not too far removed from the previous system types. They are similar to single-conversion systems in that the incoming signal is preprocessed to lower frequencies to more easily

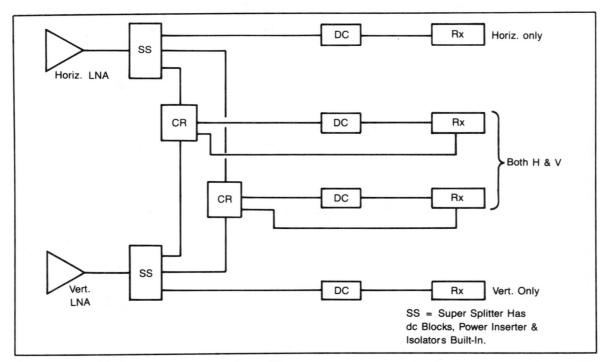

Fig. 2-15. A separate DC system using one of the new super splitters. This is the type of splitter which has integral dc blocks, isolators and amplification to make up for the splitter loss. The only auxiliary piece required is the coax relay. Compare this to Fig. 2-14's method.

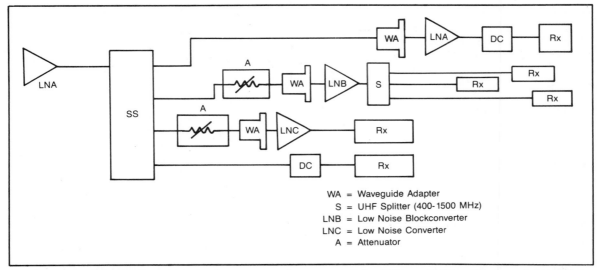

WA = Waveguide Adapter
S = UHF Splitter (400-1500 MHz)
LNB = Low Noise Blockconverter
LNC = Low Noise Converter
A = Attenuator

Fig. 2-16. Block diagram of a multiple receiver system using separate DCs, LNBs and LNCs. This is made possible by using a waveguide adapter. It converts the coax back into waveguide so that it can be fastened to an LNB or LNC. Attenuators are necessary to prevent overdriving the LNB and LNC. Notice the LNA that is used as a line amp for long cable runs.

transport them around. Once they are at this lower frequency, they are tuned in virtually the same manner as a single-conversion system.

The real advantage of block systems is in multiple receiver systems. A block-receiving system will inherently be more expensive than a single-conversion system for only one receiver. It is virtually the same situation as we had between dual and single conversion. The cost of two oscillators and mixers and filters is always more expensive then one oscillator, mixer, and filter.

But with multiple systems, block conversion takes over. It is far easier to split signals at UHF than at microwave frequencies. It is also far cheaper and more efficient. The old-fashioned single-conversion multiple system, with its dc blocks, isolators, 4 GHz splitters and 4 GHz coaxial relays was always beyond most installers capabilities. There was no way to do it easily. This has changed somewhat in the last year with the all-in-one splitters, but it still doesn't get away from the fact that the splitting is still done at 4 GHz.

Unfortunately, the first block systems on the market were low-end systems. Most used modified UHF tuners to tune in the channels, and most of them had poor performance. This was due partly to the method used and partly to the components used. The present generation of block systems have improved quality tremendously, partially because of Japanese manufacturing, but mainly

due to improved components that work well at higher i-fs like 130 and 140 MHz.

One of the most popular block systems is from Birdview. They use both the LNC, for single receiver systems, and the LNB, for multiple receiver systems. In their block system, the 20/20M, a dual LNB down converts all 24 channels down to the 500 to 1000 MHz range. Since the system uses a high side LO, the high channels are at the bottom end of the band. A separate coax cable carries each polarity's signals into the receiver. If more than one receiver is to be hooked into the system, then a dual four-way splitter is used to split both horizontal and vertical channels for up to four receivers.

The dual LNB takes the place of two LNAs and two separate BDCs. The Birdview system's LNB typically shows a conversion gain of 70 dB, which is roughly equivalent to a 40 dB gain LNA and 30 dB line amplifier. When the splitter is used, coax runs of up to 350 feet can be used to connect the four separate receivers.

Once the splitter is hooked into the system and the dual RG-6 type cables are run, the system is complete. Because all polarity switching is done in the receiver and because the +24 Vdc LNB power is sent up the horizontal cable, there are no other cables to run. Compare this system to the four-way system using single conversion.

The receiver is basically the same as the single-conversion receiver, but there are a few additional items.

One is the second down converters that are used to actually tune in the horizontal and vertical channels. The other is the polarization switching network, which chooses between the two polarities. This switching is done by a relay placed at the splitter in the separate DC system, so it is really not an addition. It is more just a shifting of position.

Once the channel has been selected, then the video demod and processing and audio demod and processing are virtually the same as other receivers. Birdview uses a discrete ratio detector to detect the video and a PLL to detect the audio.

DBS SYSTEMS

Even though I have only concentrated on C-band systems, there are actually two frequency ranges that are used to broadcast video via satellite today. The main one, C-band, extends from 3.7 GHz to 4.2 GHz. The new kid on the block is Ku-band. It ranges from 11.7 GHz to 12.2 GHz. It is the band that is being touted for DBS (Direct Broadcast Satellite). So far DBS broadcasting in the Ku-band is an on again, off again proposition. It is presently being used in Canada and to a great extent, for market test purposes, in the U.S., although its future is still up in the air because of the continuous changes that almost daily occur with the main players. The biggest usage for Ku-band at the present time is for video teleconferencing, although NBC has said it will go fully 12 GHz, and has begun installing a network of Ku-band TVROs for its affiliate stations.

BDC is a requirement for the Direct Broadcast System. This is because the Ku frequency band is 3 times higher in frequency than C-band. You already know what a bear it is to work with RG-214 at 4 GHz, so you can well imagine what happens when the frequencies are 12 GHz. RG-214's specs end at 4 GHz, so one can only guess at the loss per foot, but suffice it to say that the signal would be gone completely in only a few feet, thus in every 12 GHz DBS scheme, an LNB approach is used. The output frequencies from the LNB is typically 950 to 1450 MHz, although there is a move by some manufacturers to standardize at the lower 450 to 950 MHz band.

Since 12 GHz is 3 times higher in frequency than 4 GHz, the wavelength is 1/3rd shorter or about 1 inch. This means that the dish, the feed, and the LNB are much smaller.

Advantages and Disadvantages

Ku-band does have some advantages over C-band, the biggest being the exclusivity of the band, meaning no Terrestrial Interference (TI) caused by the telephone company's microwave system. Another advantage is that the wavelength is 1 inch versus C-band's 3 inches. Because of this, a two and a half foot dish at Ku-band performs like an eight footer at C-band. The smaller Ku-band dish must have much better surface accuracy because the wavelength is much smaller, but because it is smaller, this is more easily accomplished.

One drawback to Ku-band is that rain and moisture attenuation increase dramatically over what is experienced at C-band. This will be helped somewhat by the high powered satellites but even then unless the system is designed with what is called the *fade margin* to compensate for the signal loss, there will still be some degradation during snowstorms and thunderstorms. Drawing an analogy with 4 GHz systems, rain interference could be termed AI (Atmospheric Interference). Maybe one day we'll even see AI filters for Ku-band.

Compatibility

All the Ku-band systems that are presently being sold today use an LNB with an output of either 450 to 950 MHz or 950 to 1450 MHz. A 4 GHz BDC receiver that uses the same range would be totally compatible with the Ku-band LNB (as long as the supply voltage was compatible) since the channel bandwidth is the same between the two bands.

If a customer expresses any interest in DBS or 12 GHz compatibility, the safest bet is to sell him a receiver that uses a 450 to 950 MHz or 950 to 1450 MHz block DC. An alternative would be to sell him a separate DC receiver in which a second DC could be installed. Either way he can upgrade his system in the future.

One American company that is far-sighted and is ready for DBS is Lowrance. All of their latest receivers can be bought with a standard DC and used with an LNA, but if additional receivers are desired or if a 12 GHz system is added sometime in the future, then the receiver can be field retrofitted with a second DC inside the receiver which turns it into a block compatible receiver. The receivers can be ordered with the block system and can even be ordered with a 12 GHz LNB and dish.

Terrestrial Interference

A N UNFORTUNATE FACT OF LIFE FOR EVERYONE involved with TVRO is that we share the C-band with Ma Bell (AT&T), MCI, Sprint, and a number of other line of sight microwave communications systems. As I explained in Chapter 1, AT&T had it first, so I guess we really can't complain too loudly. But what we can do is to try to minimize the interference by using various filtering or trapping techniques.

A poll taken in 1984 of TVRO dealers found that one out of every ten installations had some kind of TI (Terrestrial Interference) problem. A similar poll taken in 1983 showed that only one out of every 15 installations had TI. Obviously it is not going to go away or get any better. In fact, as more systems are sold in suburban and urban areas and as microwave relay tower construction increases because of the deregulation of the telephone business, the future will surely bring an increase in the number of problem installations.

Terrestrial Interference, which literally means undesired Earthbound signals, is a form of crosstalk, a distortion-causing signal that enters the LNA and is amplified along with the desired satellite signals. Depending on the type of material and level being transmitted, this interference can appear as sparklies, an intermittent blanking out of the picture, as if the picture was cross-

polarized, as a solid grey screen, or as a solid black screen. In many cases, the signal strength meter will be jumping up and down in level or it may be pegged at maximum.

Often, there will be times when the interference is stronger and times when it is completely absent. The interference level depends upon the amount of information being transmitted and the type of information being transmitted. In the past, most interference has been caused by phone traffic, which is narrowband and is fairly easy to trap out. Today more and more video and digital data is being carried, which requires a much greater bandwidth to transmit. Simple notch filters will not fully trap this type of TI. Instead screening or phase cancelling techniques must be used.

TI PROFILE

Terrestrial interference is broadcast by a system of relay stations (microwave towers) that are located virtually throughout the country. There are about 10,000 legs, or two-way line of sight communications links in the system. The total number of towers is increasing all the time, especially since telephone communications has been deregulated, and organizations like MCI and Sprint are putting in their own relay equipment.

As seen in Fig. 3-1, there are 24 different frequencies that are used to transmit terrestrial communications. These channels are offset from the satellite channels by 10 MHz. Therefore, there could be two interfering carriers within the passband of any transponder. Notice that except for the carriers at 3710 and 4190 MHz, two adjacent transponders will be affected by each TI carrier.

This is more readily apparent in Fig. 3-2, which shows how the satellite channels and terrestrial channels are related. The MB indication, on lines 1, 3, 4, 6, 7, and 8, signifies that a possible Ma Bell carrier could be located at that frequency. Each transponder is 40 MHz wide, with the vertical and horizontal transponders offset 20 MHz. Each MB carrier is offset form the center frequency of each transponder by 10 MHz, and from each other by 20 MHz.

If transponder 5 is the wanted transponder, then its center frequency is 3800 MHz (line #5). It could have interference from a carrier at 3790 MHz (line #4) or at 3810 MHz (line #6), because either carrier is well within the 36 MHz passband of transponder 5. The cross-polarized transponders, in this case transponders 4 and 6, would also have problems with these same carriers, since they are also within their passbands.

Very rarely are more than 6 frequencies broadcast from one tower to another at a time. But the other tower, which is receiving those 6 carriers, will be sending back the same number of carriers toward the first tower, on different frequencies. Depending on the direction of the two towers, the TVRO installation could have interference at 12 different frequencies. If they were strong enough this could cause interference on all 24 channels.

Buildings and tall structures reflect microwaves as well as most other radio waves. This accounts for the well-known *ghosts* that haunt over-the-air TV reception in urban areas. This type of symptom does not occur with the satellite signals, but it can occur with the terrestrial microwave signals. Thus TI can be reflected by these structures and create problems by showing up in supposed TI free areas. Sometimes moving the dish a few yards, to take advantage of natural cover, will cause drastic changes in TI reception because of these reflections. Unfortunately, putting a dish on the roof or on a tall pole, especially in an urban or suburban area, is virtually asking for trouble. There is no way to effectively screen out even the weak microwave sources that normally wouldn't affect an installation, from entering the dish.

Another factor in the level of interference is that the satellite signals began 22,300 miles away from the site, while the TI began maybe 5 or 10 miles away. Both are broadcast at similar power levels, but the satellite signals lose about 196 dB of signal strength by the time they hit the dish because of signal spreading, while the TI signals, which are sent in a narrow spot beam, have minimal amounts of signal loss. This spot beam is focused so that a very narrow beam is shot directly at the next relay tower, but some spreading of the signal occurs as it travels towards the other tower. This spreading signal is what enters the TVRO system as TI.

OUT-OF-BAND TI

There are other sources of interference besides Ma Bell. These can occur at the microwave frequencies, as well as at the i-f input or rf output of the receiver.

One strong source that is very intermittent and that

SATELLITE** TRANSPONDER		POSSIBLE INTERFERING CARRIER	
1-(3720)	MHz	3710, 3730	MHz
2-(3740)		3730, 3750	
3-(3760)		3750, 3770	
4-(3780)		3770, 3790	
5-(3800)		3790, 3810	
6-(3820)		3810, 3830	
7-(3840)		3830, 3850	
8-(3860)		3850, 3870	
9-(3880)		3870, 3890	
10-(3900)		3890, 3910	
11-(3920)		3910, 3930	
12-(3940)		3930, 3950	
13-(3960)		3950, 3970	
14-(3980)		3970, 3990	
15-(4000)		3990, 4010	
16-(4020)		4010, 4030	
17-(4040)		4030, 4050	
18-(4060)		4050, 4070	
19-(4080)		4070, 4090	
20-(4100)		4090, 4110	
21-(4120)		4110, 4130	
22-(4140)		4130, 4150	
23-(4160)		4150, 4170	
24-(4180)		4170, 4190	

**Satcom/Comstar
24 transponder satellites

Fig. 3-1. Listing of the terrestrial carrier frequencies versus the satellite channels. Notice that each TI carrier is within the passband of two channels with the exception of the lowest and highest frequencies. Courtesy of Microwave Filter Company.

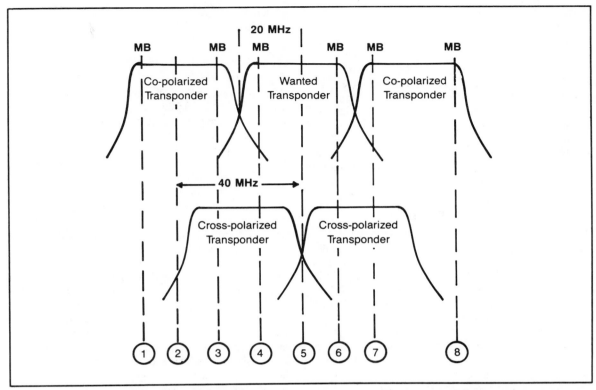

Fig. 3-2. Five satellite channels and the possible TI carriers. The co-polarized channels are offset 40 MHz, whereas the cross-polarized channels are offset 20 MHz. Dashed lines 1, 3, 4, 6, 7 and 8 represent the Ma Bell (MB) carriers that may be present.

only occurs near airports is caused by aircraft navigation and radar. This appears as a pulsing of the picture or as an intermittent blanking of the picture. It is usually accompanied by the sound of an airplane passing overhead. It is only experienced if you live under the flight path of an airport that services aircraft with instrument landing capabilities.

The interference is caused by overloading of the LNA by the aircraft's radar signals. This is done by harmonics of the signal falling into the 4 GHz passband or by out-of-band signals biasing the mixer in the DC into nonlinearity.

Other sources of interference are 3.5 GHz amateur radio and 4.4 GHz Air Force radar, which fall just outside the satellite band. Depending on the selectivity of the LNA and the strength of the signals, usually only the lower or upper channels are affected. For the most part, the waveguide is insensitive to signals below about 3.5 GHz, and the LNA is optimized to only pass signals up

to about 4.2 GHz, but the frequency roll-off is fairly gradual and the gain is still fairly high even up through 4.5 GHz.

Sun Spots

One unavoidable source of TI is the sun. Twice a year the sun will literally blank out the pictures for about 20 minutes each day as it passes behind the Clarke Belt. This occurs sometime near the spring and fall equinoxes as the sun passes the equator. The exact days and times this occurs is dependent upon your latitude and on which satellite you are looking at. Its effects can be seen every day for about a week, although complete wipeout only occurs for about 20 minutes on two or three days.

Rf Interference

Other TI sources can occur at UHF and VHF fre-

31

quencies. These are familiar to anyone that has installed TV antenna systems. Such things as CB transmitters, local TV transmitter interference and cordless telephones can all cause problems. Most often these problems will show up as distortion in the rf modulator output.

They most often appear as herringbones (small lines that wiggle diagonally through the picture) although color misregistration, poor quality pictures and ghosts can also be caused by these same sources. If the video is sent to a monitor and the picture is clean, then either the rf modulator itself is bad or there is interference entering into it. See Chapter 12 on rf modulators for more information.

OTHER INBAND TI

The biggest source of inband TI outside of Ma Bell, is LO (Local Oscillator) radiation from other satellite systems. If there is an existing installation located in close proximity to the TVRO of concern, then it should be checked to see if it uses a single-conversion DC. If it does, then chances are it is not isolated. In fact, Gould/Dexcel's model DXD-1000 down converter is the only single-conversion DC that I know of which has an integral isolator. If, in addition, the LNA is a model without an isolator then there is a good chance that the LO could be radiating out of the dish. If the DC uses a high side LO, then the interference would show up between 3 and 4 channels above the one that the interfering system is tuned to.

For instance, if your neighbor's TVRO system is a nonisolated single-conversion system, and he was tuned to channel 10 (3.900 GHz), then his LO would be radiating out at 3.970 GHz, which is in the passband of both channel 13 and 14. Usually polarization alone cannot null out the interfering LO so that both channels will be affected. Putting an isolator in between the neighbor's LNA and DC will clear up the interference.

This same problem occurs in multiple receiver systems that use single-conversion DCs. If isolators are not used between the 4 GHz splitter and the DCs, then they will talk to one another, which in a fixed channel set-up may not matter, if they are set to channels that don't interfere with the other receivers. If they are all tunable, then each one will cause interference on the 3rd and 4th channel above the one they're on.

There are two solutions. One is to put 60 dB isolators between each DC and the splitter. The second is to use a block conversion system. In the block system, all the channels are lowered in a group to a high i-f. All tuning

is done at the receiver, thus there is no LO to cause problems.

COMBATING TI

The best cure for TI is to be knowledgeable about what it is and where it is in relation to each dish you install. If you are seriously in this business, then you should get a frequency coordination map run of your surrounding area. A freq-map, as it's called, shows which carriers are active, where the local towers are and the directions that they are aiming at or receiving from. With this information, you will already know if the site may have TI, definitely will have TI, or will be TI free. If this information is known beforehand, then the customer can be advised of possible TI and the extra charges for traps or screens. That way there is no surprise installations for either you or your customer.

To get a freq-map, all you have to do is contact one of the frequency coordination companies that specialize in this task, and pay their fee, which is usually between $500 and $1,000. These companies are mainly in the research business for commercial installations, since every new installation, such as a microwave relay tower or a large commercial uplink must have a plot run to determine that they will not interfere with existing FCC licensed installations. A listing of frequency coordinators is given in Appendix D.

Once you know where and on what frequencies local TI can occur, then you are prepared to do battle. But you still need some weapons, and the largest TI weapons manufacturer is the Microwave Filter Company (MFC) of Syracuse, NY. They are specialists in dealing with TVRO TI problems. They offer courses in dealing with TI both in person and on tape.

MFC also offers two diagnostic instrument kits to track down TI on your own. One is called the Terrestrial Tracer, the other is the Sky Doc Kit. They are used to identify and isolate Ma Bell carriers so that the appropriate trap can be ordered. Both can be purchased or rented by the month. The Sky Doc Kit costs about $1300, but it can also be rented for about $200 a month.

The Terrestrial Tracer is part of the Sky Doc package. It is a variable-tune notch filter in a section of 4 GHz waveguide. It is placed in line between the LNA and DC and then tuned until the interference is eliminated. The interfering frequency is then read directly off the dial and a 4 GHz trap of that frequency is inserted in place of the Terrestrial Tracer. The Sky Doc Kit also includes a tunable C-band single-channel bandpass filter,

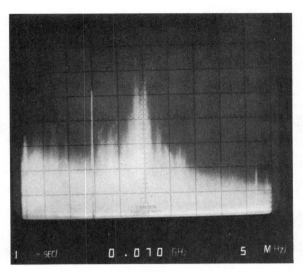

Fig. 3-3. Transponder 5 on F3-R showing a TI carrier at 60 MHz. The TI is about 5 dB below the carrier level. Since it is narrowband, it can be trapped out by a 60 MHz filter quite readily. The reference line is −50 dBm.

a 3.7 to 4.2 GHz bandpass filter, a tunable, switchable 60/80 MHz notch filter, separate 60 and 80 MHz tunable notch filters, fixed tune 60 and 80 MHz notch filters and various adapter cables, and connectors.

In addition to the above services, MFC carries the most extensive line of microwave and i-f filters available for TVRO installations. They have microwave traps as well as microwave bandpass filters, i-f bandpass, lowpass and half-transponder filters and also accessories such as dc blocks, power inserters, isolators, power dividers, terminators, and coax-to-waveguide transitions.

Many other companies make TI suppression filters for the 70 MHz i-f. Some are switchable, and others are not switchable and are left in-line all the time. This type of filter is really only useful in a fixed channel receiver, because if one filter is necessary (i.e., a 60 MHz filter), then the other one will be necessary to clean up the next upper or lower channel.

Referring back to Fig. 3-2, on channel 5, the interference that is caused by MB #4 will be at 60 MHz if a 70 MHz i-f is used. But on channel 4 that same carrier will be at 80 MHz. Thus to properly trap that one carrier out both a 60 MHz trap for channel 5 and a 80 MHz trap for channel 4 would be required. If both were in-line all the time, then needless picture degradation would occur. Thus for best performance a switchable filter is needed.

The MFC model 4616-70F is a switchable notch filter designed for 70 MHz i-fs. It contains two traps, one at 60 MHz and one at 80 MHz, with notch depths of 20 dB and a −3 dB bandwidth of 4 MHz. Either or both notch filters can be switched in or out as needed. This way channels that don't have TI are not affected by the notch filters. It is normally placed on top of the receiver and switched in or out as necessary.

To graphically show what a TI carrier looks like, Fig. 3-3 is a Spectrum Analyzer photograph of one transponder. A Spectrum Analyzer looks at signals with respect to frequency and amplitude, not time. Thus the horizontal divisions are 5 MHz apart and go from 45 MHz at the left edge to 95 MHz on the right edge. The transponder has been down converted to 70 MHz, which is the center line frequency. The spike that is located at 60 MHz (two divisions to the left of center), is a narrow deviation terrestrial carrier. This carrier can be filtered out by installing a 60 MHz notch filter in the coax cable coming from the DC or LNC. Either the Drake or the MFC filter would probably be successful in curing the interference caused by this carrier.

Cables and Connectors

THERE ARE TWO BASIC KINDS OF WIRING USED IN the typical installation. Coax cables and insulated wires. The insulated wires are most often run in pairs or groups of wires within a shielded covering. There are many brands of cabling available, some of it good, some of it bad. But to choose a cable requires more than just a name, for there is Belden cable that would be completely wrong for one job and yet it would be ideal for another. The main criteria for any cable is what type of voltages and what frequency signals it will be carrying.

All cable manufacturers sell basically similar products, at least on paper. There are many subtle differences between brands of cabling, even those with the same generic number. The type of material used in the insulation, the percentage of braid in the shield, the size of the center conductor, and the number of strands in a stranded cable all vary from one product to another. Like any product, the more you know about it, the better choice you can make.

COAX CABLING

A coax cable consists of four parts. The center conductor, which often must carry both a dc voltage and a signal, the dielectric core or insulation, the outer conductor, which is also the shield or ground, and a jacket made of PVC. A coax cable and its component parts are shown in Fig. 4-1.

Each part plays an important role in the operation of the cable. The outer jacket protects the cable from moisture, oil, oxidation, ozone, acids and abrasion. The outer conductor shields the inner conductor from other electromagnetic forces and functions as the ground return path. The dielectric core sets up the impedance of the cable and also insulates the center conductor from the shield. The center conductor is what the cable is all about. It carries the signals we wish to transport from one end to the other. It must make good mechanical contact at both ends using some type of connector. It can be stranded or it can be solid, the makeup depending on the type of signals it is to carry.

A fact of nature is that the higher the frequency the less the signal runs through the center conductor of a cable and the more it runs along the surface. This is called the "skin effect." At microwaves almost all signal is transported on the surface of the cabling. This is why the best cabling for microwaves is waveguide.

Obviously using waveguide is none too practical for the home installation due to its high cost. The exception to this is in the interface of the feedhorn to the LNA, since

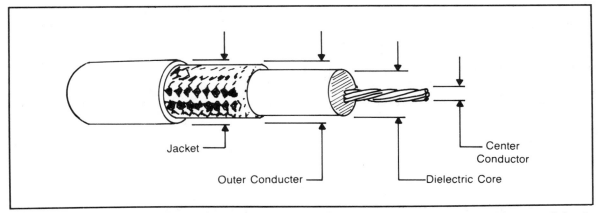

Fig. 4-1. The various parts of a coax cable. Coax cables have either a stranded or solid center conductor, a dielectric core made of various types of polyethylene, one or two braided outer conductors or shields and a PVC covering.

losses in this area far outweigh the slight added cost of waveguide. Once the signal has been boosted by the LNA, the higher losses suffered by coax cable can be lived with because of its lower cost. The coax must still have a fairly large diameter center conductor, because again the bigger the diameter, the more surface there is to carry the microwaves on. This is why RG-213, RG-214, Belden 9913 and 9914 have larger center conductors when compared with RG-59, RG-6 and RG-11. The other cables only function at lower frequencies and thus can have smaller center conductors. It is the center conductor which is primarily responsible for the signal drop in the cable.

Look at some RG-59 cabling and compare it to RG-6 and RG-11. Notice that the center conductor and hence the dielectric core gets bigger as we go from RG-59 to RG-6 to RG-11. The cable attenuation drops inversely because of the increase in conductor size. RG-59 typically attenuates 70 MHz signals at the rate of 3 dB per 100 feet. RG-6 shows about 2.5 drop per 100 feet, while RG-11 has about 2 dB per 100 feet of cabling at 70 MHz.

Coax Classifications

The coax cabling used in TVRO systems can be divided into two classifications: 50 ohm and 75 ohm. 50 ohm cabling is used between the LNA and the DC, while 75 ohm cabling is used from the DC to the receiver and from the receiver to the TV or video monitor. This ohm rating is called the cable impedance. It is a function of center conductor size, dielectric size, and the spacing between the center conductor and the shield.

The most popular type of 50 ohm cable is RG-213. It is fairly easy to work with, which is why it is popular.

A better cable is RG-214. It is a double shielded coax, meaning it has two separate braids on it to more fully shield the inner conductor. Because of the dual shield it is also stiffer and requires a wider radius to bend it in. Belden has an improved version of RG-213 and RG-214 called 9913 and 9914, which are better choices for cabling than either 213 or 214. At 4 GHz, 9913 has an 11 dB/100-feet loss while 9914 shows a 13 dB/100-feet loss while both RG-213 and 214 show loses of 21.5 dB/100-feet.

The two Belden cables are actually improved versions of RG-8, which shows an 18 dB/100-feet loss. RG-8, like 9914 uses a cellular polyethylene dielectric. This type of dielectric requires that wide bending radii be used to prevent the center conductor from cutting into the dielectric, which will disrupt the impedance of the cable. 9913 uses a semisolid polyethylene dielectric, so it is much stiffer also requiring a wide bending radii, but it is less prone to shifting of the center conductor, and the impedance changes that would occur if that happened.

On most LNAs a right angle N connector is the best choice to use because of the turn radii of 9913 and 9914. That way the cable only has to make a 90 degree turn instead of a 180 degree turn. There are LNAs on the market whose designers took this into consideration. They have their N connectors mounted on their side so that using a right angle fitting means that the cable does not need to be bent at all.

The most popular 75 ohm cable is RG-59. RG-59 is a generic title for a coax cable that is about 75 ohms in impedance and has one center conductor, an insulator, and a shield. Beyond these rather broad specs, each type of RG-59 will vary from one to the next. For instance,

Table 4-1. Comparison Chart of the Various Generic Coax Cables.

Cable Type	Nominal Impedance (In Ohms)	Jacket O.D. (In inches)	Loss (In dB/100 ft.) 70 MHz	Loss (In dB/100 ft.) 4 GHz	Dielectric	Shield Coverage (% of Braid)
RG-59	75	.242	2.2	N/A	Polyethylene	80
RG-6	75	.336	2.5	N/A	Solid Poly.	
RG-11	75	.405	1.8	N/A	Poly.	97
RG-8	50	.405	1.5	18	Cellular Poly.	97
RG-213	50	.405	1.8	21.5	Poly.	97
RG-214	50	.425	1.8	21.5	Poly.	98
9913	50	.405	.8	11.0	Semi-Solid Poly.	100
9914	50	.405	1.3	13.0	Cellular Poly.	100
9915	50	.870	.7	10.0	Solid Poly.	100

Belden carries several RG-59/U type cables. Losses at 70 MHz range from 2.6 to 3.5 dB/100 feet. This means that some of their RG-59 is equivalent to other manufacturer's RG-6. Table 4-1 lists several of the popular coax cables and compares their specifications. Table 4-2 relates Belden's and Alpha's cable numbering to the generic cable number.

RG-59 cables should have a half-hard center conductor. This ensures that they will effectively insert into the female connectors. Some RG-59 cables do not have a half-hard center conductor. This means that the center conductor can be easily flexed. This also means that it may bend as it is inserted into the female connector, since F connectors use the center conductor as the male plug. The only reason for not using a half-hard conductor is to save on manufacturing cost. If this type of coax is found in a one run cable, there has probably been other manufacturing penny pinching done to the cable as well and you'd be advised to stay away from using it.

Handling Coax Cable

The 75 ohm cables like RG-59, 6, and 11, really don't require any special precautions on use unless they are to be stapled to a wall or to the floorboards or joists. If the cable is to be cleanly run then it should be stapled, but only if the proper cable stapler and cable staples are used. If coax cable is pinched where the staple is placed, then the cable impedance is screwed up at that point. If this

is done at several points along the cable, then standing waves or impedance anomalies can be set up, even at 70 MHz. This will cause suck-outs or deep notches in the signal. In a block system, this can cause some channels to be almost completely attenuated, while leaving others untouched. Usually the cable must be replaced if this has happened, as just removing the staples does not correct the deformed dielectric shape and impedance disruption.

The biggest handling caution on 75 ohm cables is to make absolutely sure they are waterproofed. This applies equally to 50 ohm cables as well. The next section covers sealing techniques.

When burying cables, usually PVC tubing is used.

Table 4-2. Generic Cable Numbers vs. Belden and Alpha Cable Numbers.

Belden	Cable Type	Alpha
9240	RG-59	9059
8215	RG-6	9006A
8238	RG-11	9011A
8214	RG-8	9008
8267	RG-213	9213
8268	RG-214	9214
9913	RG-8	
9914	RG-8	
9915	RG-218	

In many instances, this is not necessary, but it does allow for a cleaner more professional looking installation. If PVC is used be sure that it is large enough for your connectors to go through if you are using preassembled cabling. Always tape the connectors so that they will not be bent back while they are pulled through the PVC. When routing cabling through J-hooks, the same thing applies, especially if an N connector and 213 cabling is involved. Also watch out for sharp edges which can slice into the cable covering, and which will sooner or later allow water into the cable. If PVC is used make sure that no water enters the PVC by placing an upside down U at all outside breaks. This is especially important in areas that freeze during the winter, since any water in the PVC will expand when it freezes, possibly cutting into the coax in the process. This will cause all sorts of problems when the spring thaw comes, and the water is soaked up by the foam dielectric.

The bigger cables like RG-8, 213, 214, 217, and especially heliax and hardline all require special handling. The biggest area to watch is in turn radius. Most cables should have a bend radius of at least 20 times their diameter. For the 213, 9913 families this amounts to a loop about 8 inches across. For heliax, the radius should be 3 to 4 feet!

Unfortunately, with the rise in LNA covers, the radius on RG-213 is often reduced to 2 or 3 inches. LNA covers are great for protecting from moisture intrusion and ultraviolet light, but a cover that is too small for the LNA can cause more problems than its worth. Using a flexible rubber cover is probably the best bet, that way the cable can flex out against it without being pushed down like with smaller ABS plastic covers.

When pulling cables, always cover the connectors or at least be very careful so that dirt and moisture do not enter them. This will surely cause problems down the line, like intermittent signals due to improper sealing because of a little dirt on the seal.

INSULATED WIRING

The other type of TVRO wiring is a solid or stranded conductor covered with PVC. These wires are mainly used for power supply and other low voltage dc signals. Insulated wires are used for carrying the PD's power and pulse, the DC's tuning and dc voltage supply, the coaxial relay supply, and the motor drive power and feedback signals.

The wire is rated by its gauge or size. It is typically listed by its AWG or American Wire Gauge number. The AWG number gets smaller as the wire diameter gets larger. Wire size is one factor in the current carrying capacity of the wire. Other factors are the number of stranded conductors, the maximum temperature rating of the covering, and the environmental conditions (i.e., whether enclosed in conduit, buried, or open with plenty of air flow).

Wire size also affects the amount of resistance that is built up as the cable length increases. It is always OK to use a larger gauge wire but you cannot go the other way. If the cable run calls for an 18 AWG wire, using 16 AWG is fine, but using 20 or 22 AWG is not.

CUSTOM CABLING

With the advent of motor drives and polarization controllers becoming part of the standard system, a new problem arose. That of having to run several cables out to the dish. This was addressed by Maspro in their system, by encoding the polarity and tuning signals onto carriers, so that only one coax is needed in their SR-1 and SR-2 systems. But in most TVRO systems that have a PR I, motor drive, and separate DC and LNA, there are 12 wires plus the coax cable that must be run. In addition, most of the wires should be shielded from one another.

This led to the development of the one-run type of cable. It is a cable that has all the various conductors enclosed in one PVC covering. Most of these cables are made by independent cable manufacturers, and so vary widely in quality. One one-run cable is made for Echosphere, and it is called the Echo Cable. It differs from other one-run cables in that it is like a four part siamese cable, meaning that there are four RG-6 size cables molded together along one axis. The cable can be folded to allow it to be passed through a 3/4 inch or larger hole. Its true advantage comes out when it is run in the house. It can be flattened out and run along the baseboard or through spaces where the large one-run cable will not be able to go, due to its wider turning radius or bulk.

Whenever buying cabling, especially one-run type, be sure to find out what all the various conductor sizes are and whether or not they are shielded, what the percentage of braid and foil is on the coax, and whether or not the coax has a half-hard center conductor. In most cases, the one run cables are touted as being direct buriable. If you always use PVC, then you may be paying extra for this feature. Check the motor drive manufacturer's recommended cable size vs. run to make sure the conductor size is big enough for your typical installation. Often if a long run is needed, standard UF-type direct burial 117 Vac romex cabling (for outdoor lighting) is the best bet for motor drive wire.

CABLE/CONNECTOR SEALING

Cable/connector junctions must be protected from the environment. The two biggest problems with connectors is moisture contamination and salt or chemical corrosion. To effectively seal a connector requires that the sealant adhere to the cable's PVC covering as well as to the connector's metallic surface. Before sealing any cable connector, both the cable and the connector must be cleaned. Both surfaces must be free of any oils from the fingers, dirt, solder splashes, etc. Typical light solvents like alcohol or freon can be used to give the cable and the connector a final cleaning before applying the sealant.

Such products as Coax-Seal, duct seal, and Scotchkote all work equally well, if properly applied. The popularly used clear RTV sealant, also known as bathtub caulking, is not recommended for several reasons. It takes a long time to cure, it doesn't adhere well to PVC, and it is very acidic, which can actually cause corrosion of the connector. Like cabling, all RTV sealants are not alike however. RTV is a generic word meaning Room Temperature Vulcanizing, so there are many types of sealant that are designed for different uses that all carry the name RTV. One type of RTV from Dow Corning, that I've used, is 3145. Use the grey version. It seems to be more flexible than the clear stuff, and it adheres well to coax cable. It cures on contact with humid air about 2 hours.

GE makes an RTV that is a paste-like sealant made of silicon rubber. It is called RTV-108. It was designed for CATV applications for potting F connectors, taps, and underground connections. It has superior bonding qualities over most other RTV sealants when applied to the PVC covering on coax cabling. It will withstand temperatures ranging from – 70 degrees F to + 400 degrees F without melting or becoming brittle.

Another GE product, G-635, is a dielectric compound which maintains a soft to medium consistency even when exposed to environmental extremes. It can be applied to connector threads as a lubricant to prevent seize-up caused by dissimilar metals. It can be applied to any outside connector to protect it from moisture intrusion and chemical corrosion (salt, acid rain, etc). G-635 can also be used on LNA/PD gaskets to keep them from drying out and becoming brittle. A light coating is all that is necessary to protect the gasket.

Before applying any sealant, always check to see that the cable works properly. It's a lot simpler to disassemble a connector if sealant doesn't have to be removed first. The best way to seal a coax connector is to wrap the last one or two inches of the cable and the connector with self sealing tape. Shrink tubing could also be used, if it is slipped onto the cable before the connector is put on. Cover up all the tape (or the ends of the shrink tubing) as well as another half inch of cable with the sealant. It is not necessary to ever apply any sealant to the LNA, LNB or LNC case, unless specifically instructed to do so by the manufacturer. This is especially true if N connectors are used, as they only need to be sealed on the cable entrance end.

If it becomes necessary in the future to disassemble the connector, slice down the tape, and then peel it off. The sealant will come off in one piece, leaving the connector and 1 or 2 inches of cable clean. It's much easier to work with a connector that is clean than one that is covered in RTV or Coax-Seal.

CHECKING CABLES

All coax cables should have no resistance between the center pin and the shield, if they are disconnected at both ends. If moisture intrusion in the cable has occurred, then there will be some resistance between the center conductor and the shield. If this resistance is low enough it can cause the power supply voltage to be pulled down or the tuning voltage to be pulled down depending on the cable's use. It can also cause 4 GHz signals to disappear, since water absorbs microwaves.

To check the cable, use a VOM, VTVM, or DVM set on the highest resistance scale. If there is any resistance at all, then the cable most likely has water in it, unless it is a dead short, in which case the cable's shield is contacting the center conductor. The most likely place for this to occur is at the connector or at another stress point. If a resistance reading of 2,000 ohms or more is read, then water is probably the culprit. This can often be removed by heating up the coax cable and connector, steaming the water out. This can be done by using a shrink tubing heat gun or by using an ordinary hair drier. Be sure to keep moving the heat source, as the PVC will soften and possibly melt if it is concentrated on only one area for too long of a time. Be careful of the connector as it will get very hot after the 5 minutes or so of heating that it usually takes to evaporate the water. If the ohmmeter is left hooked up, you will see the resistance drop during the first minute of heating, and then it will start to rise as the water evaporates, until it will finally be infinite, at which time the cable should be allowed to cool and then should be refastened and resealed.

To check the wiring for a PR I, turn off the receiver and check the resistance between the red and white and black wires with everything but the PR I connected. Between the red and black wires there should be about 4,000 ohms resistance (this will vary from brand to brand but should be close). If there is a capacitor at the PR I wire splice it should be disconnected before doing this reading. There should be no resistance between the white wire and ground.

SPLICING CABLES

All splices in coax cables must use connector-type splices that are referred to as couplers or female/female adapters. For 50 ohm cable using N connectors, a UG-29, female to female adapter is used. If the cable uses BNC connectors, then the female/female adapter is called a UG-643 or a UG-914 connector. On 75 ohm cabling that uses F connectors the splice is called an F-81 coupler. On cabling that uses a UHG or PL-259 connector, a PL-258 coupler would be used. All the adapters and connectors are available from most Radio Shack stores with the exception of the N connectors. Radio Shack is not quite there yet, but I wouldn't be surprised to find them carrying them before long.

The dc wiring is a bit easier to splice. For it, all that is required is a butt splice of the proper size for the wire used. All splices should be waterproofed by using shrink tubing and sealant, whether they are for coax cables or for dc wires.

CONNECTORS

The early LNAs used SMA connectors. These are small coax connectors used on miniature semirigid cable. This is a coax cable that has a solid metal cover. It was often used in dual-conversion receivers to pipe the high LO between the 1st mixer and the 2nd DC. Some LNBs still use an SMA connector with an N adapter to interconnect the LNA and the BDC. This allows the use of 4 GHz filters in the line.

There are basically seven different connectors used on present TVRO systems. The workhorses are the type-N, type-F, RCA (or phono), BNC (or type-C), UHF, Cannon (or Bendix) and Molex. The first four are for coax cables, and the last two are for insulated multiconductor wiring. The most often used connector is the F. Luckily it's also the cheapest one and the easiest one to make.

F Connectors

F connectors are simple connectors. They don't even have a center pin. All they are is a threaded shell. The center pin is the coax cable. Because of this, F connectors will only work with solid center-conductor coax cable. There are several types of F connectors, each designed

for a specific type of cabling. An F connector designed for RG-59 will not work on RG-6 or RG-11 because each cable has a different thickness of insulation. F connectors are also available with two types of crimp rings: integral and separate. The integral type is a much easier connector to work with, and it looks cleaner when you are finished.

F connectors are not waterproof, or even water resistant. Any time an F connector is exposed to the weather it should be covered with sealant. The best way to weatherproof an F connector is to tightly wrap self-adhering tape around both the female and male connectors covering an inch or so of the cabling. Cover this tape with any brand of cable sealant mentioned previously. If the sealant is clear, then the sealing tape should be wrapped first with standard black PVC or electrical tape for ultraviolet light protection.

To put an F connector onto any coaxial cable follow these steps.

1. Strip the outer jacket, shield, and insulation back 3/8". Be careful not to nick the center conductor. All of the insulation from the center conductor must be removed to ensure good electrical contact. Refer to Fig. 4-2A.

2. Strip outer jacket and shield back another 1/8". Be careful not to cut into the insulation. Refer to Fig. 4-2B.

3. If you have a separate connector and crimp ring, slip the crimp ring onto the cable. Insert the F connector fully into cable. 1/8" of the center conductor should be protruding beyond the connector end. NOTE: Each cable type requires a different F connector and crimp tool, so be sure you are using the correct type.

4. If the connector has a separate crimp ring, position it so that it is 1/8 inch behind the connector. Crimp in two places, as shown in Fig. 4-2C using a crimp tool designed for the specific cable type.

N Connectors

The N connector is probably the most widely used and abused connector this side of the garden variety F connector. This connector is mainly used on 4 GHz cabling, although it is also used on some LNCs to provide a water-resistant connector. It is typically used on the larger coax cable, like RG-217, RG-214 and Belden 9913 but there are N connectors especially made for RG-59, RG-6, and RG-11 cable.

Like F connectors, each cable will have a specific N connector. The UG-21 is designed RG-214, while the UG-204 is designed for RG-217. Specify the cable type

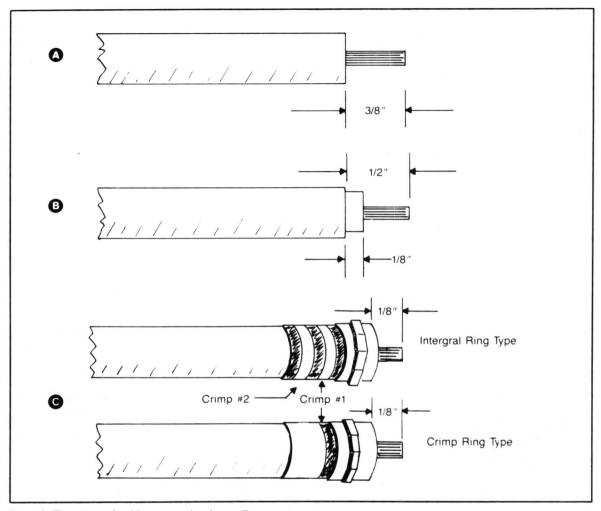

A

3/8″

B

1/2″

1/8″

1/8″

Intergral Ring Type

Crimp #2 —— Crimp #1

1/8″

C

Crimp Ring Type

Fig. 4-2. The steps of cable preparation for an F connector.

that the connector is going on when buying N connectors, or any coax connector for that matter.

A female N connector is always mounted on the LNA (except for the real old ones with SMA connectors). A mating male connector is used on the cable end. Occasionally a down converter will be found that has a male instead of a female connector. In such a case, it is designed to mount directly to the LNA connector. This is not a good mounting design unless the DC is extremely lightweight or has some kind of support because the N connector is not strong enough to support a DC by itself. Electronically it's not such a good idea either, unless a low gain (30 dB) LNA is used.

Unfortunately, with the rise in popularity of TVRO, and the attendant rise in N connector use, many imported connectors that physically look like U.S. made N connectors have made their appearance, especially on premade cables. In many cases, they do not work as well as their U.S. made counterparts. There is a big difference between brands of connectors as to materials used and quality of workmanship. Figure 4-3 is a typical N connector. Notice the thickness of the outer ring. In many foreign imports it is only one half as thick. The outer ring is used to mechanically lock the connectors together and to provide a waterproof connection. The actual shield is connected to the split ring around the center pin.

There are basically two versions of the N connector, the crimp type and the solder type. The crimp type is harder to work with, and requires a special crimping tool, but it doesn't require a soldering iron. Even if it is done correctly, a crimped connector can be a source of intermittent problems or complete loss of signals after the cable has been flexed during installation. Usually the solder type is more reliable, especially for field construction, but it too can cause problems if not assembled correctly.

Figure 4-4 shows the steps used in putting an N connector onto RG-214 cable. It is important that the braid not be cut while removing the jacket. Also take care when combing out the braid to prevent breaking off any strands. When assembling the connector hold the connector with a small crescent wrench while tightening the clamp nut with pliers or another crescent wrench. This is the only time that wrenches or pliers are used on an N connector. When it is fastened to the mating connector it is hand tight only. If a right-angle connector is used (as in step 6), be sure the cap is seated properly. If it cross-threads

then it will not seal correctly and water will enter the connector.

The N connector must not only pass rf at 4 GHz, but in the newer LNAs it must also pass the dc to the LNA. In the first generation LNAs, there were two connectors, one for rf (type-N) and another for dc (usually a Cannon connector). Because these connectors were exposed, many people thought it necessary to pack the N connectors with water repellent. This is OK (more or less) at 4 GHz but when dc is put through the connector the water repellent (or actually contaminants contained in it) can cause problems, like shorting out the dc voltage or pulling it down to where the LNA will no longer work properly. This will also cause problems in LNCs that use the coax cable as the tuning voltage line. The tuning voltage is very low current, and so can be pulled down very easily. The part of the N connector that is packed is waterproof because of the integral gasket that mates with the female connector, so packing it with anything is not necessary and will lead to more problems than it has ever avoided.

Before inserting an N connector always check the

Fig. 4-3. Close-up of a type-N connector. Notice the center pin and its depth compared to the inner ring. Also notice the thickness of the outer ring. On some import models, the outer ring is as thin as the inner ring on this model. The best type of connectors to use are ones with a gold coating on the center pin and inner shield.

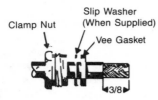

Clamp Nut

Slip Washer
(When Supplied)

Vee Gasket

3/8

1. Cut cable end square, place clamp-nut, slip washer (when supplied), and gasket over jacket. Remove jacket to dimension "A."

Contact

4. Solder contact to center conductor. For access type angle connectors omit this step and proceed to step 5.

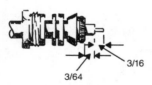

Braid Clamp

2. Comb out braid and taper forward. Place braid clamp over braid against jacket cut.

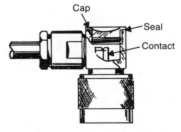

5. Thread assembly into connector, and lock securely. Vee gasket must be split by braid clamp.

3/16

3/64

3. Fold braid back over braid clamp and trim as shown above. Remove dielectric to dimension "B." Cut center conductor to dimension "C."

 When cable positioning insulators are used adjust trim code dimensions as shown below, and assemble as indicated.

Cap

Seal

Contact

6. For access type angle connectors, solder center conductor in contact groove. Close access opening.

Warning: Do NOT rotate the connector when locking assembly. Rotate ONLY the clamp nut. This will prevent short circuiting the braid to the center conductor and prevent fraying of the braid.

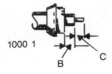

1000 1

B   C

Fig. 4-4. Instructions for installing an N connector.

center pin for the proper height and centering. The center pin must be below the inner ring, such that if you place your finger onto the inner ring, you should barely feel the end of the center pin. If it is flush or higher, then the pin is too high and the mating female connector may be damaged. The female connector will appear to be OK, but it usually will be pushed back into the LNA housing causing the solder or bond connection to the circuit board to be cracked. This is a sure cause for a future service call because of an intermittent signal.

If the pin is not centered or if care is not taken when putting on the connector, then the female center conductor, which consists of a four part circle that grasps the center pin, may be damaged. If any one of the four circle parts are broken, the connector must be replaced, which often means the LNA must be sent back to the manufacturer.

N connectors are water resistant but they are not totally waterproof, except at the point that the connectors are connected to each other. Thus water can enter the connector by running down the cable, if it has not been properly sealed. If the connector is carrying 4 GHz signals, they will disappear. If the connector is attached to an LNC and if the LNC tuning voltage is coming up the cable, this water will cause jumping or drifting channels because the water will wick up the foam dielectric causing the tuning voltage to change or to be shorted out. Note: In most cases if the cable is disconnected the connector end will still be dry, since the problem is behind the connector, where the cable enters.

To prevent water contamination of coax connectors, some type of waterproofing must be used to cover the coax cable and the upper part of the connector. The section on sealing cables covers the proper procedures and products to use.

The following list contains some do's and don'ts for the proper handling of N connectors.

☐ Treat N connectors carefully, don't allow them to be dropped in the mud or to get dirt or sand in them or to be forced through sharply bent tubing. This also equally applies to the female connectors on the LNAs.

☐ Don't pack any kind of water repellent substance into the connectors. Leave them dry. The connector is already waterproof where the repellent would go.

☐ Always check the center pin to make sure it is centered and is below the level of the inner ring.

☐ Don't use pliers or vicegrips to tighten the connector. Hand tight is good enough. If there is a problem screwing the connector on, make sure the connectors are square to each other. If they are, a very light coating of lubricant on the threads should ease attachment.

Cannon or Bendix Connector

A Cannon connector is a military style connector originally made by the Bendix Corp., hence its secondary name. Its first TVRO application was as a power supply connector for LNAs. In the early days of TVRO, all LNAs were side-powered, meaning that there was a separate connector solely for the dc voltage. A Bendix connector is shown in Fig. 4-5.

Today almost all LNAs are powered up the coax cable, thus the Bendix connector is rarely seen on new LNAs. Its use is mainly confined to LNCs, although there are some down converters that also use it. The connector is used to supply the dc voltage to the LNC or DC and, in some designs, to supply the tuning voltage.

The reason Bendix connectors are used is that they are waterproof at the male/female connection point. They are not waterproof at the backside where the cable enters the connector unless the proper precautions are taken.

The cable entrance point is every connector's weak spot and the Bendix is no exception. An improperly sealed connector will allow rainwater to get into the connector, which could short out the tuning voltage or the dc voltage. Sometimes the problem does not show up for several days, weeks, or months after it stops raining, because the water wasn't of sufficient quantity to short anything out.

Fig. 4-5. A 4 pin Cannon or Bendix connector. There are also five pin models used by some manufacturers.

However, any water at all in the connector is enough to leave a mineral deposit when it dries.

After several rainstorms there will be a conductive, chalky white, build-up inside the connector. If this build-up stretches between pins, it can short out the tuning voltage to ground or to the dc supply voltage. Likewise it could short out the dc supply voltage to ground. Since the conductivity of the build-up will change as the humidity changes, it can create intermittent problems. Symptoms such as randomly jumping or changing channels, long-term tuning instability (always rematching the LNC or DC), every channel being off by one or more numbers, or even no signals at all, can all be caused by water or by the mineral build-up.

An ohmmeter can be used to check for this build-up. Disconnect both ends of the cable, and measure the resistance using the highest scale between the active pins. If a reading other than infinity is found then the connector should be opened up and cleaned or dried out. Figure 4-6 is a male Bendix connector found on an LNC. Note the pin orientation. The pin functions are typical uses for the connector.

To disassemble and repair Bendix connectors, use the following procedure:

1. Fasten the connector onto the LNA, LNC or DC for support. Clean off any old sealant.

2. Remove the two screws and the cable clamp.

3. If a strain relief is used, slide it back up the cable out of the way. Soap or some other light lubricant can be used to allow the strain relief to slide up the cable.

4. Unscrew the upper part of the connector from the main body. Be sure to hold the cable to prevent it from turning as the connector is unscrewed.

5. Slide the terminal insulator back up the wires as far as possible.

6. Check the connector for mineral build-up or water contamination. Use a heat gun or portable hair drier to heat up the connector if any water is observed.

7. Soak an acid brush or a toothbrush in some freon, alcohol or other light solvent, and use it to clean off the terminals and the insulating material between them.

8. Slide the terminal insulator back down and check for any resistance between pins again. If there is still resistance, then the cable should be checked for shorts, or one of the leads that showed resistance could be unsoldered and checked again.

9. If the problem is solved, reassemble in a reverse manner.

10. Once the connector is back together again, make sure that it is well protected from the environment. Wrap butyl tape around the upper half of the connector until it is about 1 inch past the cable clamp. Apply duct seal, Coax-Seal, or Scotchkote to the tape and to the cable itself for another inch. It is not necessary to apply any sealant to the LNA, LNC, or DC housing or to the lower part of the Bendix connector.

In some cases, the receiver and DC or LNC will need rematching once the conductive material is cleaned out. Matching instructions for many popular receivers are covered in other sections of this book. If you have a receiver that is not covered, contact the manufacturer for the proper matching procedure.

RCA Connectors

The RCA jack is an unbalanced connector for shielded cabling used at low frequencies. It was originally designed for audio applications and because of its ease of use has been drafted for higher frequency applications. It is occasionally even used as a 70 MHz receiver con-

Fig. 4-6. The LNA, LNC, or DC male Bendix connector. Notice the numbering matches the female connector's (on the cable). The functions are typical uses for a LNC or DC.

Pin	Function
A	+12 to +15V
B	Ground
C	Ground
D	Tuning Voltage

nector, but it is a very poor connector for that application. Actually, using it at video frequencies (up to 4.5 MHz) is pushing its frequency response, but there again, its ease of use made it the winner over the BNC connector.

The RCA plug is usually soldered to the center conductor, as well as to the shield. There are solderless connectors, using a screw to fasten the center conductor and the shield, but they do not give as good a performance as a solder plug. To improve response for video, some manufacturers have started putting a thin gold coating on the connector, but it's more advertising hype than a true improvement in most uses.

Parabolic Reflectors

FROM THE SATELLITE, SIGNALS ARE BROADCAST back to Earth using two types of amplifiers: a Traveling Wave Tube Amplifier (TWTA) or a Solid State Amplifier (SSA). These amplifiers typically have power levels of 5-watts, 8.5-watts and 9-watts per channel. Even though the amplifier wattage that is used is small, the actual signal level that is sent towards the Earth is boosted many times by the satellite's parabolic reflector. A typical signal level leaving the satellite is in the neighborhood of +76 dBm.

The first part of the system that the microwaves hit after traveling the 22,300 miles from the satellite is the parabolic reflector. In everyday use, it is also called the satellite dish or satellite antenna. It is one of the most critical parts of the total system, and yet it is the part that most people don't think much about, except maybe about how to camouflage it or how to get by with a smaller dish.

As the signal travels from the satellite to the Earth, there is a considerable amount of attenuation, called free space loss, that occurs. It adds up to about 196 dB of loss. There are several long involved calculations that can be done to find the exact signal loss, and then find the exact signal level that strikes the earth at a given point from a given satellite, but suffice it to say, that in the Midwest,

a signal level of about −120 dBm is what actually strikes the surface of the dish.

Since most receivers need a minimum input of about −45 dBm, we must boost the signal about 75 dB between the dish and the receiver. There are three places where this gain takes place: the parabolic reflector or satellite dish, the Low Noise Amplifier (LNA) and the Down Converter (DC). There is also gain in the intermediate-frequency amplifier (i-f strip) in the receiver, but it is used to overcome losses in the i-f filter and to drive the video demodulator circuit.

The most critical gain is from the parabolic reflector and the LNA. Reflector gain is virtually noise free, whereas there is a penalty when getting gain from an LNA or DC. That penalty is the additional noise contributed by the LNA's active components. The down converter is used more for compensating for cable losses than for bulk gain due to its high noise figure, in relation to the LNA's noise figure. The gain and noise contributions of the DC and receiver are minor in most designs when compared to that of the LNA and dish.

SYSTEM GAIN

The initial gain in the system, and by far the most important gain, is that which happens due to purely passive physical means. By this, I mean the *gain* that occurs from the satellite dish, or to use the proper term, the parabolic reflector.

Any waveform, be it sound, light, or microwaves can be reflected to some degree by solid surfaces. In the case of microwaves, which we're concerned with, a solid metallic surface shaped into a perfect parabola, will focus all of the energy striking the parabola into one sharply defined focal point. A screen-type dish will also reflect the microwave energy, if its holes are smaller than about 1/16 of the primary wavelength. For 4 GHz reception the screen openings must be 1/8th inch or smaller to properly reflect the microwaves.

This same idea is used to pick up sound at football games. A sound engineer, holding a small parabolic reflector, points it at the line of scrimmage and every sound the players make can be heard clearly, even though there may be 100,000 people screaming in the stands. This is possible only because parabolic reflectors not only have *gain*, but are also very directional.

The same principles hold true in satellite dishes, the parabolic reflector *sees* a small fraction of the sky, which hopefully will have a satellite in the exact center of its field of view, ignoring all the rest of the sky and the earth. This directional ability is a good thing, because like the fans at the football game, both the sky and the earth are noise sources. If too much noise is picked up, the desired signal will be engulfed by the noise and will not be received.

Unfortunately, behind the Clarke Belt is the galactic plane or Milky Way, and thus, because of the radiation from the stars and the general background noise from space, the added noise temperature is around 30 degrees Kelvin. As the dish is pointed away from the galactic plane, the sky noise drops to about 4 degrees. Both temperatures are small in comparison to the earth, which has a noise temperature of about 290 degrees Kelvin, and the typical LNA, which has a noise temperature of from 75 degrees to 120 degrees Kelvin.

Kelvin Scale

Degrees Kelvin is a positive thermometric, or temperature, scale with unit steps equal to the centigrade scale. In the Kelvin scale, 0 degrees is equal to −273 degrees centigrade. This is the theoretical point at which all molecular motion stops. A perfect LNA would have a noise temperature of 0 degrees, adding no noise to the signal it is amplifying. Unfortunately, making such an

LNA work, since there is no molecular motion at zero degrees, has not quite been figured out yet.

LNAs are rated by the noise they add to the signal. Thus the lower the noise temperature the less noise is added by the LNA, so a 90 degree LNA is better than a 120 degree LNA, as far as the added noise is concerned. As can be seen from the sky temperatures, lowering the LNA temperature from 120 to 90 degrees would be roughly equivalent to moving the dish from the galactic plane to cold sky.

Since the cold sky is only about 4 degrees Kelvin, it is fairly simple to determine the noise level of a dish, feed, and LNA combination if the proper test gear is available. By reading the cold sky temperature of the system with a noise meter, and then subtracting 4 degrees K (or about .05 dB), we have the noise level of the system or N. If a static satellite signal, like color bars, is observed and if the galactic noise level (30 degrees or .4 dB) is subtracted from the noise meter reading, then the C + N/N or Carrier and Noise to Noise ratio can be directly calculated. If this total is above 8 dB, then the system most likely will give good or excellent pictures. If it is below this level, then the dish may be too small, or the LNA may be too noisy.

As the dish is pointed closer to the horizon, more and more of the Earth will be seen by the dish and the Earth noise (290 degrees K) will be increased to such an extent that the satellite signals can no longer be demodulated. This is why a practical limit to viewing satellites is about 5 degrees above the horizon. The exact point will vary depending on many factors, the most important being dish size, feed design, and the surrounding terrain.

Calculating the G/T

G/T or Gain over Temperature, is a rating of antenna, feed, and LNA as a system. It uses the antenna gain, the antenna noise temperature and the LNA noise temperature. The G/T formula is given in Appendix B (equation #14). It is rated in dB/K or dB per degree Kelvin. It is used in the calculation for C/N, thus it must be figured before the C/N can be figured.

For example, the G/T for a 10 foot dish with a 45 degree look angle and a 120 degree LNA is 17.5 dB/K. A 10 foot dish has a 35 degree K noise figure at 45 degrees. If the LNA noise temperature is lowered or if the dish size is raised, then the G/T goes up. An equivalent increase can be obtained by using a 12 foot dish or by using a 70 degree LNA. If either is done, then the G/T would be raised to about 19 dB/K.

The C/N Ratio

As was seen, the gain and noise of both the LNA and the dish play an important part in providing sparklie free pictures. But in addition, the signal level from the satellite and the bandwidth of the receiver also play important parts.

To calculate their effect on the total system, you must consider the carrier to noise ratio or C/N. The C/N is a reference number that relates the signal level coming from the satellite (EIRP or Effective Isotropic Radiated Power), the G/T, some mathematical constants, and the receiver's i-f bandwidth. When all these factors are calculated using the C/N formula (listed in Appendix B, #13), what is derived is a number in dB which is called the C/N.

A typical C/N for adequate reception with most receivers is 9 dB. The system's C/N must be above the threshold point of the receiver, or the picture will be nonexistent or at best have a lot of sparklies. *Sparklies* is the name given to noise pulses in the video that result in the demodulation circuit losing the signal for a fraction of a second. They appear as black or white spots or streaks on the screen.

If a system has a C/N of 12 dB, and the receiver that is being used has a threshold point of 9 dB, then there is a 3 dB fade margin. This means that if the signal drops by 3 dB then the receiver will be at threshold. This can easily occur with slight dish mispointing, feed focus offsets, and polarization offsets.

The C/N will vary for each satellite transponder, as the signal level being broadcast varies. Thus to achieve perfect reception on all satellites on all transponders, the weakest transponder to be viewed should be used in the calculations. If the system's C/N is still above threshold, then a sparklie free picture should be obtained. Another factor is what the receiver's threshold is based on. A common threshold figure is 8 dB. But the signal that is used to measure this number can be static (color bars) or moving (standard video). Using standard video will cover up the true threshold, extending it by about 1 dB. Thus there are very few real 7 dB static threshold receivers. Most receivers actually fall between 8 and 10 dB.

Dish Gain

The most critical gain and hence the most critical losses in a system occur before the LNA. The dish and the feed system must work together to extract the maximum signal from the dish. A typical gain for a 10 foot dish is 39 dB. If we use the figure of -120 dBm as the signal level that actually strikes the dish, then the signal that enters the feedhorn, will be about -81 dBm (-120 dBm + 39 dB gain). That is, if the feedhorn is matched to the dish and if it is perfectly centered and focused.

Table 5-1 is a listing of dish size vs. several typical parameters, such as gain, noise, sidelobes, and first null location. These are averaged from specifications listed for several dishes of each size catagory. Most dishes will fall very close to these parameters.

Dish gain is basically dependent on three factors: the parabolic surface accuracy, the dish pointing accuracy, and the match of the feedhorn to the parabolic surface. If any of these are off, the gain will drop rapidly as the error increases.

If the parabolic surface is not accurate, or if it is badly rippled, or sags in one quadrant, then the focal point will no longer be a sharp point, but will be more like a fuzzy ball of microwave energy that will not be effectively transferred into the feedhorn. This will be most readily apparent by the amount of movement that the feedhorn can make before the signal strength is affected. Normally a movement of about one inch up, down or to the side will be reflected in a drastic drop in signal strength. If the feed can be moved 4 or 5 inches before the signal drops off and if the signal level seems low to begin with, then the surface is not even close to being a true parabolic.

Checking the Dish. Checking the surface accuracy can be done by simply running two taut strings across the dish at right angles to one another. Their starting and ending points must be equidistant around the rim of the dish. They should cross at the center of the dish and should be touching each other where they cross. If there is a gap between the strings then one or more quadrants is low or high in relation to the others. If the dish comes in sections try rotating the strings to the next section to see if more than one section is off.

If a Focal Finder is used, in addition to the strings, the centering of the feedhorn can also be checked. The *Focal Finder* is a device which is inserted into a Polarotor. It has an extendable rod which is used to check the feed centering and focal length. This rod should point exactly at the string's crossing point. If it doesn't, then the feedhorn is off and must be centered up.

If the surface itself varies by more than 1/8 to 1/4 inch from a true parabolic curve, then gain will be lost because of the defocusing effect on the signal. As a general rule, if your eye can see the variance, then there is too much surface inaccuracy.

Dish Mispointing. Probably the most common problem with satellite systems is mistracking of the satellite belt. Each dish has a specified main beam. It is between about one degree and about four degrees wide, depending on dish size. This main beam is where the dish

Table 5-1. Comparison of Dish Parameters to Dish Size.

Dish Size	1 Mid-Band Main Lobe Gain	2 Dish Noise	3 Mid-Band Beamwidth @-3 dB	3 Mid-Band Beamwidth @-12 dB	1st Sidelobe Location	1st Null Location	4 G/T	5 C/N
6 Ft.	35.1 dB	60° K	3°	6°	4.9°	3.9°	12.4	4.7
8 Ft.	37.6 dB	50° K	2.3°	4.5°	3.7°	2.9°	15.5	7.8
10 Ft.	39.5 dB	40° K	1.8°	3.6°	2.9°	2.3°	18.0	10.2
12 Ft.	41.1 dB	33° K	1.5°	3.0°	2.5°	1.9°	19.7	12.1
14 Ft.	42.3 dB	27° K	1.3°	2.6°	2.1°	1.7°	21.4	13.5

1 55% Efficient dish @ 3.950 GHz
2 Measured @ 30° elevation
3 Using 3.950 GHz
4 Using 85° LNA and 42° look angle
5 35 dBw EIRP and 26 MHz receiver i-f

gain is. If the satellite falls on the side of the main beam, then the gain that the dish gives to the satellite will be much less than the published dish specifications. Figure 5-1 is a typical performance chart for a satellite dish.

Figure 5-1 is a typical antenna pattern plot of a 9 foot dish. It is plotted as antenna gain vs. on and off-axis response. It shows the main lobe gain and beamwidth, the −3 dB points, the −12 dB points, the first null, and the sidelobes. These are the most important parameters used in determining dish performance. Unfortunately antenna plots such as this must be taken with the proverbial grain of salt. This is especially true for dishes that are built for home TVRO use, as the repeatability of the pattern from one dish to the next is dependent upon many factors that occur both before and after the dish has left the manufacturer.

If it is more than about two degrees off from where it should be pointing, then the satellite will not be on the main lobe and thus will have less gain. If the satellite falls into the 1st null, then it will usually be down 25 to 30 dB. If it is sitting in the 1st sidelobe, then it will be between 15 and 25 dB down.

Mispointing can be caused by several factors. The biggest problem is with actuators. All actuators are not suitable for all dishes. Torque, cable runs, and jack size all affect actuator action. Some positioning systems start and stop the motor at full power. This causes a jerking motion as the dish begins to move and as it stops, which can shake the dish and cause the screws and bolts to loosen. In screen dishes, it can actually warp the parabolic surface because the dish will not move together (i.e., the half with the actuator's arm attached starts to move first, then the other half catches up to it).

Not tightening all the bolts and screws properly during assembly will usually cause problems a week or two after installation, or during the first windstorm that hits the dish. Manufacturer's literature that reads, "This dish will withstand 100 mph winds" is usually referring to a test dish that has been properly assembled with the proper torque applied to all the bolts and screws and that is probably hit from a certain direction for a specified amount of time. Actual field installation is a different matter entirely, a loose bolt may mean the difference between survival and destruction, even in moderate winds below 100 mph.

LNA and DC Gain

The LNA receives the signal from the feedhorn. It typically boosts the gain 50 dB. Thus the signal level that leaves the LNA is about −31 dBm (−81 dBM from the dish + 50 dB gain). If the down converter is placed behind

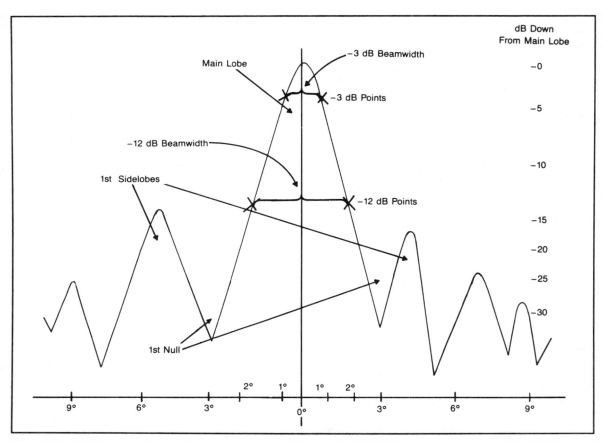

Fig. 5-1. An antenna on and off axis chart. It shows the main lobe's beamwidth and gain, the dish's sidelobe position and strength and null position. The important criteria for 2 degree spacing is beamwidth and sidelobe location.

the dish, there will be a maximum of about 20 feet of RG-214 between it and the DC. RG-214 shows a typical loss of about 3 dB per 10 feet. Each type-N coax connector has about .5 dB loss. Thus the 20 feet of cable and the two connectors show a 7 dB loss. This results in a level of −38 dBm that enters the down converter.

A typical down converter will show a conversion gain of about 20 dB and an output frequency of 70 MHz. Using a 150 foot run of RG-59 will result in about 6 dB of loss between the down converter and the receiver. The 6 dB cable loss coupled with the 20 dB conversion gain and −38 dBm input signal means that the signal entering the receiver is about −24 dBm, which is well above our −45 dBm minimum.

SATELLITE SPACING

There has been a lot of controversy surrounding the

FCC's decision to move the satellites closer together, to a mixture of 2- and 3-degree spacings. This plan is necessary to increase the number of satellite parking spots to accommodate all the new applicants. This ruling has led to many different claims regarding dish performance vs. dish size when the satellites are moved to two degree spacing.

The most important dish specifications under 2- and even 3-degree spacing is the width of the main beam, the first null position, and the sidelobe performance of the dish. Table 5-1 listed the typical beamwidth, sidelobe, and gain to be expected in various size dishes. As can be seen, dishes 10 feet and up should have no problem with 2-degree spacing. Smaller dishes may have problems with rejecting signals from the other satellite because their mainbeam widens up to include the other satellite within it.

Figure 5-2A shows how a 6-foot dish would react to 2-degree spacing. In a 6 footer, the beamwidth is about 3 degrees at the 3 dB point. This would put the 2-degree satellite at about −5 dB from the desired satellite. This would cause interference and an unwatchable picture. But in Fig. 5-2B, the dish is pointed so that the desired satellite is actually off-boresight. This would decrease the gain slightly (about 1 dB), but it would put the interfering satellite about 12 to 15 dB down, making its interference liveable. Thus the smaller dish could survive in 2 degree spacing with a slight decrease in gain, by positioning the interfering satellite farther down the main lobe, or within the 1st null. Of course, if the system was close to threshold the 1 to 2 dB drop could send it over the edge. The result would be worse than what the interfering carrier would cause.

Another important fact is the location of the antenna in relation to the closely spaced satellites. At this time, the FCC has relegated 2-degree satellites to the ends of the useable U.S. Clarke Belt, namely, east of 80 degrees W and west of 130 degrees W. This means that a dish in the western U.S., looking towards a pair of 2-degree spaced eastern satellites, would actually see them as being spaced 2 degrees apart, whereas that same dish looking south or slightly southwest would see a pair of 2 degree satellites as if they were 2.5 degrees apart. This, of course, would work in reverse for a dish on the east coast.

A well-designed 8 or 9 foot prime-focus dish, designed with an eye towards 2-degree spacing, should have no more trouble coping with 2-degree spacing than any well-designed 10 or 12 foot dish. The key word is well-designed. I've seen a 10 foot dish that had interference (cross-polarization problems) between F3-R and Galaxy I, which are 3 degrees apart.

As long as the signals hitting the antenna from off-boresight (i.e., coming into the skirts of the mainbeam or into the sidelobes) are at least 12 dB down from the desired signal, then there should be no interference noted while watching standard video programming. For private cable, teletex and other digital information transmissions, the sidelobes should be 18 dB down. Almost every dish ten feet or larger in diameter should easily meet the 12 dB down figure. The 18 dB down specification is harder to meet and unless the dish is carefully designed for low sidelobe performance, it will be hard to meet even using a prime focus feed.

Another criteria beyond the dish, is the type of demodulator used in the receiver. The most insensitive receiver to outside interference is the discrete Phase Lock Loop (PLL). The ratio detector or discriminator, delay line, and the divide-by-two (or four) PLL circuits progressively become more sensitive to interference.

Of course, all this could be academic if the FCC does its homework and places the 12 GHz and 4 GHz satellites next to each other in 2-degree spacings. Even if two 4 GHz satellites are parked at 2 degrees, if they use opposite polarization formats, then cross-polarization would keep most interference from occurring.

Currently there are two satellites that are parked 2.5 degrees from one another, and yet are using the same polarity. They are Westar 5 at 122.5 degrees W and Spacenet 1 at 120 degrees W. They are a nice test pair for viewing 2-degree spacing compatibility on smaller dishes, since most of the time TR 11 on W5 has static color bars, while TR 11 on S1 has standard video. Thus two-degree compatibility can be easily checked while viewing the color bars.

DISH MATERIALS

There are four main types of dishes being made today. Some are almost regional in their appeal, while others are seen all across the country. Each has their own plusses and minuses. Choosing which type of dish to use is more an individual preference than a true performance decision, as any variety of dish can be made accurately. A reputable company, a quick and easy installation, and good weatherability are the primary dish buying factors which should be considered by the TVRO dealer. On the other hand, the customer will be looking for appearance, price, and size more than anything else.

Screen Mesh

The most popular type of dish is made of screen mesh. As a whole, mesh dishes have evolved from assembly and performance nightmares, into some of the most efficient and easy to assemble products on the market. There are even a few that are UPS shippable.

Most of the newer mesh designs feature precut panels that slide into tracks in the ribs. Assembly time is often less than two hours, a far cry from the four to six hours that earlier J-hook type of designs took to assemble. Most don't require any power tools, which was another drawback to the earlier designs.

Besides using screening material, mesh dishes are also made out of perforated aluminum. This type of dish has been popularized by Winegard in their SC-1018 satellite TV antenna, shown in Fig. 5-3. It has the see-through advantages of screen, but the structural stability

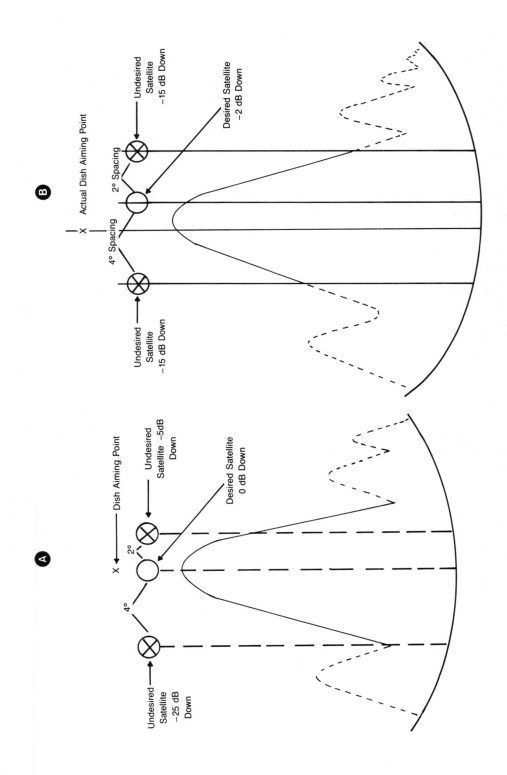

Fig. 5-2. When two satellites are two degrees apart and a small dish is aimed at one of them the second may interfere. Notice that at 4 degrees there would be no problem. A way around the 2-degree spacing if there is not another satellite 2 degrees in the other direction) is shown in B. By intentionally mispointing the dish, you can achieve 13 dB separation, which is about the minimum required to yield good rejection. Unfortunately, this does knock the desired satellite signal down 1 or 2 dB, which may be just as bad as having the interfering carrier, since it could send the signal level below threshold.

Fig. 5-3. An example of a 'mesh dish' is the Winegard perforated sectional dish. Winegard has been known for many years for their quality TV antennas, and they've carried on with that tradition with their perforated dishes. Courtesy of the Winegard Co.

of a solid aluminum dish. It also has the advantage that it holds its shape well and can be shaped in both axis. Most panels are curved from the center out, but most do not have a side to side curve. This limits the efficiency as the dish is not a true parabola except in the center to edge direction.

One fallacy about mesh dishes is that they let the wind pass right through. To some extent this is true in light breezes, but once the wind gets to gusts of 40 to 50 miles per hour and above, the wind loading is not much different than that of solid dishes. The mount can be easily destroyed if it is too light for the wind loading. If you install a mesh dish in an area that experiences high winds, it is best to examine the mount with an eye towards overkill, which is not often found in mesh dishes.

Unfortunately, most mesh dishes will not perform well at 12 GHz. If you are selling a system that may be upgraded to both 4 and 12 GHz in the future, a mesh dish is not the way to go, as the surface accuracy will not be tight enough for 12 GHz nor will the hole size be small enough. Remember the higher the frequency the smaller the wavelength. This is one reason why those 2-foot, stamped 12 GHz dishes work even with medium power satellites. They have just about as much gain at 12 GHz as an 8 foot dish does at 4 GHz.

Another often overlooked item is the weatherability of the dish hardware and the structural strength of the entire assembly. Screen mesh dishes can be made extremely lightweight, which if coupled with the wrong motor drive could mean the dish will get shook to death. The system will lose gain due to dish warping, feed skewing, or overall sloppyness of the mount and reflector structure.

Warping will also be more noticeable with screen dishes due to snow and ice build-up. Excessive snow and ice build-up will cause the lower half of most screen dishes to sag, cutting gain by as much as 2 or 3 dB. Plus as water freezes it expands, causing panels to loosen up and ripple.

Fiberglass

There are three types of reflector surfaces applied to fiberglass. One is a fine mesh screen much like the standard mesh dishes. The second is a flame sprayed metallic coating which looks similar to a spun dish until the top layer of fiberglass is applied. The third is a fine aluminum foil that is applied to the fiberglass with a thin protective layer over it.

Fiberglass dishes are most popular in the south and southeast mainly because most fiberglass dishes came out of the many boat manufacturers that are located in these areas. The early dishes were extremely heavy as hand layering of the fiberglass was mostly used. Today injection molding and resin transfer molding are often used to increase production and to ensure quality and repeatability; both are essential for easy assembly and good performance in the field.

Fiberglass dishes are usually easier to assemble than mesh dishes, but they require an extra person or two to lift the dish onto the mount, as they are usually much heavier than mesh on a size for size basis. One drawback to fiberglass is that the individual layers of fiberglass may separate and cause warping of the reflective surface. Often this won't show up for several months after installation.

Another problem with fiberglass is sagging. Because the fiberglass is usually the heaviest type of dish, it is also the most prone to changing its shape from a parabola to more of a circular shape due to gravity. In some cases, nonmetallic supporting wires that run across the dish can be used to parabolize a sagging dish. In some cases, just tightening the bolts and manually adjusting the panels will do the trick. The fiberglass dish is one type where the strings stretched across the dish can really show up sagging or warping in a particular panel.

Spun Steel or Aluminum

The spun as well as the stamped dishes seem to be concentrated mainly around the Great Lakes/Midwest region. This is due to the many sheet metal fabricators catering to the auto industry in the area. Of course, as dish size comes down and ease of shipping improves, both the fiberglass and the stamped or spun dishes will see increased acceptance outside of geographic boundaries.

The spun dish is one of the most accurate dishes if the spinning process is carefully controlled. Short term repeatability is good with most reflector surface tolerances being held better than .05 inches. Wear and tear on the mold or improper rolling pressure can cause tolerances to slip to .1 inches or more if the manufacturer is sloppy. Shipping can also cause warping of the dish if it isn't packed properly. Birdview ships all their systems using their own trucks to prevent this from happening. They use an 8 1/2 foot spun dish shown in Fig. 5-4.

Hydroformed or Stamped

A hydroformed dish is formed from a flat piece of metal that is pressed into a parabolic shape using a large hydraulic press. Repeatability is somewhat improved over spinning, and the long term accuracy of the press is very good. The most well-known stamped dish is from Kaultronics.

Fig. 5-4. An example of a spun dish. Spun dishes are formed by applying pressure on a sheet of metal that is against
a wooden or teflon parabolic form. The metal is turned against a wheel which presses the metal against the form, creating
the parabolic shape. Courtesy of Birdview Satellite Communications.

Fig. 5-5. An example of a stamped dish. The Janeil Junior is made by Alcoa. It is a six foot dish whose panels are stamped by a hydraulic press. Repeatability from panel to panel is very high. Courtesy of Janeil.

Another type of stamped dish is sectional. Here individual panels are stamped out of steel or aluminum. The Janeil Junior, made by Alcoa, is one type of stamped panel dish. It is shown in Fig. 5-5. The accuracy is high due to the stamping process, which will allow dual axis shaping so that each section is parabolic in both directions.

Polarization
and Polarity Control

THERE ARE FOUR POLARIZATIONS THAT ARE USED to broadcast satellite signals. The familiar linear horizontal and linear vertical are broadcast by domestic satellites and are what most home TVRO systems are set up for. Horizontal polarization means the transmitted waveform's electric field is perpendicular, or 90 degrees, from the satellite's north/south axis. A vertically polarized waveform has its electric field aligned with the satellite's north/south axis.

Notice that the signals are aligned with the satellite's axis, not the Earth's axis. This is why F3-R has such a *skew* to it, the satellite s tilted about 20 degrees in orbit, and so are the horizontal and vertical polarizations. The reason for the tilt is to supply more EIRPs to Alaska.

The other two polarizations are right hand circular (RHC) and left hand circular (LHC), which are used by the International satellites (Intelsats). At present, all Intelsat video is broadcast using RHC polarization. Circular polarization is like a corkscrew, as the signal travels through space, the electric field rotates clockwise, for RHC and ccw for LHC, at the rate of 90 degrees for every 1/4 wavelength of forward movement.

Because of this rotation, a standard linear Polarization Device (PD), like the Polarotor I, will not be able to tell the difference between a RHC and a LHC signal. It will receive something, but it will lose 3 dB or half the signal in the process.

To overcome this loss the feedhorn must be modified by putting a bi-refringent element in the waveguide. This element changes the phasing of the RHC or LHC waveform so that it appears as a linear signal to the probe. A commercial version of a bi-refringent element is the Dielectric Plate from Chaparral. It is constructed out of G-10 fiberglass, which is cut so that the dielectric constant is optimized for the 3.7 to 4.2 GHz wavelengths. Unfortunately it must be removed to receive linear polarizations again, or else they will be cut down by 3 dB.

POLARIZATION DEVICES

Before 1978, there was no reason to change polarization because all video was broadcast horizontally polarized, even on the dual polarity satellites. This changed when both polarities of Satcom 1 began to be used to broadcast video.

Antenna Rotors

The first semiautomatic polarization device was an antenna rotor. To be specific, it was a Channel Master antenna rotor, and it was installed in early 1978 on Bob

Cooper's 20 foot CATJ Lab dish. Mechanical antenna rotors, using the standard indoor controller, marked North, South, East, West, were quickly adapted by other early TVRO experimenters, and became a standard part of the first home TVRO systems.

Of course, one had to adapt the LNA and feed to use a rotor. This was usually done by jerry-rigging the motor to the feed support and then fastening a pipe onto the LNA and sticking it into the rotor. It worked after a fashion. Polarization change only took 30 seconds to one minute to accomplish and could be done in your easy chair. This was far better than 15 minutes to half hour that it would take if the feed was rotated by hand, especially if it was raining or snowing outside!

There are two drawbacks to this type of switching method, beyond the time involved to change polarities. The feedhorn was hard to keep perfectly aligned as it was changed from horizontal to vertical, and the cable on the LNA had to also move slightly every time the feed was rotated. This movement caused the cable to flex slightly which meant that bad connections and water entrance into the N connector would occur after long usage of the rotor to move the entire feed around.

But the antenna rotor was the only way to remotely change polarity until 1981. In 1981, at the SPTS show in Omaha, two quantum leaps in polarization rotation schemes were unveiled. One was designed by Bob Luly. It used a split piece of ferrite and a ±15 Vdc supply to switch polarities. The other, introduced by Gene Augustin, used a small model-airplane control motor to rotate the pick-up probe. Two versions of the probe rotating design would be popularized by Chaparral Communications in their Polarotor I and II, introduced in late 1981. The ferrite switching scheme has just recently begun to become popular, mainly due to improved ferrite materials and switching schemes which have lessened the loss inherent in the ferrite material.

Polarotor I

The Polarotor I is by far the most popular feed designed to date. Its popularity is due in part because it was based on the Super Feed (also by Chaparral) and because it was the first automatically adjustable probe system to get into production. It is controlled by a relatively simple circuit that is available in a hand-held controller, or built into many popular receivers. The receiver manufacturers found that built-in Polarotor I control circuits were essential to sales beginning in 1982. Today virtually every receiver has built-in polarization control circuits. The Polarotor I and II are shown in Fig. 6-1.

The Polarotor I's operating principle is pretty simple. A small servo motor drives a probe which actually changes the polarization of the signals that are passed on to the LNA. This probe can freely travel back and forth through 140 degrees. The motor's position is controlled by a driver circuit, which is controlled by a PWM detector circuit, which is controlled by a PWM generator in the receiver. What's PWM? It's short for Pulse Width Modulation. This is a method of sending information by varying the length of a stream of pulses. In this case, the probe's location must be controlled.

To control the PR I and compatible units, the motor control circuit is fed a stream of pulses that are between .8 ms (milliseconds) and 2.2 ms long. The PWM detector translates the pulse length into a driving signal for the motor. It uses a potentiometer to get feedback from the motor, so that it can tell where the motor and hence the probe are at. To position the probe in the center of its range, a pulse length of 1.5 ms must be detected. If the pulses are longer or shorter, then the motor is moved cw or ccw until the pulse length and measured motor position match.

Unfortunately, the motor can only move so far before it comes up against a mechanical stop. This is translated, in pulse widths, to a pulse that is .7 ms long for one stop and 2.3 ms for the opposite stop. If pulse lengths shorter than this or longer than this are sent to the PR I, then the motor control circuit will try to move the motor beyond its stops, which will cause the motor to overheat. This can burn out the motor.

In the Polarotor I and other compatible PDs, the control pulses are standard TTL logic levels (+5 Vdc and zero volts) with a repetition rate of 17 to 21 milliseconds, and a pulse length of .8 to 2.2 milliseconds. Figure 6-2 is an oscilloscope picture of the minimum and maximum pulse lengths that are sent to the PD. To measure the pulse length set the oscope's vertical attenuator to 5 volts/div and set the sweep rate to .5 ms/div. The upper trace is set to .8 ms, while the lower trace is set to 2.3 ms. The signal is found on the white wire that goes to the Polarotor I.

At the PD, these pulses are detected and compared with the probe position. If they aren't equal, then the circuit turns on the motor which rotates the probe clockwise or counterclockwise until the pot feedback from the motor indicates its position is equal to the position determined by the incoming pulse length.

Figure 6-3 is the polarization control circuit from a

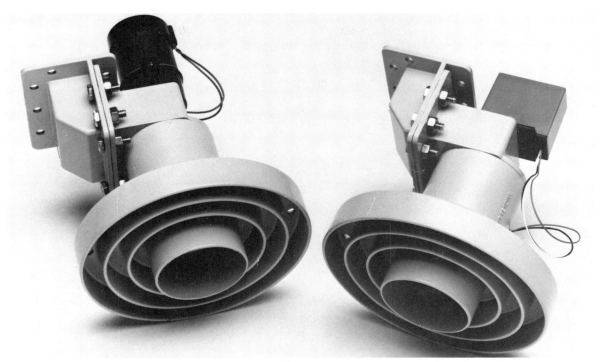

Fig. 6-1. The Chaparral Polarotor I (on the right) and Polarotor II (on the left) are two of the most popular feedhorns in the TVRO world. Courtesy of Chaparral Communications.

Dexcel DXR 1200. It is a typical *smart* receiver control circuit. As the channel is changed from an odd number to an even number or vice versa, the pulse length automatically changes to maintain the correct polarization.

The heart of just about every polarization control circuit is a NE555 timer chip (IC2) which is configured as an astable oscillator with two adjustable pulse length outputs. The pulse length is controlled by the selected channel (whether odd or even), the front panel polarization or *skew* control setting, the rear panel format (Satcom-Westar) switch, and the two rear panel installation set-up adjustments. The 4 switches are parts of a CMOS chip, a CD4066, which consists of 4 SPST (single pole single throw) solid-state switches.

VR1 and VR2 set the maximum and minimum pulse length to 2.2 ms and .8 ms respectively. Pulse widths above or below these limits can drive the motor against its physical limit points. Because the feedback from the motor will never equal the desired pulse width, the motor is kept running. This causes the motor to heat up, which may burn up the motor and/or melt the surrounding plastic housing.

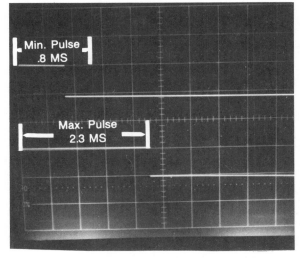

Fig. 6-2. Oscilloscope photo of the minimum and maximum pulse lengths used for controlling a Polarotor I. The upper trace is the minimum length and the bottom trace is the maximum pulse length. A stream of these pulses are sent to the Polarotor to control its position. The scope is set for 5 volts/div vert. and .5 ms/div horiz.

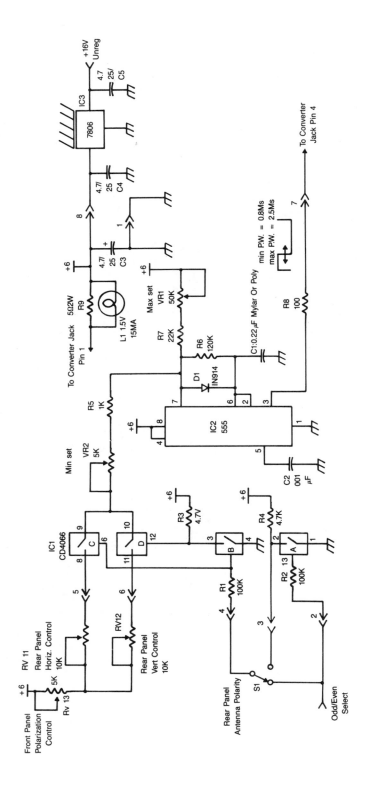

Fig. 6-3. Typical polarization control circuit used to create the pulses shown in Fig. 6-2. The output pulse width is determined by IC1's switches, RV11 or RV12's position, RV 13's position, and S1's position. It is from a DXR 1200. Courtesy of Gould/Dexcel.

Table 6-1. Polarization Skew Across the Clarke Belt.

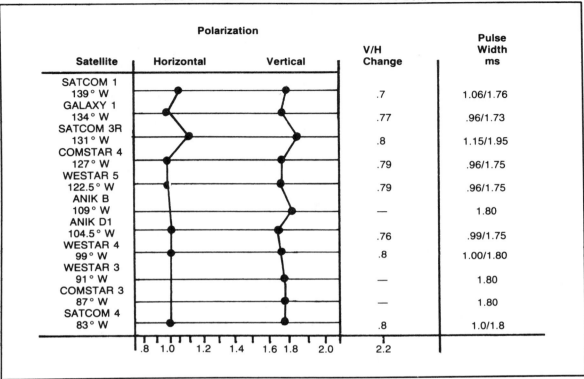

Satellite	Polarization		V/H Change	Pulse Width ms
	Horizontal	Vertical		
SATCOM 1 139° W			.7	1.06/1.76
GALAXY 1 134° W			.77	.96/1.73
SATCOM 3R 131° W			.8	1.15/1.95
COMSTAR 4 127° W			.79	.96/1.75
WESTAR 5 122.5° W			.79	.96/1.75
ANIK B 109° W			—	1.80
ANIK D1 104.5° W			.76	.99/1.75
WESTAR 4 99° W			.8	1.00/1.80
WESTAR 3 91° W			—	1.80
COMSTAR 3 87° W			—	1.80
SATCOM 4 83° W			.8	1.0/1.8

There are two pots (RV11 and RV12) that set up the actual horizontal and vertical probe positions in the Polarotor. These set-up pots are located on the back panel of the receiver and are set during installation, peaking the polarization on the satellite closest to due-south with the polarization fine tune control centered. RV13 is the skew, or the fine polarity adjust. The end user uses the skew control to adjust for satellite polarity differences. Table 6-1 is a chart of the polarization differences or skew between satellites as seen from the West Coast. It is plotted in pulse length timing.

Polarotor I Compatibles

There are many other feeds that have been designed to be compatible with the Polarotor I's control signals. Units by Boman, M/A Com, and Seavey are but a few. Most of them have almost identical specs as far as controlling pulse width and timing, power supply voltage, and current draw are concerned.

One of the main differences between designs is in the probe. Chaparral has their probe design patented, and it appears to have the most repeatable performance from unit to unit. One unique feature of M/A Com's Omni Rotor is the adjustable scalar ring, which is used to match the f/D ratio of the dish. Figure 6-4 is a drawing of the adjustable scalar feed from M/A Com. The distance A determines the beamwidth of the feed and should be flush with a .44 f/D dish, .38 inches with a .4 dish, .62 with a .35 dish and .75 with a .30 f/D dish.

Table 6-2 is a comparison chart between the Chaparral Polarotor I and the M/A Com Omni-rotor (also sold as the Uniden Unirotor). It compares the electrical ratings between the two.

Other Polarization Devices

There are three other PDs that are popular. These are the dc motor (Polarotor II), the ferrite (The Twister), and the pin diode switch (M/A Com Polarizer). A fairly new model from Chaparral is the Polarotor III. It uses the same control signals as the Polarotor I, but instead

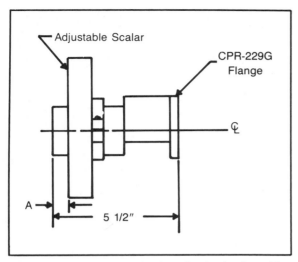

Fig. 6-4. Drawing of an adjustable scalar ring feedhorn. The ring is loosened and moved up or down the throat to match the f/D ratio of the dish. As dimension A is changed, the effective aperture of the feedhorn is changed.

of a movable probe, it uses a thin metal serpentine plate that is twisted to induce a skew and change the polarity of the signals. Chaparral claims a noise decrease of 1 degree vs. the Polarotor I, and a slight increase in efficiency.

Ferrite PDs. Even though it was the first type of polarization switching device developed, the ferrite idea has taken some time to come to fruition. There was a very wide range of insertion loss and performance variances in the first models produced. This is mainly due to ferrite's temperature and frequency sensitivity.

The original type of ferrite used changed its skew when the frequency or temperature was varied, but new ferrite materials and manufacturing techniques have been developed since Bob Luly first showed off his original design in 1981. Most of the problems have been overcome or at least minimized in the many models that are currently on the market. Chaparral, Hytek, Seavey, and M/A Com all manufacture ferrite models, most of which can be controlled by Polarotor I control signals if an interface like that shown in Fig. 6-5 is used.

The Twister from Chaparral uses the interface to change the pulse width into a variable +15 to −15 Vdc. To hook it into a PR I system requires that the black and white wires that normally would go to the PR I from the receiver, be instead spliced to the black and white wires that extend from the Twister interface box. Two wires are run to the Twister from the two terminal screws on the interface box. These are marked 1 and 2. They are hooked up to the same numbered terminals on the Twister. The red wire which carries +5 Vdc to the PR I is not used.

DC Motor Controllers. Figure 6-6 is a blow-up of the various subassemblies used in the Polarotor I and II. As can be seen, the only difference between the I and

Table 6-2. Polarotor I and Omni Rotor Comparison Table.

	Polarotor I	Omni Rotor
Stall current	160 mA	170 mA
Standby current	25 mA	12.5 mA
Min. pulse	.8 ms	.9 ms
Max. pulse	2.2 ms	4.0 ms
Pulse period	17 to 21 ms	16 to 22 ms
f/D ratio	.33 to .45*	.28 to .44
Probe rotation	160 deg	180 deg
VSWR	1.25:1	1.25:1
Insertion loss	.08 dB	.1 dB

* to .28 with Gold Ring

Fig. 6-5. The Twister is the ferrite feedhorn from Chaparral. It can be controlled by PR I signals when used with the interface box on the left, or it can be controlled directly by using the hand controller on the right. Courtesy of Chaparral Communications.

II, is the motor. The II is controlled by a momentary double-pole, double-throw switch which reverses the polarity of the voltage to the motor to either drive it clockwise or ccw. Its current draw is minimal, since the motor is only on during polarization changes. It can even be supplied from a nine-volt battery. Be sure to use an alkaline or rechargeable type if you change polarization a lot.

MATCHING FEED TO DISH

Most dishes have a f/D ratio between .27 and .45. Those with .27 to .32 f/D ratios are termed *deep dishes*. Those from .33 to .45 are termed *shallow dishes*. Using a 10 foot dish as an example, the focal length goes from 32 inches on a .27 f/D dish, to 54 inches on a .45 f/D dish. The curve of the dish gets shallower and the focal length gets longer as the f/D ratio goes up.

Ideally the dish will have a sharply focused *hot spot* at the focal length, regardless of the f/D ratio. This hot spot is where the reflected signal coming off the dish will be concentrated. The feedhorn is used to direct this energy into the rectangular waveguide on the LNA. If the feed is not placed in the hot spot (placed too far from the dish, too close to the dish, or off to one side) then the G/T of the dish will suffer.

Illumination

To properly collect the reflected energy, the feed must see all of the dish. If the feed is not matched to the f/D ratio of the dish, then it will either see beyond the edge of the dish and allow earth noise to enter the feed and lower the G/T; or the feed will not see all of the dish and the gain will again drop, which also lowers the G/T. When the f/D ratio and the feed are properly matched and focused, then the dish is fully illuminated and the G/T will be at maximum.

A dish with an f/D of .30 needs a feed that has a beamwidth of 159 degrees, to properly see all the dish, while a dish with an f/D of .44 will need a beamwidth of only 121 degrees to see the dish. If the wide beamwidth feed designed for a deep dish is used on a shallow dish, overillumination will result. Going the other way, using a shallow feed on a deep dish will result in under illumination.

Figure 6-7 is a plot of the pattern of a Polarotor I. As the f/D changes the edge illumination level changes. As can be seen, the feed works best on shallower dishes. If the Gold Ring (which was shown in Fig. 6-6) is added to the feedhorn, then edge illumination on a deep dish, f/D of .30 and below, is increased. Effectively it's like moving up the chart to a higher f/D reading, making the

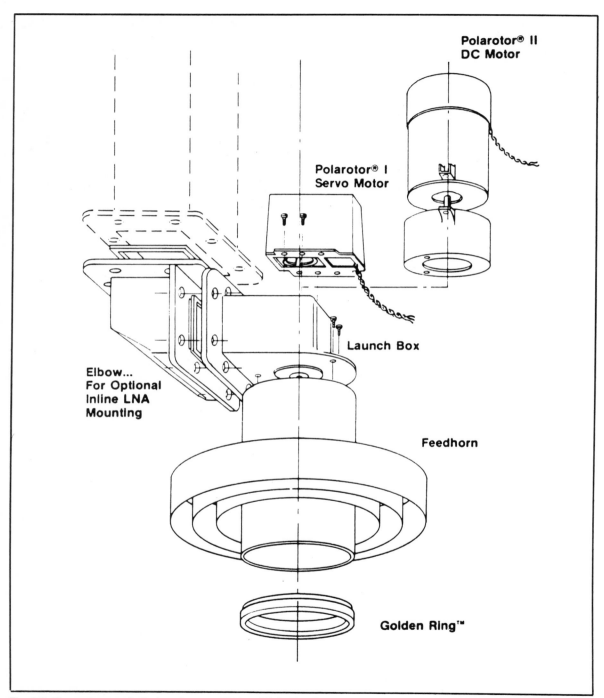

Fig. 6-6. The various parts that make up a PR I or PR II. The Golden Ring is used to adapt the nonadjustable feedhorn to deep dishes. The feedhorn is similar to the Super Feed, minus the WR-229 flange. The probe is located in the Launch Box and is driven by either the PRI I servo motor or the PR II dc motor.

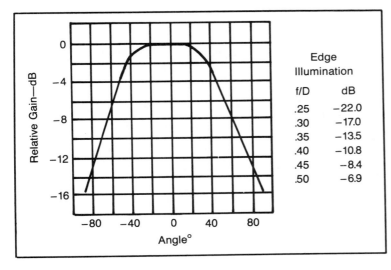

Edge Illumination	
f/D	dB
.25	−22.0
.30	−17.0
.35	−13.5
.40	−10.8
.45	−8.4
.50	−6.9

Fig. 6-7. The gain and edge illumination chart from a Polarotor I. As can be seen, it is optimized for shallower dishes. The Pattern is the PR I's response at 3.9 GHz.

feedhorn about .9 dB more efficient.

Another way of looking at illumination is to imagine a flashlight and an adjustable lens positioned where the feed would be. In a deep dish, the feed is closer to the dish and so a wide-angle lens is needed to spread the light out all the way to the edges of the dish. In a shallow dish, we need a narrow lens, since the feed is farther from the dish and the dish is flatter. If the light is spread beyond the dish then the feed is over-illuminating the dish. If the dish edge is dark, then the feed is said to be under-illuminating the dish. In either case, a lowering of the G/T will be the result.

Some feedhorns like the Omnirotor, Unirotor, and Omni Polarizer have an adjustable scalar feed, which can be used to match the feed to any .28 to .40 f/D dish. Older Chaparral feeds, designed for f/D ratios of .33 to .45, must use a throat extension called a Gold Ring to optimize the feed for use with f/D ratios of .32 and below. The latest ones are available with an adjustable scalar ring.

Measuring the f/D

Mismatching the f/D between the dish and feed can cause losses of 2 dB or more, thus checking the actual f/D of the dish is an important part of each installation. The first step is to measure the actual dish diameter. Do this from two directions at right angles to each other. If there is a large discrepancy, then the dish is not parabolic and may be sagging in one quadrant, or it may be cupped, as if the dish was being squeezed from opposite sides. Figure 6-8 shows the various measurements that are used to correctly set the feedhorn for optimum focus.

If the diameter measurements are exactly equal or very close, then by squaring the diameter and dividing by 16 times the depth, the focal length can be calculated. Then by dividing the focal length by the diameter the f/D ratio can be determined. Of course it would be easier to read the manufacturers specifications, but what if they changed the dish and didn't change the instructions or specs? The two equations are written out in Fig. 6-9A.

On a good dish, an inch or two change in the focal length will make a big difference in performance. If the f/D is calculated beforehand, then you know the exact focal length without the trial and error method of watching the signal strength meer and moving the feed in and out. Plus it looks much more professional to the customer if you calculate the focal length than to do it by trial and error. And once you've done the calculations, you'll realize how easy they are.

For instance, let's figure out the focal length and f/D ratio of a ten foot dish. The calculations are shown in Fig. 6-9B. Measuring the dish I find it's really 9 feet, 10 inches, or 118 inches in diameter. Measuring from the center of the dish to a straightedge (like a taut string) held across the dish gives us the depth. In this case 19 inches. So squaring 118 gives 13,924. While 16 times 19 equals 304. Now taking the diameter squared and dividing by 16 times the depth gives me a focal point of 45.8″. Now dividing this number by the diameter, I get the f/D, which is .39. Thus, this is a shallow dish.

In a deep dish, the depth is much greater. Therefore, the same size dish (118″ with a depth of 26″) will have a focal length of only 33.47″, and the f/D ratio will be .28. The calculations are listed in Fig. 6-9C. On this dish,

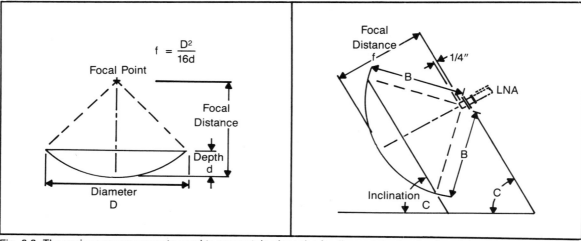

Fig. 6-8. The various measurements used to accurately place the feedhorn at the focal point of the dish. These include the focal distance, the feedhorn inclination and the feedhorn's centering.

A

$$\frac{D^2}{16d} = F$$

Where = Dish Diameter
  d = Depth Of The Dish
  F = Focal Distance

$$\frac{f}{D} =$$ Ratio Of Focal Distance To Dish Diameter. Also Called F To D Ratio.

C

Figuring F & f/D For Deep 10′ Dish Of Same Diameter.

D = 118″ $\frac{f}{D} = \frac{33.5''}{118''} = .28$
d = 26″

$$\frac{D^2}{16d} = \frac{13,924''}{416''} = 33.47''$$

Focal Point Is 33.5″ From Center Of Dish.

B

Figuring F and F/D For A Shallow 10′ Dish.

D = 9′10″ = 118″ $\frac{D^2}{16d} = F$
d = 19″
D² = 13,924″
16d = 304″

$$\frac{13,924''}{304''} = 45.8'' = \text{Focal Distance}$$

$$\frac{F}{D} = .39$$

Fig. 6-9. The two calculations that are necessary to accurately compute the focal distance and f/D ratio for any dish are given in A. B shows the calculations for a 9′10″ dish with a depth of 19″. This would be considered a shallow dish since the f/D is .39. C shows calculations for a 9′10″ dish with a depth of 26″. It is classified as a deep dish since its f/D ratio is .28. If a standard PR I is used with this dish, then a Golden Ring would be used.

if you plan to use a Polarotor, you better have a Gold Ring on hand in order to properly illuminate the dish.

Centering the Feed

The focal distance is measured from the center of the dish to 1/4 inch inside the circular waveguide. To check proper feed centering, measure from the edge of the dish to the edge of the circular waveguide on the feedhorn from three or four equidistant points around the dish. All measurements should be identical, if not then the feed is not centered and will need to be biased toward the higher measurement. If the feed is supported by a J-hook that has guy wires, adjusting their lengths will center up the feed. If the J-hook does not have guys, they should be installed. Guy wires cut down on feed oscillation as the dish is rotated or as the wind blows. It also assures that the feed is held in the center of the dish at the focal point.

Another, often overlooked fine tuning measurement is to determine whether the feed is square with the dish. If the feed is not pointed at the exact center of the dish, then gain will be lost and tracking of the dish will be false.

One way to check this is to take two strings and stretch them across the dish, as in the method used to determine whether the dish is a true parabolic. They should be touching where they cross, which should be right below the center of the feedhorn. Mark the strings 4 inches from where they cross and then measure between the strings and the feedhorn's scalar plate. All four measurements should be exactly the same. Be careful not to apply any pressure on the strings or dish or else the measurements will be in error.

Another method is to use an inclinometer to measure the scalar ring's inclination when the dish is set due south. Subtract the declination angle setting from this reading. Measure the polar axis angle. The two should equal each other, if they don't then the feed is off and will have to be adjusted by shortening or lengthening one of the legs if it is a tripod mount, or by shimming or tweaking it if it is a J-hook type of mount. There are also several feed centering devices that can be used to assist in centering and focal length measurements.

Attaching the PD to the LNA

There are two factors which are often overlooked when installing systems. These are that water absorbs microwaves (which is why a microwave oven cooks food), and that microwaves only travel along the surface of waveguides. If these two facts are kept in mind, then it will be readily apparent why a gasket must be used between the PD and the LNA and why there must be metal to metal contact between the two surfaces.

The gasket is to keep water from getting into the waveguide, while the carefully machined surfaces of the LNA and PD waveguide openings are to ensure the most efficient signal transfer between the two devices. Never install an LNA without a gasket. Never use RTV or any other sealant on the PD or LNA. And always use at least six bolts to hold the LNA in place.

If there is a groove on the PD's waveguide opening, then an oversize gasket will be required. If an oversize gasket is not available, then one for the PD's groove and one for the LNA's groove will be required. In such a case, carefully attach the two together to ensure that the gaskets line up properly.

Once the LNA and PD are together, they can be mounted at the focal point. Their position in relation to the incoming signals is important if the PD is one in which the probe has limited travel. Table 6-2 listed the probe rotations for the Omni Rotor and the Polarotor I. Notice that the Polarotor I rotates 140 degrees, while the Omni Rotor has 180 degree of rotation. Thus it is possible to set the feed so that only one polarization can be received properly, which is not very desirable.

Figure 6-10A shows what happens when the Polarotor is not rotated correctly in relation to the incoming signals. The vertical dashed line is one polarization, which will be received perfect. Notice that from this position the probe can only travel ± 70 degrees, thus the dashed line or horizontal polarity will never be properly tuned in. Figure 6-10B shows what the correct setting is. The feed has been rotated 45 degrees in either direction from Fig. 6-10A. Now both polarizations can be received properly, with about 25 degrees of movement on each side so that there is plenty of room for skew adjustment before the motor reaches its travel limit.

The feedhorn should be set on the dish so that the LNA is tilted in relation to the earth. In the Omni Rotor, this should be 15 to 20 degrees, while in the Polarotor this setting should be 45 degrees. This assures that the probe will be able to rotate through both polarizations. Figure 6-11 shows this setting.

Another way of determining proper probe rotation in a Polarotor I is to hook up the feedhorn to the receiver and run the probe to both ends of its travel. Mark these on the edge of the scalar ring in pencil. Then split the difference between the two marks. When attaching the feedhorn to the dish, set it so that the mark is setting at 45 degrees to the polar axis.

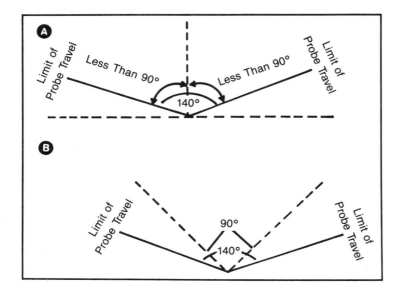

Fig. 6-10. If the PR I or other pulse controlled feedhorn is not set up properly, both polarizations will not be received as in A. The PR I must be physically moved so that it falls within range, as in B. Both polarizations are received and there is plenty of room for skew adjustment in either direction.

When setting up the receiver, the skew control must be able to be adjusted through the polarization peak. The signal strength meter should have a distinct peak and then should decrease as the control is rotated cw and ccw. If the polarization can be peaked but does not show a decline with further adjustment, then the feedhorn is being driven right to its stop. This can cause the motor to be continuously running on that polarization. If this is experienced, then the feedhorn should be physically rotated 10 to 20 degrees. If after rotation, the polarization is worse, then rotate the feed in the opposite direction about 40 degrees (20 of it to make up the first change). As was seen in Fig. 6-10B, there is ample movement of the probe

beyond the 90 degree polarization difference if the feedhorn is set up correctly.

TROUBLESHOOTING GUIDE

This section describes some very common troubles that can occur with the feedhorn and polarization device. In each case, a number of possible causes and suggested cures for each are given.

Problem. Hum bars, white or black horizontal lines, in the picture. Unplug the red wire to the PD, if the hum bars stop then it is definitely related to the PD. Hum bars are usually caused by using incorrect wire size for the length of the cable run. If the cable is less than 75 feet,

Fig. 6-11. LNA orientation for the the Omni Rotor. It should be 15 to 20 degrees from horizontal. The PR I should be about 45 degrees from horizontal because of its reduced probe travel.

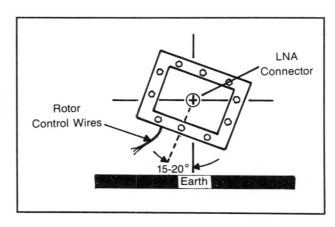

then 22 gauge wires are OK, but if the run is from 75 to 125 feet, use 20 gauge, if it is up to 200 feet, use 18 gauge, while runs beyond this length should use 16 gauge. Instead of ripping up the cables, this can sometimes be cured by attaching a filter capacitor at the PD. Chaparral recommends a 1,000 μF electrolytic rated at 10 or more volts. The positive lead is attached to the red (for the PR I) + 5 Vdc wire and the negative terminal is connected to the ground wire (black for the PR I).

Hum bars can also be an indication of a motor that is against its stop and is still running. If the skew control can be adjusted or if the polarization changed and it stops, then the feedhorn should be rotated to center the probe travel.

If the hum bars do not stop when the PD is unplugged, then chances are there is either a ground loop between receiver and dish, or an oscillation in the LNA. Lift the ground at the receiver by using a three prong to two prong wall plug adapter. Substitute the LNA and/or cabling.

Problem. Polarization does not change. If the PD is a moving probe device and if the weather is below freezing, then the probe could be frozen to the housing. To prevent this from occurring, use a feedhorn weather cover and, if necessary, heat tape to keep the internal temperature above freezing. Another way to prevent this from occurring is to use a ferrite PD, which has no moving parts.

The power or the pulse may not be getting to the PD. If it is a movable probe type PD, then there should be a slight voltage on the control wire (white wire for a PR I). This voltage can be measured by a voltmeter. It should read from about .4 to .8 Vdc as the receiver changes from odd to even channels. The red wire should have + 5 Vdc. Alternately, the pulses could be checked for proper voltage and pulse length by using an oscope. Also check for loose or corroded connections, especially where wires are spliced outdoors.

Problem. Servo motor oscillation. This is usually caused by the motor being driven to its physical limits. If the motor is in the center of its travel and it is still oscillating, then the control pulses should be checked for stability, by viewing them with an oscope. In some receivers, there is a relay that is used to change the pulse length, this can cause the polarization to change if the contacts are intermittent.

Cable to driving circuit impedance can also cause servo motor oscillation. If the other possible causes have been checked out, then this may be the real cause. Put a 100 to 500 ohm resistor in series with the control wire

(white in the PR I) going to the PD.

Problem. Inequalities between polarizations. If one polarization is good and the other is poor, look first to skew settings at the receiver. Usually there are two controls for setting the vertical and horizontal positions of the probe or ferrite PD.

If it is a PD that uses pin diodes, then check for the proper voltage at the terminals. One diode may not be turning on or may be burned out. This can also be caused by ground loops which can partially turn on the diode that is supposed to be turned off. Try temporarily using the reverse bias supply shown in Fig. 6-12. If this allows both polarizations to be received equally well, then a ground loop is probably the source of your problem. If the dish is properly grounded, then try lifting the receiver's ground.

An oscillation may also be occurring in the LNA as the polarization is changed. Often this is due to the placement of the coax cable in relation to the feed. They will also occur if the LNA does not have metal to metal waveguide contact with the feedhorn. This is especially true of nonisolated LNAs. Substitute with an isolated LNA if this is suspected to be the problem.

TI can also show up on one polarization and not the other. If the polarizations are both OK on some satellites while others are poor, then TI is probably the cause.

Problem. DC motor-type PD not rotating correctly. If the motor only turns in one direction, then the polarity switch is defective or has a cold solder joint.

If the probe operates intermittently or sluggishly, then the voltage may be too low from the battery (if so equipped) or there may be a cold solder joint or bad connection at the PD. Replace the battery if it is below 7 Vdc.

Problem. Poor performance, or gradual decline in performance. This is usually due to shifting of the feedhorn away from the focal point, foreign matter in the waveguide, moisture in the waveguide, degradation of the LNA, or bad cables or connectors. Also check the centering and focus of the feedhorn as stated earlier in this chapter.

Foreign matter in the waveguide is usually of an organic nature. Spiders building webs, wasps building nests, even hummingbirds moving in and calling the waveguide home have been noted. This can be prevented by gluing a thin piece of mylar or flexible plastic across the waveguide opening. Be sure to check the signal meter reading before and after adding the cover, as some types of material will cause some signal loss and should not be used.

Remove the LNA from the feed horn and check for

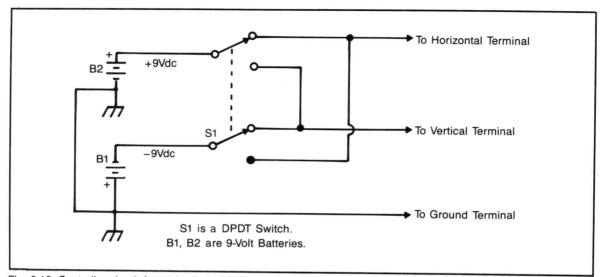

Fig. 6-12. Controller circuit for a pin diode PD. It can be used to check for ground loop problems that may be biasing the opposite polarization diode back on. By using dual voltages, the opposite diode is hard off and cannot be affected by ground loops.

moisture. Moisture absorbs microwaves so it must be kept out of the waveguide as well as the LNA and DC connectors and cabling. Dry out the moisture and check the gasket. If it is cracked or damaged in any way, replace it. If there was none, be sure to put one in before reassembling the LNA/feedhorn. Check that the flange is clean and smooth on both PD and LNA before reassembly.

If after trying all the above and the signals are still poorer than they were previously, then substitution of the electronics is the next step. Try a different LNA, DC, or receiver. Don't forget the cables as they could be a source of problem also. Refer to the particular chapter on the other electronic components in the system for more information.

LNAs, LNBs, LNCs, and LNFs

E VER SINCE SOMEONE CALLED THE LOW NOISE AM-
plifier an LNA, there has arisen a multitude of
TVRO abbreviations. Most of these concern microwave
down conversion. LNAs, LNBs, LNCs, and DCs are explained. I'll leave the PDs, APs, i-f (although there's a
little i-f stuff here), rf, UHF, VHF, TI, agc, afc, PLL,
EIRP, and all the rest for other chapters.

LOW—NOISE AMPLIFIERS

An LNA is the basic building block of every system. If
the LNA is poor, then even a 15 foot dish on Galaxy 1
won't give you good pictures. It can be considered the
heart of the system. But to work properly, it must be
situated correctly on the dish and attached properly to
the feedhorn or PD (Polarization Device).

The LNA in Fig. 7-1 is really a pretty straightforward
piece of equipment. Its purpose is to raise the level of
some very weak signals that have been reflected off a
dish located about three or four feet in front of its
waveguide opening. The physics of this interface is pretty
esoteric. Suffice it to say that the ideal dish and feed
presents 50 ohms of impedance to the LNA's waveguide.
Of course, in the real world, this rarely happens, and this
is one of the reasons for using an isolator on the input

of the LNA. The isolator shows a fixed impedance to the
first stage GaAs-FET, making it act more predictably,
no matter what the dish impedance actually is.

Without an isolator, the LNA is prone to oscillate if
everything's not right. Some nonisolated LNAs will even
start to oscillate everytime the polarity is changed. I once
pointed this problem out to a local TVRO dealer whose
response was, "Oh yeah, sometimes they do that, here
. . .," and he proceeded to turn off the receiver and turn
it back on. "See, that fixed it." I bet he even told his
customers that that was normal.

Once the signals have been focused by the dish into
the PD and one polarization separated out and passed on
through the feed, the LNA takes over. It has a 500 MHz
wide band of signals that it must amplify about 50 dB
or 100,000 times, in order to have enough signal level to
drive the RG-214 cable that connects it to the DC. The
kicker is that these signals are at a level of about – 120
dBm, which is just barely above the level of noise that
is coming in with the signals from outer space.

To properly do this, the LNA must have an across
the band evenness to its response. Such factors as gain
flatness, group delay, noise figure, and VSWR are all important specifications, that are often expressed as one
number. For example, the noise figure is almost always

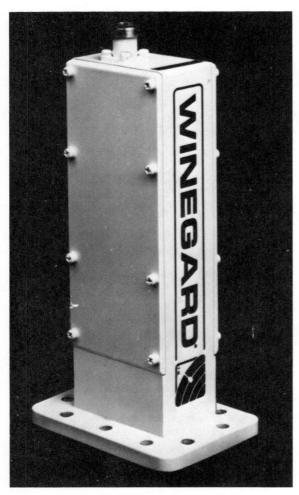

Fig. 7-1. Typical LNA from the new generation which has the power supply sent up the rf cable. This LNA is made in Japan by JRC. It is virtually the same as that sold by Gould/Dexcel in their GT-series and by Norsat. Courtesy of Winegard Satellite Systems.

usually better, again due to the matching and tuning that is done to the GaAs-FET circuits.

Waveguides

To route the signals through the PD and into the LNA requires two types of waveguides. A circular waveguide is nonpolarized and is used to gather the signals from the dish. A rectangular waveguide is a polarized transmission line. A *waveguide* is like a hollow piece of coax to microwaves. It is the most efficient mode of transporting them around. They just love to travel down the surface of a waveguide if it's the right size. Of course it is a mighty expensive way to transport signals, which is one reason why the home TVRO system has the LNA up at the focal point of the dish instead of down behind it like on commercial systems that use a J-hook shaped waveguide to feed the signals back behind the dish to where the LNA is located.

The waveguide is of a specific size and shape that matches the frequency band we are interested in. Since we are dealing with the C-band, we must use a C-band waveguide. This waveguide opening is shown in Fig. 7-3; it is 2.29 inches wide and 1.145 inches high. You might have noticed the first measurement, 2.29″, just happens to be the name of this waveguide, which is technically referred to as a WR-229 or CPR 229 waveguide. The flange or face must make metal to metal contact with the mating flange to ensure the most efficient signal transfer between them. There is also a groove around the flange that is used to hold a gasket to keep water out of the waveguide. Water absorbs microwaves, which is why a microwave oven works, and it will drastically attenuate the satellite signals as well as cause corrosion to the waveguide.

The reason that a waveguide of a certain size and shape is needed is so that the microwave signals will propagate, or travel down its surface to our antenna. The real antenna in a satellite receiving system is located near the end of the LNA's waveguide. It is a short piece of metal, of a variety of shapes and sizes, all designed to effectively transfer the microwaves from the focal point of the waveguide into the first stage amplifier or isolator in the LNA.

Once the signals enter the LNA, they are amplified, which sounds simple enough, except that they must be amplified with the least amount of added noise that is possible. If this is done correctly, then the signals can be manipulated by the RG-214, the DC, and the receiver, and still come out looking good. If the LNA adds excessive noise, or it doesn't add enough gain, or if the dish doesn't have enough gain, or if the feed is off center, then

better at midband than at the ends. This is mainly due to the bandpass characteristics of the LNA. Thus, the LNA could be called several different noise temperatures just by changing the specified frequency that it is measured at. Most manufacturers specify worst case or maximum noise temperature.

The LNA whose specifications are shown in Fig. 7-2 would have a maximum noise temperature of about 95 degrees, an average noise temperature of about 87 degrees and about a 85 degree midband noise temperature. Notice also that the gain at midband is

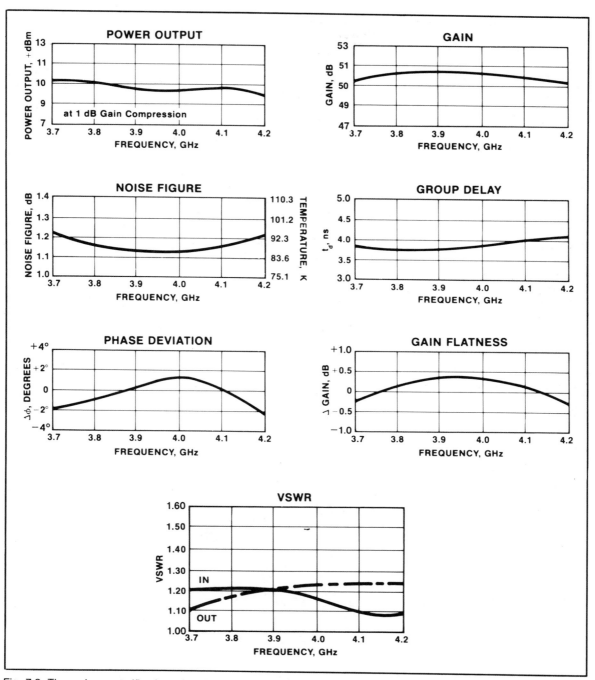

Fig. 7-2. The various specifications that determine the quality of an LNA. The most important being noise figure, gain and gain flatness. VSWR (Voltage Standing Wave Ratio) is a measure of the match of the input and output to an ideal impedance. If an isolator is used, then the input VSWR should be close to that listed when the LNA is put on a dish. LNAs without isolators typically show a much worse VSWR when actually put on a dish.

no matter what happens to the pictures after the LNA they still won't make it. They'll probably be grainy, streaky, and loaded with sparklies.

Handling and Installing

Back when LNAs were $3,000 a piece, they were treated with reverence. They were handled as if they were made of gold, which pound for pound they might as well have been. Another more important reason was that they were very fragile. If one was to drop an LNA while attaching it to a feedhorn, your heart would stop as it fell to the ground because you knew that it would never amplify another microwave again without an expensive return visit to the factory.

Today it's a different story. LNAs cost $99 or less now, and they are imported by the boatload from Japan. They can just about be dropped from a plane before they'll die on you. But LNAs should still not be treated too lightly. They are still a sensitive piece of electronics that can be damaged by the improper handling. So to prevent that from happening, here's a list of what not to do with an LNA, LNB, or LNC.

☐ Don't ever adjust the monopole transition. That's the official name for the small probe sitting somewhere in the waveguide. As the Avantek installation guide so aptly states, "Any attempt to touch, adjust, or alter the waveguide probe in any way will result in the unit malfunctioning, causing all warranties to become null and void." And you'd best believe them.

☐ Never ever use a pair of gas pliers, a pipe wrench, or any other kind of tool on the N connector. Your hand is the only tool that should be used. Hand tight is sufficient to prevent any moisture from entering the connector from the mating end. The cable end is a different story. See Chapter 4 for more information on N connectors and coax cables.

☐ Never ever use anything on the flange of either the PD or the LNA. They should be inspected to ensure they are clean and then assembled dry. The only thing that should come between the LNA and the PD is the rubber gasket (or two rubber gaskets, or one oversize double gasket if there is a gasket groove in the PD's waveguide flange as well as the one on the LNA's flange).

☐ Always inspect the center pin of an N connector before connecting it to ensure that it is centered and that it is the right height. Again, see Chapter 4's section on N connector do's and don'ts.

☐ Always properly ground the dish. It is your best protection against lightning as well as possible electrocution should the motor drive's voltage short to the dish. And remember, lightning damage is not covered by anybody's warranty.

☐ Make sure the power is off before connecting or disconnecting the LNA, LNB, or LNC. It's a minor point, and it is probably more important from the standpoint of the power supply, but it's better to be safe than sorry.

☐ Do not fill up the N connector with silicon grease

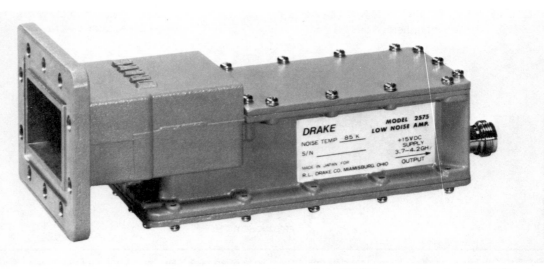

Fig. 7-3. A Drake LNA showing the waveguide opening. The opening is called a WR 229 waveguide, since it is 2.29 inches across. When the LNA is mounted, always make sure a gasket is used and that the inner flange makes mechanical contact with the feedhorn's waveguide. Use a minimum of six 1/4″ × 20 bolts to mount the LNA.

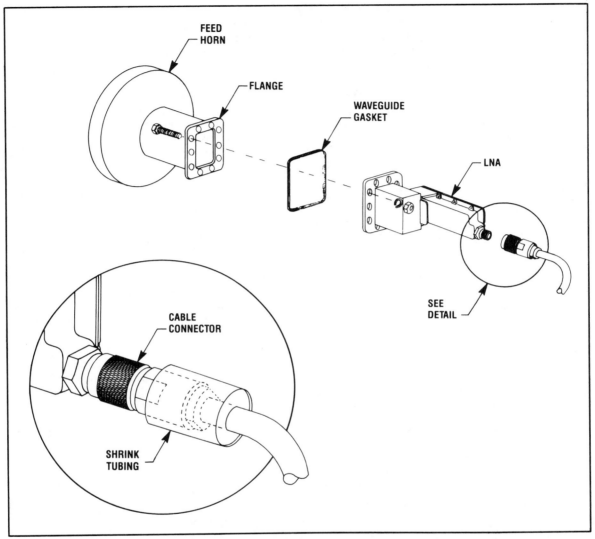

Fig. 7-4. Method of mounting the LNA to the feedhorn. Always use the gasket (which is typically included with the feedhorn) and six 1/4″ × 20 bolts. The insert shows how to seal the connector by using shrink tubing. It is a good idea to cover the shrink tubing with a coating of sealant, since it is better to be safe than to be called back by the customer.

or any other type of water repellent, as it will cause more problems than it's worth. Properly seal the cable entrance point on the connector and water won't get into the inside of the connector!

Figure 7-4 shows the parts used to install an LNA, LNB, or LNC. The rectangular waveguide gasket fits into a groove on the mating flange surface. Both surfaces, LNA and feedhorn, must be smooth and must make mechanical contact. Never put anything else on the flange

in an attempt to weatherproof it, especially RTV. The gasket by itself will do just fine. If using a PD that has a right angle elbow, be sure to use it. It does not cause any change in the signal, and is necessary to enable a cover to fit over the LNA and feed. Be sure to use two gaskets if its surface also has a groove.

The insert in Fig. 7-4 shows the use of shrink tubing over the N connector. This is important, for all N connectors are only watertight where they fasten to their mating connector. The cable entrance end of every connector must be sealed. If shrink tubing is used, be sure to apply water-repellent to it after the cable is in place. Using duct seal, Coax Seal, Scotchkote, and RTV are popular sealing methods. Two minutes spent sealing a cable will prevent a two hour service call later on.

I also strongly suggest that you use an LNA cover. It is the easiest and cheapest way to prevent moisture damage to the LNA and PD. Use a light colored cover, as the dark covers absorb too much sunshine and raise the temperature below the cover to the point that the noise figure increases in the LNA and drift occurs in the LNC.

Troubleshooting

LNAs are not field serviceable, but they can still be troubleshot in the field. Basically there are three things that happen in the LNA that cause it to go bad. A cold solder joint or intermittent connection develops somewhere in the LNA, one of the GaAs-FETs dies or degrades, or the regulator and probably one or two GaAs-FETs have died together. GaAs-FETs are touchy little parts that don't like static electricity or too much voltage being applied to them. They also don't like to be left without proper bias applied to all points, so, if the regulator goes, they all usually get nailed at the same time.

As to bad connections, there are many little jumpers and solder joints (or bonds) that can go bad. These go between the regulator and the various other parts of the

LNA, and are often the cause of temperature related failures and intermittents. However, the number one cause of problems is improper handling of the N connector. All of these type of symptoms can be caused by the female connecting pin being pushed into the LNA by a male pin that's too high, that's off center, or that's applied with too much pressure.

With GaAs-FETs it's usually an either/or situation. Either they are good and draw the right current, or they open up and draw no current. So to tell if something is amiss in the LNA, measure the amount of current being drawn by it. Figure 7-5 shows an easy way to measure the current when using an external test supply. Of course if you have a fancy supply that gives you a voltage and current readout, then the current meter isn't necessary.

You could also check the system's supply voltage, by using the set-up shown in Fig. 7-6. It consists of two power inserters hooked up back to back with an N barrel adapter. Be sure to put the rf/dc side of each towards the DC and the LNA. Most LNAs draw around 100 mA. This current is usually divided into about 25 mA per amplifier stage, with another 20 to 25 mA being used by the bias circuit and dissipated as heat by the regulator and dropping resistors.

If a customer calls with a symptom like noisy pictures or low gain (typical LNA caused symptoms), then the LNA's current draw should be checked first. If it's correct, then it's a good chance that the LNA is OK and it's the dish or the DC that's not right.

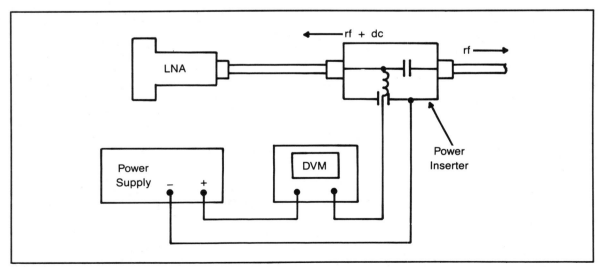

Fig. 7-5. Diagram of a test bench set-up for checking LNA current draw. Typically an LNA will draw about 80 to 120 mA at +15 to +18 Vdc.

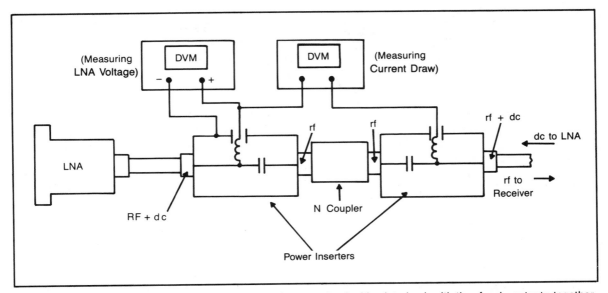

Fig. 7-6. To check the LNA in the field, use two power inserters hooked back to back with the rf only outputs together. A single DVM (or voltmeter) could be used if a jumper was installed for the voltage test. This should be done at the LNA so that any cable voltage drop is measured.

If it is drawing only 75 mA, then it can be assumed one stage is gone and that the signal is capacitatively being coupled through. In this case, the LNA should be changed. If the problem is solved then send the LNA in for repair. If the current is found to be 200 mA or more, then the regulator may be bad or there may be a shorted stage, and it too should be sent in for repair. One point though about high current, be sure to check the LNA connector and jumper cable, as they could have an internal corrosion build-up which is causing a varying resistance short between the center pin and ground. This will act as a voltage divider, drawing some current itself as it pulls the voltage down. So for excessive current, either check it at the input to the LNA or substitute a known good cable.

There's not much hope of fixing an LNA in the field unless you're a combination of Mr. Spock and a brain surgeon who just happens to carry a Bonder, high power microscope, and a few spare GaAs-FETs in the old tool box. In fact, with every LNA, if it's opened, the warranty becomes null and void. Of course, if the warranty has expired and it's dead already, why not, right? The only problem is, if you're planning on getting it fixed, you'll probably cause more damage by opening it up and poking around, and hence the repair bill will be even larger. This is especially true with MIC (Microwave Integrated Circuitry), a little grease off the fingers and the whole

substrate is gone, which means another $50 or more added to the bill.

Every LNA should have a sticker on the side saying, "There are no user-serviceable parts inside, refer servicing to a qualified service center." It's truly one place that deserves the label.

DOWN CONVERTERS

Since microwave carriers are pretty hard to detect and turn directly into video, it is first necessary to lower the carrier to a more reasonable frequency at which to do the demodulation. Typically this lower frequency has been 70 MHz. To get to 70 MHz there are basically three methods. Use an LNA and a separate DC, use an LNC, or use an LNB or BDC and a second tunable DC. In practice, only the first two use 70 MHz. Most BDC systems use 130 or 140 MHz as their i-f.

Separate DC Methods

When the first home-brew satellite receivers appeared on the scene, the most common approach was a direct line at 4 GHz from the LNA to the receiver via hard-line or heliax. This was the way it was done by the commercial systems, so it must be the best way, right? It became apparent right away that this was not the way to do things for the home receiver system. Even a fairly short run of cable was expensive, attenuated lots of signal at 4 GHz and was a real bear to install. Those who have had to put N connectors on hard-line will know what I'm talking about. If you weren't extremely careful and the cable should kink, all you could do was scrap it and start over.

In receivers that had an internal DC, cable loss could be compensated for by line amplifiers. The SA-50 Signal Purifier from ICM could cut or boost the gain from − 10 dB to + 5 dB. It also includes a 5-pole bandpass filter to clean-up out of band TI from overdriving the DC. It will operate on any voltage from + 10 to + 24 Vdc.

The first improvement that came along, at least as far as cabling and installations are concerned, was to separate the receiver and the down converter. This meant that RG-214 could carry the 4 GHz signals a few feet to the down converter (DC), and from there all that was needed was RG-59, or RG-6 for longer runs. This is still the most popular arrangement of system blocks, although BDC is rapidly making inroads.

Moving the DC outside led to a few problems though. The biggest one being the weather. Most DC manufacturers didn't have the slightest idea how to waterproof a down converter, and so there was water contamination problems. Another little item that wasn't noticed when the DC was inside the receiver, and hence inside the fairly constant temperature of the house, was temperature drift. This is caused by the VTO (Voltage Tuned Oscillator) or VCO (Voltage Controlled Oscillator) being sensitive to changes in temperature. What happens is the frequency to voltage curve changes with temperature and unless there is some kind of atc (automatic temperature compensation) network, in addition to the normal afc (automatic frequency control), the channels are no longer tuned at the right setting.

Low cost receiver manufacturers approached this problem by using a pot to tune the channels. This way no none could really tell how far the receiver had drifted. You just changed the dial until you got the channel you wanted.

Figure 7-7 is a plot of the tuning-to-voltage characteristics of a Dexcel VTO. When the temperature drops this curve lowers, designated by the dashed line. If no compensation was added, then the channels would all be off the same amount, because the tuning voltage remains constant and the change is fairly linear across the band. By adding in the right amount of shift, this temperature change can be compensated for. The afc will supply some of this change automatically, but what is really needed is something that is temperature sensitive in the opposite direction from the VTO. Germanium

diodes have just the sort of temperature change that is needed.

By putting three germanium diodes in series with the tuning voltage line inside the DC or LNC, they will raise or lower the tuning voltage slightly as the temperature shifts. The germanium diodes have a .3 volt drop across them at room temperature, but as the temperature gets hotter or colder this drop changes slightly. Their voltage drop changes in just the right fashion to compensate for the shift in frequency that occurs in the VTO when its output changes because of temperature. Without the diodes, we would be off almost one channel, but with the counter-change to the tuning voltage, we have compensated for the VTO's change in frequency, thus the DC is virtually free of temperature related drifting.

Image Reject Mixer

Single conversion DCs are only possible because of a circuit called an image reject mixer. A mixer is a circuit that combines two frequencies together to yield a third frequency which is either the difference or the sum of the two frequencies. This is how up and down conversion take place.

If two frequencies are combined in a mixer, the mixer outputs the original two frequencies plus another signal equal to the sum and a fourth signal equal to the difference between them. By using tuned circuits, you can isolate out only the desired frequency.

To illustrate this, lets take transponder 15, at 4.00 GHz and mix it with a 4.070 GHz signal. This signal is called the Local Oscillator or LO. The mixer output would contain the original signals at 4.00 and 4.070 GHz, the sum signal at 8.070 GHz, and the difference signal at 70 MHz. Filter out everything above 100 MHz and you are left with channel 15 centered at 70 MHz.

Unfortunately, in the real world we don't only have transponder 15 coming into the mixer. We have the entire 3.7 to 4.2 GHz band. Which for the most part is no problem, except for the frequency that is 70 MHz above our LO frequency.

Why? Because like all the signals entering the mixer, it will be added and subtracted with the LO frequency. Because it is 70 MHz higher, the difference signal is a second undesired 70 MHz signal. When tuning in channel 15 (4.00 GHz), the image will be from channel 22 (4.140 GHz). The image is always 7 channels or 140 MHz above the one you are tuned to. Notice that it is always on the opposite polarization, which helps us somewhat, but does not completely solve the problem of how to stop

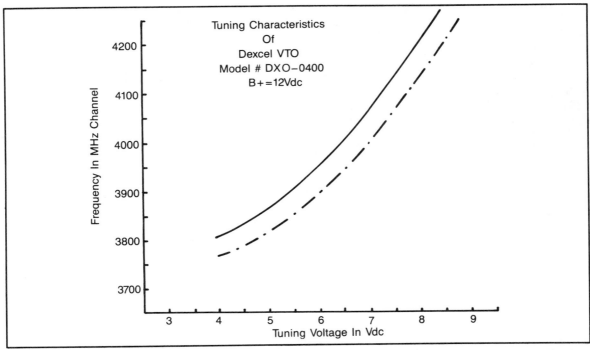

Fig. 7-7. Typical tuning characteristics of a VTO or VCO. The dashed line indicates the difference in tuning voltage to output that would occur without temperature compensation. This accounts for the drift that occurs in some DCs and LNCs when the temperature changes.

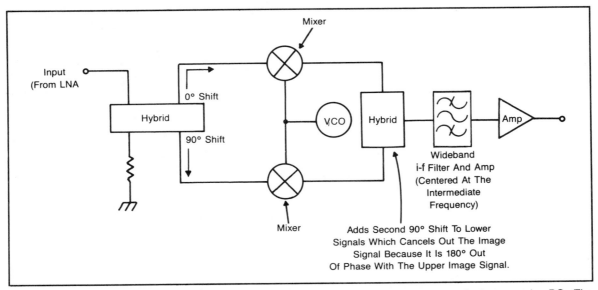

Fig. 7-8. Block diagram of an image reject mixer down converter, which is also known as a single-conversion DC. The two hybrids cause two 90 degree phase shifts to occur to the lower signals. This causes the image frequency to be cancelled out, leaving only the desired channel.

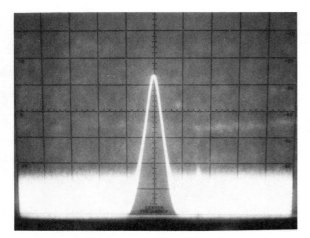

Fig. 7-9. Spectrum analyzer photograph of the LO leakage from an STS DC-102 down converter. It is measured without an input signal and is 3.80 GHz. The peak level is −36 dBm.

oscillator that typically is operating 70 MHz above the desired channel, there are several possibilities for interference. This is because the LO signal will leak out the input of the DC. Figure 7-9 shows what this LO leakage looks like. It is a spectrum analyzer photo of the LO leakage coming out of the input connector on a DC-102 down converter from STS. It is at a −36 dBm level, which is about 10 to 20 dB above the level of the incoming signals.

This leakage can do two things. It can black out the picture on another receiver tuned 7 channels above the one the STS receiver is tuned to if it is used in a multiple receiver installation without an isolator installed between the DC and the splitter. If the DC is connected directly to a nonisolated LNA, then the LO signal could be broadcast out into the dish, and be radiated into neighboring dishes, causing the same effect.

One way around this problem is to use a half-frequency mixer. This approach is used by Sat-Tec in their R-5000 DC. As Fig. 7-10 shows, the in-band leakage is at a level of −60 dBm. This is because the LO is tuned to one-half the frequency that the STS would be tuned to. Thus to tune in channel 15 (4.00 GHz), the oscillator would be running at 2.35 GHz, instead of 4.70 GHz. This half-frequency LO is harmonically mixed with the satellite signals to yield the proper channel at 70 MHz. The second harmonic of 2.35 GHz is 4.70 GHz. The half-

the undesired 70 MHz signal while passing the desired 70 MHz signal.

That is where the image reject mixer comes into play. It does just what its name implies. It stops the undesired 70 MHz signal, the *image*, by cancelling it out, or rejecting it, while reinforcing the desired signal. It does this by using two 90 degree phase shifts, two mixers, and some high school algebra.

Figure 7-8 is a block diagram of an image reject mixer. The input signal if from the LNA. All the signals go through the hybrid circuit which outputs two signals, one of which is delayed by 90 degrees. These signals go to the two mixers, which have the tunable LO equally applied to them. The image frequency, because it is above the LO in frequency in inverted in the mixer, thus when it is again shifted 90 degrees it ends up being 180 degrees out of phase with the upper path signal. Thus when they are recombined in the second hybrid, it cancels itself out.

Single down conversion is feasible without an image reject mixer if the LO frequency is more than 240 MHz. That way the image is always above the satellite channels and will not cause any problems. In this case, the i-f output would also be higher than 240 MHz, instead of at 70 MHz. With new designs in demodulation circuits and i-f filters at higher frequencies, this is becoming a much more feasible approach.

LO Leakage

Because the single conversion approach uses an

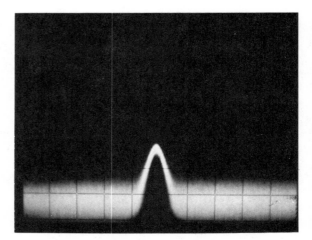

Fig. 7-10. Spectrum analyzer photograph of the 4 GHz leakage from an R-2000 DC from Sat-Tec. It uses a half frequency mixer circuit, so the LO is actually at about 2 GHz. The leakage is the second harmonic and so is very low in level (−60 dBm) at 4 GHz. The leakage is much higher at 2 GHz and the DC must be used with an isolator if other half-frequency mixer DCs are used in a multiple receiver system.

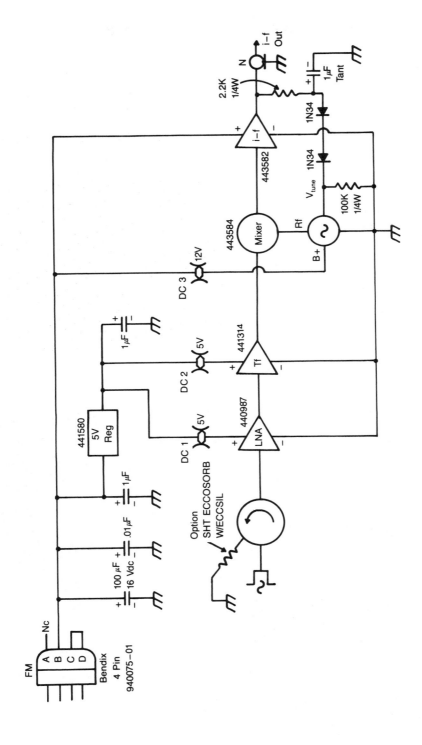

Fig. 7-11. Block diagram of a DXS 1200 LNC. It uses a +12 volt Vdc power supply and a +4 to +9 volt tuning range. The tuning voltage is applied up the coax cable. It consists of an isolator, a low noise module and bulk gain amplifier (which comprise the LNA), and an image reject mixer-type DC, made up of a mixer (with integral 90 degree hybrid), a VTO and an i-f amp (with the second hybrid). Courtesy of Gould/Dexcel.

frequency mixer's advantage is a lower-frequency oscillator and much less in-band leakage.

LOW-NOISE CONVERTERS

The next down conversion idea to emerge was to put the DC into the same housing as the LNA. Dexcel was the first manufacturer to come out with an LNC or Low Noise Converter. This meant that now there was no RG-214 to mess with, no second box to weatherproof, and only two connectors; one for the i-f signal and one for Vcc and tuning voltage.

The LNC was thought to be the most logical approach to down conversion for some time, but it never really reached its potential. Dexcel and M/A Com are the only companies that ever made a substantial number of 70 MHz LNCs. The LNC never took over the market as many had predicted. This was due to several things. First the lack of standardized tuning voltages and power supply voltages from various manufacturer's receivers meant that several LNCs would be necessary to cover all the receivers. Another problem was the lack of flexibility. The LNC was a one dish/one receiver type of installation. At the same time that the LNC was introduced, most receiver manufacturers began making separate DCs that matched their own receivers. The final problem was the cost. As LNA prices nose dived, it put the LNC at a disadvantage, especially the Dexcel version, as its inherent manufacturing cost was much higher than most LNAs and DCs due to its MIC circuitry.

Figure 7-11 is the block diagram for a DXS 1200 LNC made by Dexcel. It uses a +12 Vdc power supply voltage, and a tuning voltage that ranges from +5 Vdc for channel 1 to +9 Vdc for channel 24. The tuning voltage is applied up the coax cable.

LOW-NOISE BLOCKCONVERTERS

The last type of system to become popular uses an LNA and a DC, but the DC is special. It doesn't tune in one transponder, but rather tunes the whole band. By mixing the 4 GHz signals with a single-frequency oscillator, all 12 transponders are transferred to a lower band of frequencies. This approach is called Block Down Conversion or BDC. There are several frequency ranges which are popular, most of which fall into the upper end of the

Table 7-1. Listing of the Common Block Frequencies and the Companies that Use the Frequencies.

Manufacturer	Block Frequency
Arcom Microdyne S-A	270-770 MHz
Anderson Janeil	440-940 MHz
Uniden Dexcel Locom (LNB)	450-950 MHz
M/A Com GI Scannar Gensat Lowrance DX Luxor Sed Systems Amplica	950-1450 MHz
Winegard	1140-1640 MHz

UHF spectrum, usually around 1 GHz. Table 7-1 lists the most popular frequencies and which ones the various manufacturers use.

When the Block DC is combined with an LNA, the combination is called an LNB or Low Noise Blockconverter. The Birdview LNB is unique in that there are two LNBs, one for each polarization. They are active at all times, thus any receiver in a multiple receiver installation can view either vertical or horizontal polarizations without the need for any external switching set-ups.

LOW-NOISE FEED

Another LNA combination is the LNF or Low Noise Feed. It is an integral feedhorn/LNA. Many companies are using this approach, including Birdview. Their feed could actually be classified as a dual LNF, since it does include the feed and LNBs in a single housing. Some of the other companies that have LNFs are Syntronics, Chaparral, and Gamma-Feed.

Power Supplies

THE POWER SUPPLIES FOUND IN TVRO RECEIVERS are similar to those found in FM broadcast receivers and other stereo components. The current draw is minimal, being under 1 amp in most receivers. The most common type of supply consists of a full-wave bridge rectifier with one or more IC voltage regulators. Because of the very low current used by some receivers, a wall-plug type power supply is sometimes used with the final voltage regulation taking place in the receiver.

Antenna actuator power supplies are similar except that they must supply up to 6 amps of current, and so are much beefier. They too use full-wave bridge rectifier circuits, but for the most part are nonregulated.

RECEIVERS

Figure 8-1 shows the power supply from a Luxor 9550. It's a typical multiple output power supply that uses IC voltage regulators. There is an unregulated + 55 Vdc output for the antenna actuator controller and motor drive as well as an unregulated + 23 Vdc supply. This is in addition to the three regulated supplies of + 5 Vdc, + 12 Vdc and + 18 Vdc. It uses three taps on the secondary of the transformer, two of which go into bridge rectifier circuits, for better current and voltage stability. The 22

nF (.0022 μF) capacitors are used for rf by-passing and power supply decoupling.

Almost every TVRO receiver has a + 12 volt regulator. This voltage is used for powering most transistor circuits and linear ICs. There are also several families of digital ICs used in TVRO receivers. A + 5 Vdc is needed to supply voltage for TTL ICs (7400 series) and for ECL ICs (MC10000 series). CMOS chips 94000 and 4500 series) can be powered from about + 5 to + 15 Vdc, depending on their application.

Most receivers supply a voltage of from + 12 to + 28 Vdc to power a down converter and an LNA, or an LNB, or LNC. These components usually draw about 100 to 200 mA (milliamps) each. Almost all LNAs, LNBs, LNCs, and DCs have on-board regulation. Therefore, this supply may or may not be regulated. In the Luxor 9550, the 7818 (+ 18 V regulator) is used to supply the LNC or LNA/separate down converter with power.

In receivers with a built-in Polarotor I-type control circuit, there will also be a regulated + 5 or + 6 Vdc. This voltage is regulated by an IC regulator like the LM317, 7805 or 7806, and an output filter capacitor and decoupling capacitor. A current limiter on the output is also important, as shorting the dc line is a common occurrence which will wipe out the IC unless the current is limited.

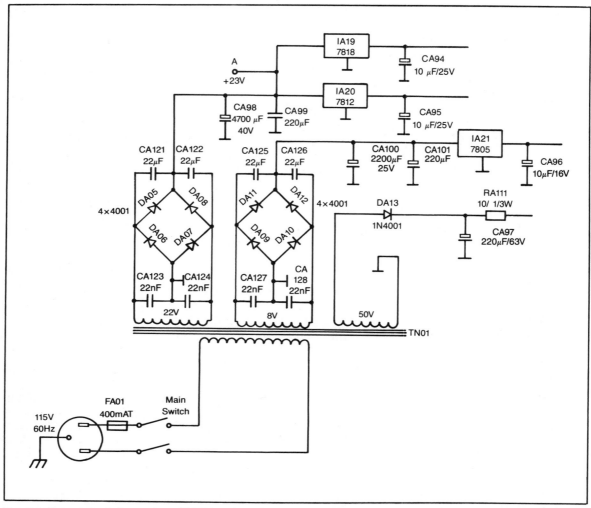

Fig. 8-1. Power supply from a Luxor 9550 receiver. It is typical of the multiple voltage supplies that use IC regulators. It has a +18, a +12, and a +5 Vdc regulated output. There is also an unregulated +55 Vdc for use by the 9534 Antenna Actuator Controller. Courtesy of Luxor North America.

Usually this is done by using a current limiting resistor, although a small incandescent lamp can also be used. The lamp turns on as the current increases preventing excessive current to flow. This same idea is used in the test bench current limiter.

ANTENNA ACTUATORS

Figure 8-2 is the schematic for the Polar-Trak by KLM. It is typical of the simple up/down type of actuator. It consists of a power transformer that steps the voltage down to 36 Vac, a full-wave bridge rectifier, which puts

out +18 and −18 Vdc, and a double-throw, single-pole momentary switch, which selects east or west movement.

A ten-turn potentiometer is used to supply positional feedback to drive a simple voltmeter so that relative dish position can be determined. A very basic control circuit for a Polarotor 2-type polarity changer is also included. Notice the use of the LM317 adjustable regulator to supply about 10 Vdc to the polarizer.

The addition of a Fast/Slow switch is really the only frill. It determines whether the motor will get 18 Vdc (slow) or 36 Vdc (fast). There are no Up/Down limit

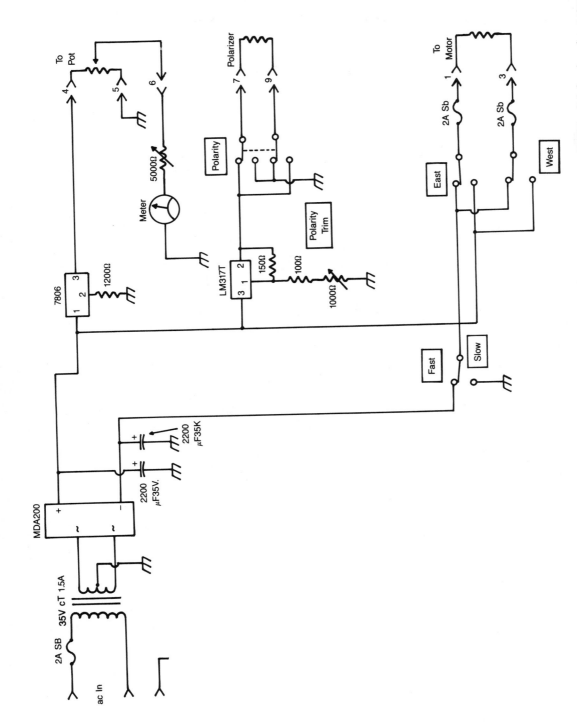

Fig. 8-2. Schematic of a basic motor controller, The Polar-Trak by KLM. It consists of a bridge rectifier and a momentary DPDT switch for changing motor directions. The Fast/Slow switch changes the motor voltage from 18 to 36 volts. There is also a dish position indication meter and a PR II-type controller. Courtesy of KLM.

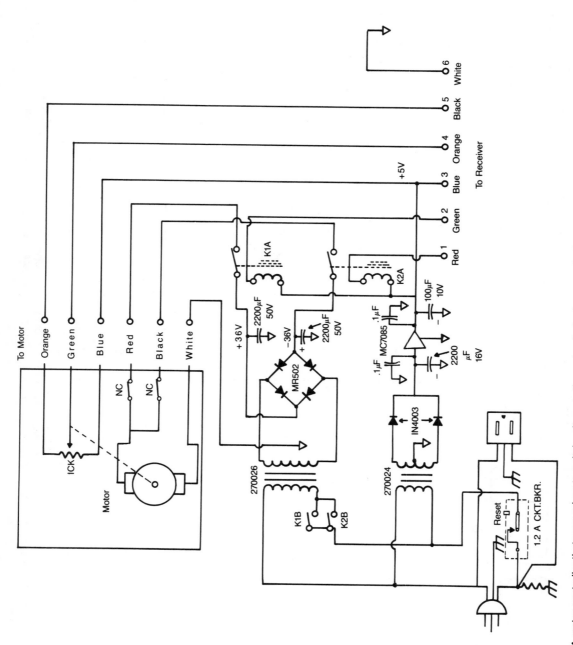

Fig. 8-3. A motor controller that uses relays to switch voltage to the motor. It also has a pot for feedback, which functions as a voltage divider. Courtesy of Winegard.

switches shown, but they are usually included in series with the motor. three 2 amp slo-blo fuses complete the circuit.

Figure 8-3 is the Winegard Polar Drive power supply schematic from their original actuator system. It required 110 Vac at the dish. It's very similar to the KLM circuit in basic operation. It uses + 36 or – 36 Vdc to drive the motor either east or west. There is no fast/slow switch as in the KLM. Positional feedback is almost identical except that it's used to drive a digital display instead of an analog meter.

The biggest change is in the use of relays to switch the motor on and off. Not only is the dc voltage turned on and off by relay, but so is the input ac to the transformer. This causes the voltage to ramp up as the 2200 μF capacitors charge so that there is a smoother motor start-up. Thus the dish is not jerked when the motor drive relay comes on (unless the capacitors have not discharged). When the driving voltage is removed, the motor slows down without being braked, also giving a smoother stop.

To improve the performance, several key elements could be added to the circuit. Voltage backswing protection diodes (IN4002 or equivalent) should be added across the relay coils. A 50 volt varistor should be added across each relay output. And a voltage protection diode (IN4002 or equivalent) should be added to the 7805 regulator. See Fig. 8-4, for placement.

Protection diodes should be installed on every IC regulator and relay. They will prevent damage to the components in the power supply by stopping harmful voltage and surges from going back into unprotected parts of the circuit. The diode across the IC regulator protects the regulator from the capacitor discharging into the regulator. The diode across the relay limits the spike that is produced when the relay is deenergized. This spike can be several hundred volts, which will eventually damage the driving transistor if not swamped out by a diode.

IC REGULATORS

Figure 8-5 shows pinouts of some of the most common IC regulators. The 7800 and LM340 series of positive voltage regulators can supply up to 1 amp with a heat sink. Often, positive regulators are mounted directly to the chassis, which is used as the heat sink.

The 7900 and LM320 negative-voltage regulator series can also supply up to 1 amp of current, with a heat sink. A note of caution on negative regulators, the body of the regulator which mounts to the heat sink is not at ground potential and thus must be isolated from chassis ground. A mica insulator and insulated screw must be used to mount them. On all 7800 and 7900 regulators, the output voltage is stated by the last two digits, (i.e., 7912 is a negative 12 volt regulator). Commonly used voltages range from 5 to 18 volts.

Two other regulators show up quite often. These are the LM317T and LM723. Both are adjustable regulators. The LM317T comes in a TO-220 package, and the LM723 is available in either the standard 14-pin DIP package or a TO-5 metal can package. The TO definition defines the case package, as to size and shape.

The LM723, by itself can only supply about 150 mA of current, but by adding an external pass transistor, it can control up to 10 amps of current. The pass transistor is what actually passes the bulk of the current to the load, while the LM723 monitors the output voltage and continually adjusts the bias on the pass transistor to only allow the proper voltage and current to reach the load at all times.

Figure 8-6 is a block diagram of a 723 regulator. The heart of the regulator is the reference voltage amplifier, which is typically connected to pin 5 of the error amplifier through a voltage divider. Pin 4 monitors the output voltage and the error amp (or comparator) outputs a voltage that is equal to the difference between the reference and the output voltages. This voltage drives an internal pass transistor which is connected to the base of the external pass transistor, thus it is constantly varying the base bias to maintain the correct output voltage. Pin 2 senses the output current by measuring a bias voltage, which is usually 0 volts. If the voltage rises, the current limiter transistor turns on and limits the drive to the internal and external series pass transistors.

Figure 8-7 shows the power supply schematic from a Dexcel DXR 1000 which uses a 723 regulator to obtain a regulated + 28 Vdc to supply the tuning circuit and the LNC. RV7 adjusts the + 28 Vdc, by adjusting the reference voltage feeding the error amp. Pin 4 of the error amp monitors the voltage at the junction of R96 and R97, which should be about 6 Vdc for a 28 Vdc output. If this voltage changes or if RV7 is adjusted, then the bias on Q8 will be changed, raising or lowering the output voltage. The main supply current passes through Q8 and R93.

The LM317T is used when an adjustable regulated supply of up to 1.5 amps is needed. Figure 8-2 showed a LM317T being used to supply current to operate a dc motor polarization device.

SAFETY

Almost all receivers and actuators use a three prong

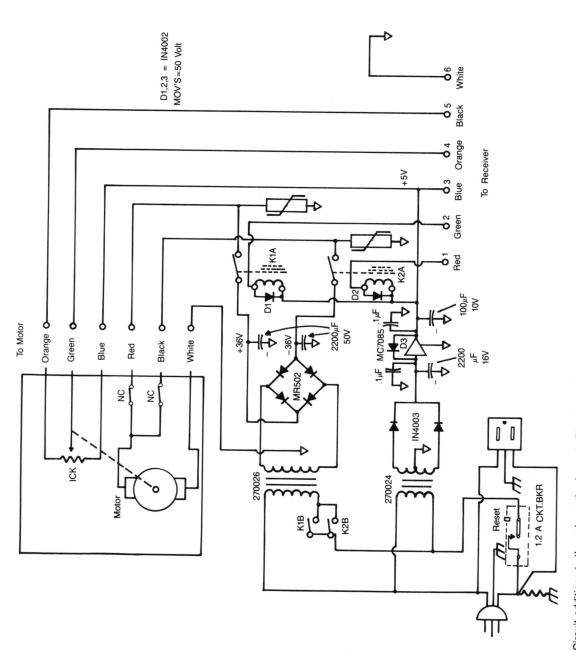

Fig. 8-4. Circuit additions to the schematic shown in Fig. 8-3. Adding the protection diodes to the relays will prevent backswing voltage damage to the regulator. The protection diode across the regulator prevents damage to the regulator from capacitor discharge. The varistors shunt out any lightning voltage that could damage the controller.

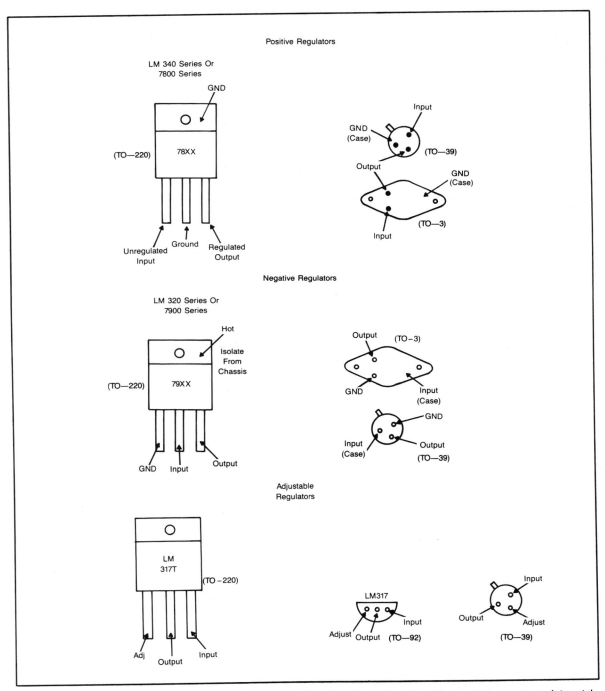

Fig. 8-5. Pinouts for three types of regulators in the three most popular body styles. The most common regulator style is the TO-220, although the TO-39 and TO-3 style components will also be found. There are adjustable regulators (like the 723) that come in a standard DIP (Dual Inline Package).

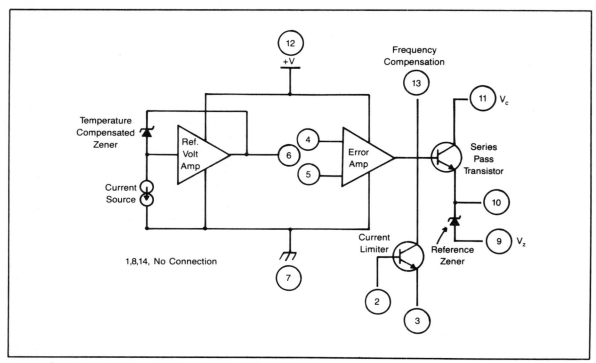

Fig. 8-6. Block diagram of a 723 regulator.

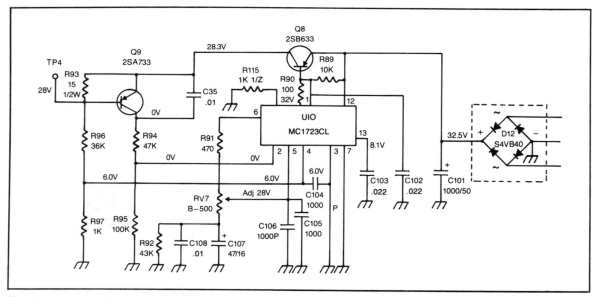

Fig. 8-7. A typical 723 regulator circuit. Q8 actually passes all the current. Q9 is used to sense the current and voltage. This is fed back to the 723 which adjusts the base bias on Q8 to maintain a +28 volt output. The output voltage is adjusted by RV7. Compare this to the block diagram of the 723 shown in Fig. 8-6. Courtesy of Gould/Dexcel.

power cord. If the system is to be installed where there are only two prong outlets, use a ground relief adapter. Never cut off the ground pin on a three prong plug hoping to save the .89 that a ground relief adapter costs. The two remaining prongs are usually not polarized, which means there's a 50/50 chance ground loops or even a shock hazard could be caused.

Before plugging in any electronic or electrical equipment into an unknown wall plug, it is always best to test the output for the proper polarity and voltage, especially if the equipment is to be run outdoors. A simple plug-in tester is available from most electronics and hardware stores. It tests for proper neutral, common and hot, and tells if any are reversed or open.

Most manufacturers tie a large value resistor (typically 4.7 Meg) between the common (white wire) and the chassis to ensure that there is not a voltage differential between common and ground. The chassis is always tied directly to the ground (green) wire in a three wire system.

For maximum safety, in most installations all components are grounded together through the wall plug. This type of installation is shown in Fig. 8-8A. If the cable run between the receiver and the dish is fairly short and the proper cabling is used then the resistance difference between grounds should be less than 5 ohms. Typically they are less than 1 ohm.

If a longer run is used, it may be necessary to "float" the indoor components from the wall socket ground and instead only use the ground at the dish. This is because of the resistance between the earth ground at the dish and the receiver, and the interconnecting cable's ground resistance are different. This resistance difference causes a ground loop. A ground loop can cause hum bars in the video, poor polarization control if a pin diode polarity controller is used, a low-level audio hum, or actuator positioning errors. If the problem is caused by a ground loop, then lifting the ground at one end will stop such problems.

To lift the ground, a ground relief adapter is used. A ground adapter has two polarized prongs for hot and neutral, the ground (normally the third prong) is a separate green wire, which is normally connected to the cover plate screw or a cold water pipe. To lift the ground means that the green wire is left unconnected.

A system that has the dish as the only ground is shown in Fig. 8-8B. This method of grounding gives maximum protection from lightning damage to the LNA, DC, PD, and motor drive. Consult a local electrical contractor for the ground rod depth in your area.

Antenna actuators require the most current and voltage of any component in a satellite system. Most of the motors use either 36 or 24 volts dc and draw from 1.5 to 6 amps of current.

Working with actuators is potentially the most dangerous part of repairing a TVRO system because of the relatively high voltage and current. Anytime you are working outside, even if you're standing on dry cement or bare ground, your body may be at ground potential, depending on the type and condition of shoes you have on. Thus merely touching a live circuit with only one hand can complete the circuit and cause a shock.

If the ground is wet, you will definitely be at ground potential and extreme care must be taken then, as an electric shock, even at a low 36 Vdc, will not only be very dangerous, but may also be deadly. Remember it takes less than 1 mA of current to override your heart's electrical pulses. To minimize the danger, carry an ac power strip with an integral ground fault interrupter (GFI) circuit. Connect the TVRO receiver, actuator and TV or monitor into this power box before working on the equipment. That way, if there is a ground problem, like a short through your body, the GFI will sense current flow and will kick the circuit breaker before possible harm is done.

Carrying a rubber mat on which to stand while working on actuators is also a good habit to get into. The rubber isolates you from being at ground potential, and if you follow the old electrician's adage, "on live circuits always keep one hand in your back pocket," the chance of receiving a fatal shock will be reduced to almost zero.

And although obvious, never work on an actuator or a dish if there's an electrical storm in the vicinity. A dose of 60,000 volts does not do good things to your body!

PREVENTING FAILURES

Probably the weakest area in most satellite receivers (and actuator controllers) is in the power supply section. Almost every manufacturer has put out at least one model that is deficient in some aspect of power supply design. It may not be readily apparent by just looking at the schematic, but if you look through an application book on using regulators or on designing power supplies in general, some things stick out like a sore thumb.

Regulator Protection

The most often overlooked potential problem area is in IC regulator protection. For the lack of a 2-cent protection diode, the IC regulator may fail due to capacitor discharge, which in turn could damage the receiver, LNA, or down converter because of excess voltage. Even the dish or motor drive could be damaged if the motor drive was to turn on by itself and drive the dish into the ground.

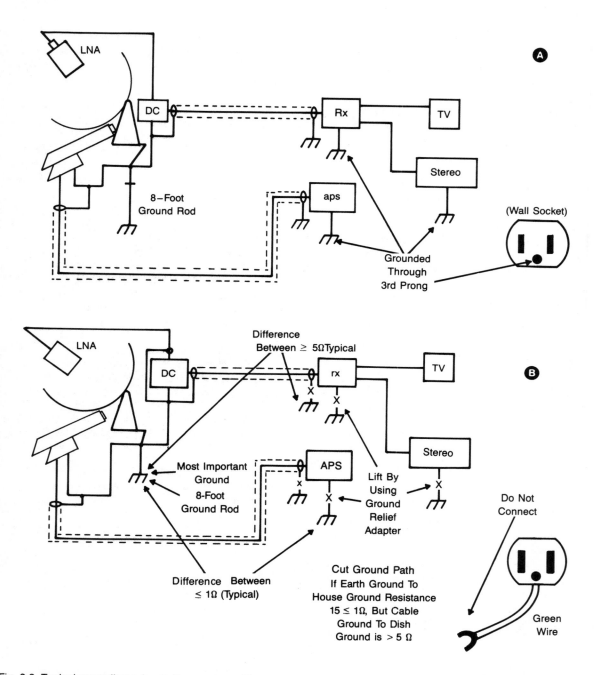

Fig. 8-8. Typical grounding scheme for a system with a short run of cabling is shown in A. The dish is grounded through ground rods and the receiver, APS and stereo are grounded through the wall plug. If a longer cable run is required, as in B, then a voltage drop can be formed on the shields if there is a 5 ohm or larger difference in resistance. Thus a ground loop is formed. To break the loop, lift the receiver, APS, and stereo above ground by using a ground-relief adapter.

92

To quote from National Semiconductor's Applications Notes on using IC regulators, "When external capacitors are used with any IC regulator, it is sometimes necessary to add protection diodes to prevent the capacitors from discharging into the regulator . . . Discharge occurs when either the input or output is shorted."

To add a protection diode onto a regulator is very simple, especially if you are in the process of replacing the regulator. Figure 8-9 shows a LM7805, and a LM340T with protection diodes installed.

Filtering

Another area that causes problems is marginal power supply filtering and bypassing. Filtering is usually accomplished by putting large filter capacitors at the output of the rectifier bridge. Typical values range from 500 μF to 8,000 μF. Bypass capacitors are usually disc type with values of .01 μF or .1 μF.

Symptoms of filtering problems are hum bars in the picture, a hum or distortion in the audio, or erratic picture lock-in. A shorted bypass capacitor can cause regulator damage. Not having enough bypass capacitors in the circuit can cause erratic motor drive or remote control functioning by allowing spikes and pulses to get into the wrong circuits.

Varistors

Another item that every receiver, DC, and LNA should have is a varistor. A varistor is like a combination zener diode and spark gap in that it allows voltages to pass that are up to a certain value. Any voltage beyond that value is shunted to ground. Some varistors can handle 50,000 volt spikes and respond in microseconds. They shunt the spike to ground while continuing to pass their rated voltage right on through.

If lightning storms are prevalent in your area, you've probably seen what happens when lightning strikes near a dish or near a power pole. Varistors are small, inexpensive, and will usually keep your equipment up and running even during an electrical storm.

Voltage Fluctuations

Most satellite receivers are being used in rural areas. Rural areas typically have power fluctuations of 10 to 20 volts or more on a regular basis. This fluctuation can increase to 30 or 40 volts during winter storms and during peak summer heat. This will be most apparent if the installation is at or near the end of a power line, or if there are other high-current loads down the line. Because of

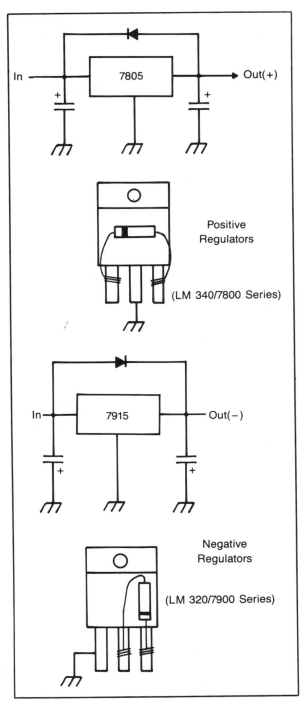

Fig. 8-9. Adding protection diodes onto positive and negative regulators.

this, most receiver manufacturers design in an operation margin to compensate for low power-line voltages.

Some receivers may still function at 90 Vac, while others will drop out of regulation at 105 Vac. In many cases, this extra margin can cause receivers to become very warm when operated at normal voltages of 117 Vac or higher. The warmth is from the power supply, which is dissipating the extra voltage being dropped as heat.

Overheating

Covering up the vent openings on the receiver, placing it on a thick pile carpet, installing it in a nonventilated wall unit, or even placing it where dust can accumulate on internal parts, can all lead to premature power supply failure. Typically, the receiver should be placed on a hard surface on its own four feet. A two-inch air space above and to the sides and back of a receiver should be suffi-

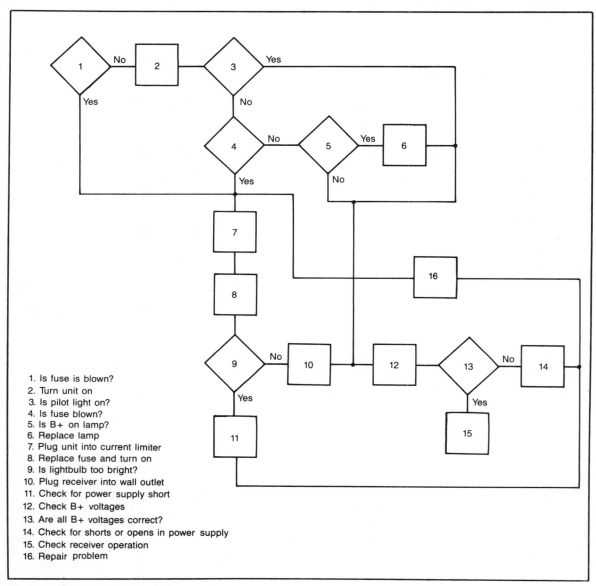

1. Is fuse is blown?
2. Turn unit on
3. Is pilot light on?
4. Is fuse blown?
5. Is B+ on lamp?
6. Replace lamp
7. Plug unit into current limiter
8. Replace fuse and turn on
9. Is lightbulb too bright?
10. Plug receiver into wall outlet
11. Check for power supply short
12. Check B+ voltages
13. Are all B+ voltages correct?
14. Check for shorts or opens in power supply
15. Check receiver operation
16. Repair problem

Fig. 8-10. Power supply troubleshooting flowchart.

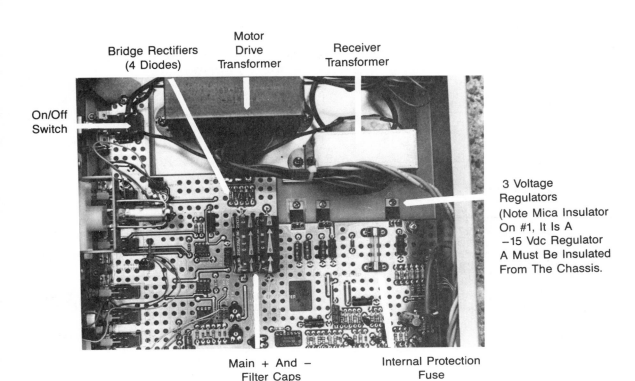

Motor
Drive
Transformer

Bridge Rectifiers
(4 Diodes)

Receiver
Transformer

On/Off
Switch

3 Voltage
Regulators
(Note Mica Insulator
On #1, It Is A
−15 Vdc Regulator
A Must Be Insulated
From The Chassis.

Main + And −
Filter Caps

Internal Protection
Fuse

Fig. 8-11. The power supply section from a Conifer receiver made by Drake. It shows the power transformers, rectifiers, filter capacitors, and voltage regulators. Courtesy of Conifer Corp.

cient to keep the receiver from overheating.

If it should overheat and shut down, the line voltage should be checked. If it is consistently above 117 Vac, or below 100 Vac, then a constant voltage transformer may be necessary. One with an output of 110 Vac would probably be best. The same transformer could also be used on the TV or monitor to improve their performance. Be sure to add in all the component's power draw when deciding what amperage constant-voltage transformer to use. Constant-voltage transformers are available from most local electrical supply houses. They can be ordered by mail through Newark, Grainger, and other suppliers.

Spike Suppression

Another very important power supply protection device is the spike suppressor or surge protector. This device consists of a varistor and several filters which swamp out surges, spikes, and rf energy that enters the house wiring on the power lines. It is an absolute necessity on any microprocessor controlled receiver or antenna posi-

tioner, as they are like home computers. Their memory can be scrambled, their directions to the receiver or positioner can be interrupted or changed causing the channels to change, or the dish can move by itself, or any number of strange things can occur. Suffice it to say that TVRO equipment is much more sensitive to line disturbances than the TV set is.

If you are using a Maspro receiver, don't use spike suppressors which have rf filtering, as it will interfere with the remote control's signals. Suppressors with rf filtering are designed to stop all emi-rfi noise, which means anything above 60 Hz is effectively by-passed to ground. Since the remote control from Maspro uses a 38 KHz carrier to transmit the information back to the receiver, this too will be shunted out.

One drawback to spike protectors, especially the $9.95 house special type, is that after a few large spikes, they lose their effectiveness. Thus they will not really be doing their job. And unfortunately, there's no easy way to tell that this has occurred until the equipment is damaged and the customer says, "but I had a spike pro-

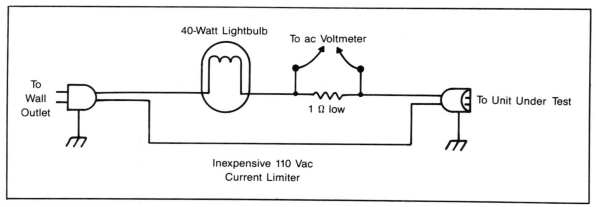

Fig. 8-12. Inexpensive current limiter for protecting units under test (UUT's) on the bench. It consists of a 40-watt lamp in series with the power cord to the UUT. If there is a short then the lamp will soak up the current. Under normal use the lamp would barely glow indicating the unit was not drawing excessive current. If a voltmeter is hooked across the 1 ohm resistor then the total current draw of the UUT can be measured.

tector in the line." Your best protection is to go with a name brand computer-type suppressor which has multiple outlets so that all the components in the system can be hooked into it.

TROUBLESHOOTING

Figure 8-10 is a power supply troubleshooting flowchart for determining the cause of failure in a receiver. By following the steps in the flowchart, you can cut down on initial receiver troubleshooting to ten minutes or so from the time the top is removed to the time the defective component is found.

On any receiver or antenna positioner, as well as just about all electronic equipment, suspect the power supply first. Always check to see that all the voltages are present. Such things are no picture or sound could be due to no +5 Vdc if ECL chips are used as the i-f limiter. The same symptoms could occur if an active video demodulator was used and the +12 Vdc was missing.

Figure 8-11 is a photo of the Conifer 2001 receiver showing the two power transformers (one for the receiver proper and one for the motor drive), the four diodes that make up the bridge rectifier, the large filter caps, and three of the voltage regulators that use the chassis as a heat sink. The easiest place to measure voltages is on the capacitors. Hook your voltmeter's negative lead to the chassis and the positive lead to the various electrolytic

capacitors. If there is no voltage, check that you aren't on the grounded end of the cap. The Conifer circuit uses +15, −15, +12, and +6 volts.

Chapter 6 shows how to check LNA and down converter current and voltages. Chapter 14 does the same for antenna actuators. Chapter 18 covers various receivers in more detail.

When working on a receiver which is blowing fuses, you usually go through several fuses before isolating the problem. If you have a variac and a current meter, you can slowly bring the voltage up while watching the meter, stopping if the current starts to rise. An easier and, I think, better way is to use a test bench current limiter. It's probably the simplest piece of test gear around, even easier to use than a variac.

As shown in Fig. 8-12, it consists of a standard light bulb in series with the ac line going to the receiver or actuator under test. The bulb acts as a current limiter, if there's a direct short, then the light bulb gets very bright and soaks up the current. An unhealthy overbias condition shows up as a brighter than normal light. A 40-watt lamp is good for most receivers and actuator controllers (without the motor attached).

By adding a 1 ohm, 10 watt resistor in series with the lamp (between it and the receiver), current draw can be measured as a function of the voltage drop across the 1 ohm resistor. A clip-on lamp and a short three wire extension cord can be modified to make up the current limiter.

Intermediate
Frequency Circuits

THE I-F OR INTERMEDIATE FREQUENCY IN MOST home TVRO receivers is centered around 70 MHz. This frequency became the defacto i-f standard for home TVRO mainly because of the phone company. The phone company's predominant i-f for their microwave down converters was 70 MHz. Since the earliest homebrew receivers consisted of surplus telephone company and military microwave equipment, it logically follows that TVRO development would start off using the same frequency.

70 MHz wasn't just picked out of the blue by the phone company. It turns out that 70 MHz is a good frequency to use in single-down conversion systems. Its image frequency and upper harmonics fall between channels and any subharmonics would be 35 MHz or below, which is out of band. 70 MHz is also about the lowest frequency that can be used to easily filter and detect the information that's carried on it. Plus there are many components designed to work in this frequency range, so designing inexpensive circuits is fairly easy.

Figure 9-1 shows that the satellite transponder's effective maximum bandwidth is 36 MHz. This means that there are signals as far as 18 MHz above and 18 MHz below the center frequency. At 70 MHz, the maximum i-f range that is encountered is from 52 MHz to 88 MHz, which stretches from just below VHF channel 2 to the top of VHF channel 6. This range is the best compromise between cost of amplification and cable losses and for component size versus wavelength.

But modern technology marches on, and new i-fs in the UHF band are becoming popular thanks to newly developed UHF SAW filters and UHF PLL circuits. So far, most of this development is being done in commercial receivers, but as prices come down for components, the next generation of home receivers will be moving upwards in i-f. Already, most BDC systems are using 130 or 140 MHz as their i-f frequency. Some of the UHF components are discussed in Chapter 17.

70 MHZ AMPLIFIERS

Amplifying a wideband signal at 70 MHz can be accomplished in two ways. One method is to amplify the entire band at one time. The second is to break it up into frequency sections and amplify each one separately while passing the other frequencies without gain. The second approach leads to phase and gain errors across the band and is not used very often in home receivers. Wideband amplifiers are predominantly used with peaking circuits centered at the i-f midpoint.

97

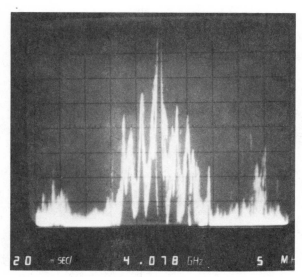

Fig. 9-1. Spectrum analyzer photograph of one transponder showing the main energy distribution. Each horizontal division is 5 MHz wide. The feed is slightly cross-polarized to show the opposite polarity at + and − 20 MHz (the spikes at both sides of the photo). Notice that the main concentration of energy is within 20 MHz of the center frequency of the transponder.

There are several devices that are typically used for 70 MHz amplification. TVRO receivers use transistors, ICs, and Hybrid modules. Each device ideally does the same thing. It boosts the signal without adding any objectionable noise. Most circuit differences between receivers in the area of i-f amplification have more to do with cost and design preference, than actual end performance, since all amplification is really done by semiconductors whether it is in discrete transistor form, hybrid or encapsulated circuit form or in monolithic IC form (since ICs are basically just a whole bunch of transistors stuck in one package).

Amplification is usually found in two places. The initial 70 MHz amplifier is right after the mixer or second hybrid in the DC. This is sometimes called the bulk gain stage, since its purpose is to provide enough output power to drive the 100-plus feet of coax cable. The second amplifier is found in the receiver's i-f strip. It usually consists of several stages of gain with bandpass filtering in between. There is a final gain stage which drives the limiter circuit. There is often another amplifier that is used to drive the signal-strength meter.

On extremely long runs of coax cable, a line amplifier may be necessary to properly drive the receiver. This is simply a wideband amplifier centered at 70 MHz. It

typically consists of a hybrid module like the NEC-5157 or a transistor like the MWA120.

An ideal amplifier would add zero noise and phase shift and would raise the amplitude equally over the entire passband. Unfortunately, in the real world there will always be some phase shift between high and low frequency components as they go through the amplifier. Gain variations will always occur as well across the band, but they must be kept to a minimum. Poor differential gain and phase will degrade performance, and will often be the reason a descrambler won't work properly with the receiver.

The differential gain and phase of most standard, low-cost VHF antenna amplifiers, if measured from 55 to 85 MHz, is not too good. This is why they make lousy 70 MHz line amps. They're designed to pass a 6 MHz channel with good differential phase and gain, but a TVRO channel is about 30 Mhz wide, which is fine if your design takes that into consideration from the outset. That's why, if a line amplifier is necessary, choose one that is designed specifically for TVRO usage.

70 MHZ BANDPASS FILTERS

The purpose of a bandpass filter is to allow only a certain frequency range of signals to get through. The frequency range of interest starts out 36 MHz wide in the

Unfiltered 70 MHz
Signals @ Output
Of The D/C

30 MHz   50 MHz   70 MHz   90 MHz   110 MHz

Fig. 9-2. The unfiltered 70 MHz output from a down converter. The desired channel is centered at 70 MHz. The adjacent channels (again cross-polarized) are at 50 and 90 MHz, while the adjacent copolarized channels are centered at 30 and 110 MHz.

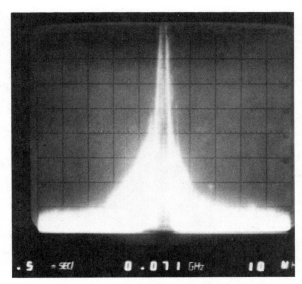

Fig. 9-3. The same setting as Fig. 9-2, but taken at the output of the i-f filter. Notice that all the other signals have been attenuated and that the only signal remaining is the desired channel's energy.

uplink and is maintained that wide in the downlink signals. Because home TVRO systems have to compromise on price, the size of the dish, and on the specifications of LNAs and DCs, the real bandwidth that is used by most TVRO receivers is only 22 MHz to 28 MHz wide.

To detect video and audio, the i-f bandpass must be properly sized to allow the necessary information to pass without allowing excessive noise or unwanted signals to get amplified. This filtering predominantly takes place in the receiver's i-f strip.

When the desired channel is down converted to 70 MHz, all the other channels are likewise down converted. Figure 9-2 is a spectrum analyzer picture of what is on a 70 MHz cable. It is centered on TR 7 on F-3R. In the photo, TR 5 is centered at 30 MHz and TR 9 is centered at 110 MHz, TR 11 is at 150 MHz and so on.

Figure 9-3 is another analyzer picture, this time taken at the input to the limiter circuit. What is now left after filtering is just 13 MHz above and below 70 MHz. All the other transponder signals have been filtered out. The upper and lower 5 MHz of the desired transponder is rolled off to maximize the signal to noise ratio.

LIMITER CIRCUITS

Since most types of FM demodulators will also detect AM signals and since noise is an AM signal riding on the FM signal, then before the receiver demodulates the video it needs to strip off the AM noise to prevent it from being detected. This is done in a circuit called a limiter. The limiter takes the FM signals which look something like Fig. 9-4A, and more or less chop the top and bottom off the sinewaves. This results in a signal like that shown in Fig. 9-4B. Thus the input to the limiter is a sinewave, and the output of the limiter is a squarewave which is identical in frequency changes to the original sinewave.

If the output looks like a squashed down version of the input signal (i.e., still a sinewave), then the limiter is said to be a soft limiter. If the output is a clean squarewave, then it is a hard limiter. Basically all that we are interested in, is the zero-crossing transition points. If we know where the signal crosses through zero then we can use that information to determine its frequency, and in turn, create a replica of the broadcast signal. If the signal is still a sinewave, then there will also be phase changes taking place which will cause spurious level changes which distort the true video or audio reproduction.

If our limiter circuit is not driven hard enough, because of design or because of low signal level to the

Fig. 9-4. (A). Typical input signal to the limiter. It is a frequency modulated sinewave. Ideally the amplitude would be the same, but because noise is amplitude modulated and because the signal is fairly noisy due to atmospheric noise, LNA noise, and 70 MHz noise, amplitude variations result. These are stripped away in the limiter. (B). Output of a hard limiter for the input signal in Fig. 9-4. The video detector should only see the frequency changes as the signal crosses zero. If the output of the limiter is not a squarewave, then phase modulation will occur and cause a streakiness to the video.

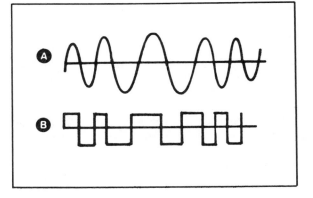

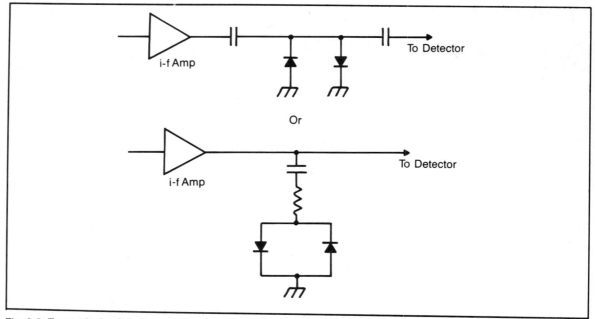

Fig. 9-5. Two typical soft limiters. As the voltage goes above the turn-on point of the diodes, then they conduct shunting the remaining positive or negative signal to ground. The output is close to a squarewave, but there will always be some tilt left due to the characteristics of diodes. Thus there will be some phase modulation of the signals left.

receiver, then it may not go into hard limiting. The result will be a soft limited signal that hits the detector circuit. A receiver with soft limiting will have streaky and noisy video due to the signal noise that is still present, and also due to the fact that the detector is trying to decode an FM signal and a phase modulated signal.

Types of Limiters

The most basic limiter circuit consists of two diodes hooked between ground and the signal line. The diodes will conduct or shunt the signal once they are forward biased. In the case of germanium diodes, this is .3 volts, for silicon diodes this is .6 volts. If a much higher level signal is applied to the diodes, the output will be a clipped waveform which will be .6 volts or 1.2 volts in level. This is shown in Fig. 9-5. If the signal is much higher in level than 1.2 volts, then a fairly good approximation of a squarewave is obtained. If it is too low, then there will be too much phase modulation to contend with and the detector circuit will not give an accurate representation of the original signal. Many receivers use this type of limiting, including the DXR 1100 from Dexcel, the 9530, 9540, and earlier 9550's from Luxor, and the 20/20 from Birdview.

A more useful limiter can be made from a number of digital ICs. The ECL (Emitter Coupled Logic) family is a series of fast digital ICs designed for high speed operation. Thus they are perfectly suited for 70 MHz operation. They basically take the incoming signals and respond to the zero-crossing point. The output is a cleanly clipped squarewave. The 10114, 10115, and 10116 are used by STS, Dexcel, Gillaspie, ICM, and Amplica in most of their receivers. The 10107 is used by Toki in their receivers.

Drake favors the 74S00 quad NAND gate chip as a limiter. It is used in the early ESR 24, the ESR 240, the ESR 324, and in the Drake-manufactured Conifer RC-2001. Drake also uses an overdriven MWA130 as a limiter in their later ESR 24's, as well as for the receivers made for Winegard (models 7032, 7035, and 7037).

Sweeping the I-F

To properly adjust the bandpass filter of a receiver requires a sweep generator to supply a signal that sweeps through the passband of the receiver, and that also syncs an oscilloscope so that a meaningful picture is obtained. The sweep range should be from about 50 MHz to about 90 MHz. Figure 9-6 shows the typical hook-up for

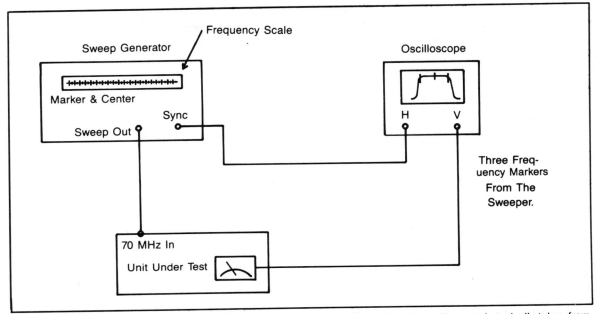

Fig. 9-6. Sweep oscillator hook-up for sweeping the i-f strip. The vertical input for the oscilloscope is typically taken from the signal-strength meter.

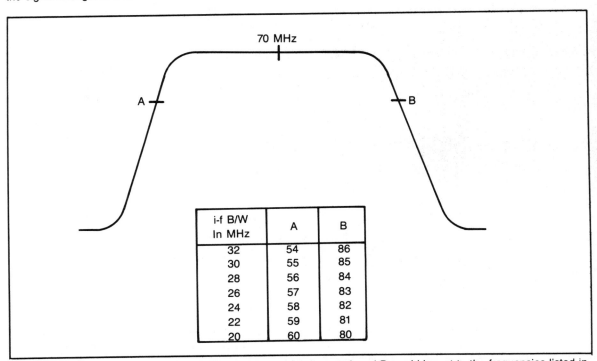

i-f B/W In MHz	A	B
32	54	86
30	55	85
28	56	84
26	57	83
24	58	82
22	59	81
20	60	80

Fig. 9-7. Ideal i-f filter output from a 70 MHz i-f strip. The markers at A and B would be set to the frequencies listed in the table according to the desired i-f bandwidth (B/W).

sweeping a receiver. The sweeper output is hooked into the 70 MHz input jack on the receiver. The scope is hooked into the receiver just before the limiter. Often the signal strength meter is the easiest place to tap off the signal since it has a detected signal. There is also a connection between the sweeper and the oscilloscope for horizontal sync.

The signal must not overdrive the i-f strip or else the filter adjustments will be incorrect for the much lower-level satellite signals. There are two good ways to tell if the signal level is correct. First, never go beyond the half way indication on the signal strength meter. Second, watch the response curve as the input signal is adjusted. There will be a point at which it will flatten out at the top and saturate the i-f amplifier. Back the level back down and raise the gain on the oscilloscope until a trace similar to that shown in Fig. 9-7 is obtained.

Markers are used to tell where you are in regard to frequency. These are generated in the sweeper and should be adjustable so that different bandwidths can be properly swept. Typically these markers are set at the -3 dB points and also at 70 MHz. The lower marker is usually between 52 MHz, for a 36 MHz bandwidth and 59 MHz for a 22 MHz bandwidth. The upper marker is set between 81 MHz and 86 MHz as the bandpass goes from 22 MHz to 36 MHz. The markers are the blips on the skirts of the bandpass trace shown in Fig. 9-7. A frequency counter is useful for setting the exact marker frequencies.

TROUBLESHOOTING

There are basically three symptoms that are indicative of possible i-f problems in a TVRO receiver: no picture or sound, poor picture and some sound, and a negative-image picture. These same symptoms can also be caused by the DC, so before you start tweaking the i-f coils, be sure to check out the DC.

If you have a 20 MHz or faster oscilloscope, then the 70 MHz signal can be seen by using a detector probe. If you don't have a detector probe, then a germanium diode will suffice. Hook one end of the probe to the diode and probe the circuit with the other end of the diode. The diode's polarity is not critical, although it does affect the video polarity as seen on the scope. What you are doing is looking at the video components in the 70 MHz signal. If you were to look at them directly (without the detector probe), then all that will be seen is a fuzzy fat trace, usually fairly low in level. This by itself may be enough to tell you that something is getting through.

If we take a typical i-f circuit that uses a MWA120 transistor for gain, then we should see some signal increase from pin 1 (the input pin) to pin 2 (the output pin). Most MWA110, MWA120, and MWA130 circuits have about 15 dB to 25 dB gain at 70 MHz. The easiest way to check them is to find the bias resistor at the output. It is usually about a 330 ohm resistor. A typical schematic of an MWA120 is shown in Fig. 9-8.

If one of the stages in the i-f strip fails, there is usually enough signal leakage that some signal will bypass the dead stage. Unless the dead stage is an emitter follower, a negative image will be formed because the signal should have been inverted by the dead stage, but of course wasn't. Again probe the circuit to find either where the signal dies, or where there is no gain.

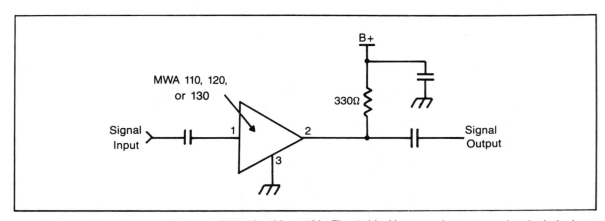

Fig. 9-8. Typical circuitry surrounding an MWA110, 120, or 130. The dc blocking capacitors are used on both the input and the output. The resistor is typically 330 ohms.

If someone has been playing with the i-f filter coils, then a full alignment is necessary. In most cases, the coils would have to be broken or have had the ferrite removed to cause complete loss of signal. The symptoms will be excessive sparklies or, in the case of a PLL, loss of lock and the subsequent tearing on black/white transitions. Usually poor picture quality, loss of audio or colors, or low signal strength readings will indicate possible i-f misalignment. Of course, broken ferrite or ferrites all the way in or out are also a good indication that something is amiss in the i-f strip.

One other problem that sometimes occurs, especially in KLMs that use a ceramic filter in the i-f, is a temperature sensitivity. This can cause loss of signal strength, picture, colors, sound, etc. In some of the KLM receivers, the ceramic filter they use has a tendency to become heat sensitive. Replacing it will usually clear up the complaints like it works fine for the customer for only an hour or two, and yet on the bench (without the cover on) it plays fine for days and days. In heat sensitivity cases, I use a variac and set the voltage to about 125 Vac. Usually this is enough to bring those on-the-edge components out in the open after about a half hour or so. Adding some heat with a heat gun will speed up the process.

70 MHZ I-F CIRCUITRY

Figure 9-9 is the i-f strip from a Gould/Dexcel DXR 900. It uses a quad-section ECL chip, a MC10115, for both 70 MHz amplification and limiting. RV-1 is the input level control, and five adjustable coils and one trimmer cap are used to adjust the bandpass shape. Its −3 dB bandwidth is about 24 MHz.

The filter circuit is really two filters, one tuned above 70 MHz and one tuned below 70 MHz. L6 and C48 make up the upper filter, while L8 and C49 make up the lower filter. CT-2 controls the coupling between the two filters.

L1 is used to adjust out any tilt across the whole bandwidth. L6 adjusts the centering of the 78 MHz bandpass filter, while L8 adjusts the centering of the 62 MHz bandpass filter. CT-2 adjusts the coupling between the filters and will cause frequencies between about 62 MHz and 78 MHz to raise or dip. L9 adjusts the upper side (82 MHz) response skirt, while L10 adjusts the lower side (58 MHz) skirt. The proper i-f response is shown in Fig. 9-7.

U2A is used as a buffer amp, and its output is split to feed the limiter and to feed another amplifier section to drive an analog signal-strength meter.

KLM I-F

The KLM Sky Eye X takes a different approach. It uses 2SC2498 and 2SC1674 npn transistors to supply amplification and two hybrid modules with fixed component values to filter the 70 MHz signals. The input gain is selectable for 0 dB and -5 dB attenuation. It is changed by moving a jumper on the main board. VR-006, which is on the display board, is listed as a signal-strength adjustment, but it strictly sets the voltage level going to the LED display. All KLM receivers employ a similar style of i-f filtering and amplification.

Birdview I-F

Birdview's approach, shown in Fig. 9-10, uses four MWA120's supply i-f gain. They use a similar filter to Dexcel's, in that it is a split filter, whose coupling is controlled by L4. The higher filter is controlled by L3 while the lower-side filter is adjusted by L5. L6 is a 70 MHz peaking filter while L7 couples the 70 MHz signal into the next MWA120. I3 and I4 are used to drive their limiter, which consists of CR1 through CR4.

Luxor I-F

The Luxor i-f, shown in Fig. 9-11 is out of the 9540 and early 9550's. It differs from the Dexcel, Birdview, and KLM i-fs in that it has an agc (automatic gain control) circuit. The agc circuit takes a split from the amplified i-f signal via CC27 and rectifies it (DC06 and DC07). The derived dc voltage drives transistor TC02, which in turn drives TC01, which is common base amplifier. As its bias changes, a varying dc voltage is sent through LC01, an rf blocking inductor. This dc voltage biases the TDA1053 diode block either on or off, adjusting the input level to TC03.

The i-f filter response is set by adjusting LC03, LC04, LC06, and LC07. The limiter circuit is composed of DC02 and DC03. Unfortunately, these diodes had a tendency to only soft limit. This was the cause of the generally poor quality pictures that plagued the 9530, 9540, and early 9550s. Figure 9-12 is the limiter from the later 9550s. All the rest of the i-f circuit is the same as the earlier receivers. The big difference in performance is due to the hard limiting that is found in the CA3102 circuit. The CA3102 is a dual differential amplifier, which is biased so that it clips the tops of the signals. This results in a close approximation of a square wave.

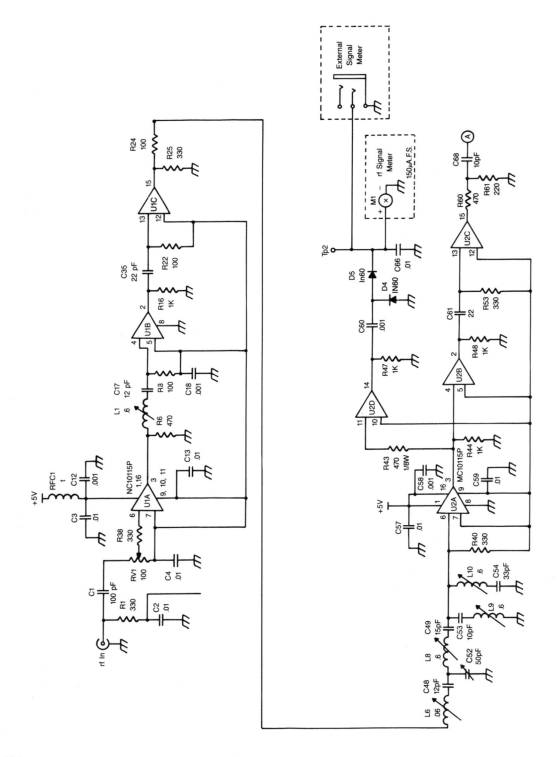

Fig. 9-9. An i-f strip that uses the MC 10115 ECL chip to both amplify and limit the i-f signal. Courtesy of Gould/Dexcel.

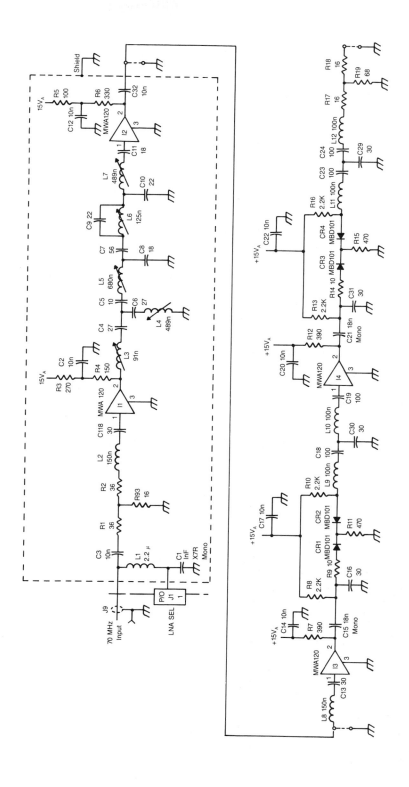

Fig. 9-10. The i-f schematic from a Birdview 20/20 receiver. The top section is similar to a Drake filter in that it uses two MWA 120 transistors and a five coil filter. The lower part is the remaining MWA120 amplifiers and the two section limiter (CR1 through CR4). Courtesy of Birdview Satellite Communications.

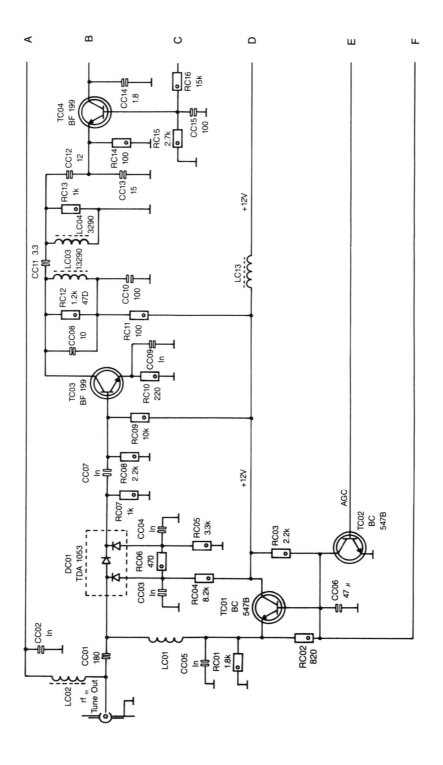

Fig. 9-11. The i-f strip and detector from a Luxor 9540. It features an agc circuit as well as an afc circuit. The limiter in the 9540 (DC02 and DC03) was a soft limiter, so the pictures are not as good as those of the 9550 which used a hard limiter. Courtesy of Luxor North America.

(Continued from page 106.)

107

(Continued from page 107.)

108

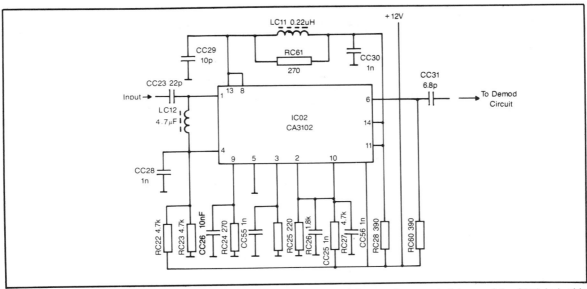

Fig. 9-12. The difference between the streaky pictures of the 9540 and the new improved pictures of the 9550 is in this circuit. It is the hard limiter circuit which replaced the diode limiter shown in Fig. 9-11.

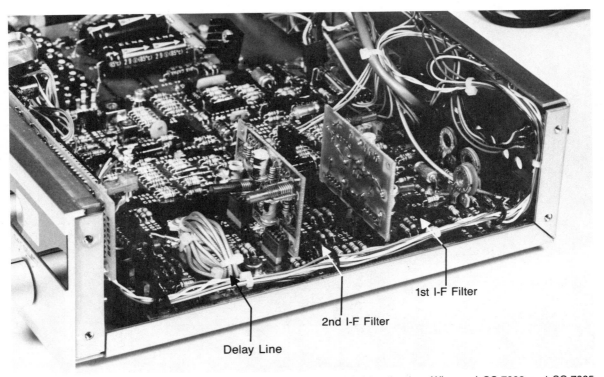

Fig. 9-13. Photograph showing the location of the i-f filter boards and delay line in a Winegard SC-7032 and SC-7035 receiver. Courtesy of Winegard.

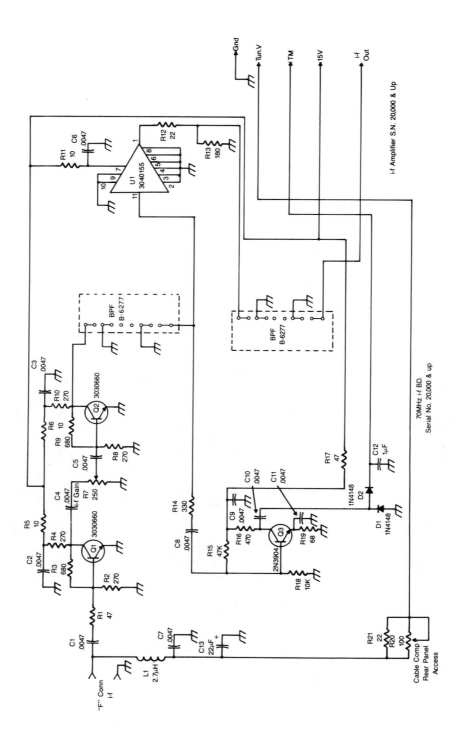

Fig. 9-14. Schematic of the newer Drake i-f filter and amplifier. It also includes the meter amp and detector (Q3 and D1 and D2). The BPF (bandpass filter) modules consist of the 5 coils and 7 capacitors shown in Fig. 9-15 between the two MWA 120 transistors. Courtesy of R. L. Drake.

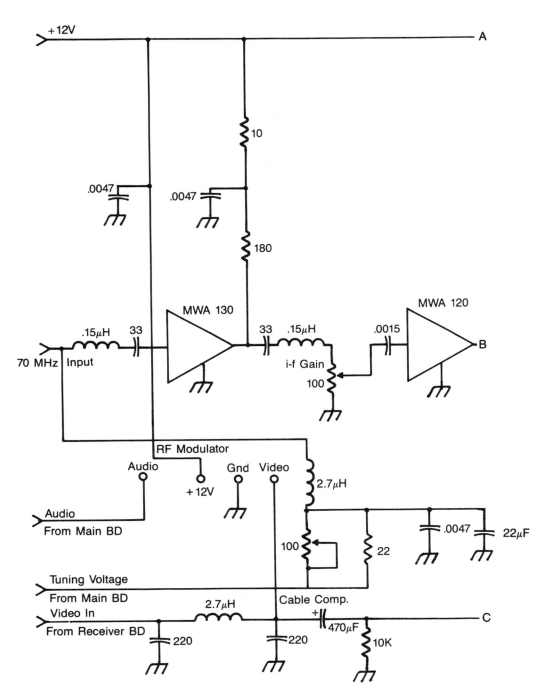

Fig. 9-15. An earlier version of the i-f strip used in many Drake, Winegard, Conifer, and Channel Master receivers. In some models, the MWA120 was replaced by a 5157 amp module. The 74S00 NAND gate is the limiter. Courtesy of R. L. Drake. (Continued on page 112.)

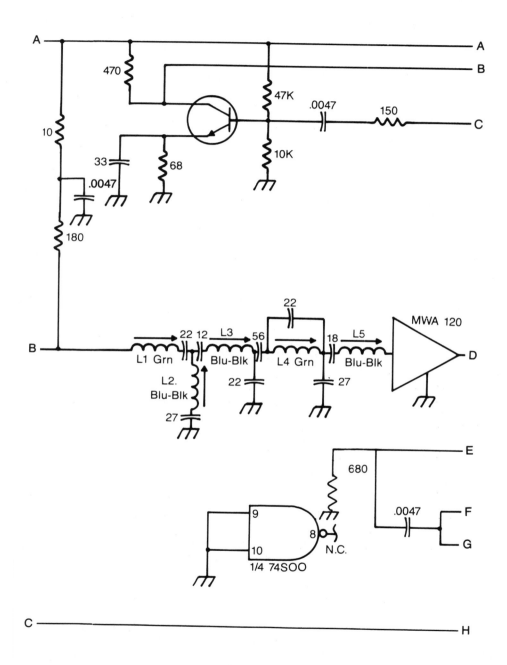

(Continued from page 111.)

112

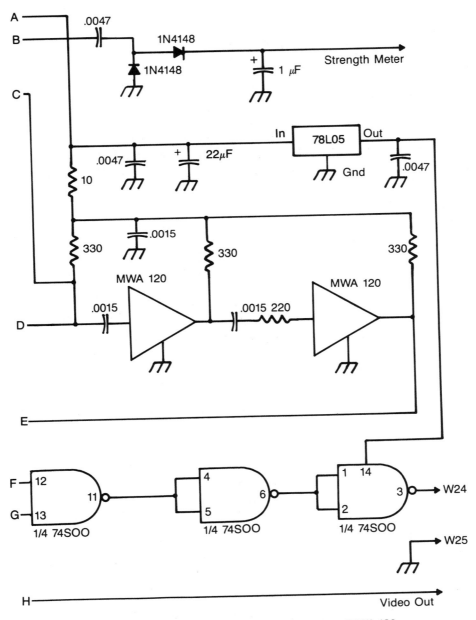

Schematic 70 MHz i-f Amp/MWA 120

DWG.#C-6107

i-f Amplifier
Below S.N. 20,000

(Continued from page 112.)

Drake I-F

Since Drake has manufactured receivers for Conifer, Channel Master, and Winegard, it follows that the circuitry would be very similar to those used in Drake's own receivers. To identify a Drake-made receiver, look at the back panel, if the terminal strips and labeling looks like Drake's, then the receiver was probably made by Drake.

All the various private label receivers are basically the same as the Drake receivers, especially in regards to the i-f amplifier, filter, and limiting circuits. Figure 9-13 is a photograph from a Winegard SC-7032 showing the two i-f filter modules. These are identical to just about every other Drake manufactured receiver. The plug-in board contains five bandpass-shaping coils and all the resistors and capacitors used in the i-f filter.

Figures 9-14 and 9-15 show two versions of the i-f amplifier and filter that are used in Drake's own receivers (model ESR 24). The first version (Fig. 9-14) consists of two npn transistors that are used to drive the first i-f bandpass filter module (BPF). The output from the first BPF is split to drive the meter amplifier circuit as well as the next i-f amp. The meter amp (Q3) and detector (D1 and D2) in turn drive the tuning meter (TM). An MC5157 (U1) hybrid amplifier drives a second i-f filter module. The output drives an MWA130 transistor (that's not shown) which clips or limits the signal.

In Fig. 9-15, MWA130 and MWA120 transistors are used to boost the signal enough to drive the filter module. The .15 μH inductor and 33 pF capacitor make up a broadband series-resonant circuit centered at 70 MHz. The filter module's various components are shown in this schematic. They are the five adjustable coils and eight capacitors between the two MWA120's in the center of the schematic. The coils are adjusted while sweeping the i-f for a bandwidth of 27 MHz at the -3 dB points. The meter is tapped off right after the first MWA120 after the filter. The 2N3904 amplifies the 70 MHz signal to drive the meter through the two 1N4148 diodes. The three MWA120's after the filter are used to drive the limiter, which is made up of three sections of a 74S00 quad NAND gate IC. The early ESR24's (below serial #20,000) used this circuit. The Conifer RC-2001's i-f strip is very similar, except that the MWA120's are replaced by two MC5157's.

Video Processing

ONCE THE SIGNAL HAS BEEN BANDPASS FILTERED to the desired bandwidth and the amplitude-modulated noise stripped off in the limiter, the next section of the receiver that the signal comes to is the demodulator. The demodulator circuit separates the carrier from the information that it is carrying. The carrier is eliminated leaving what is called baseband video.

In an AM radio or TV, this type of circuit is called a detector and in an FM radio it is called a demodulator or discriminator, depending on the type of circuit that is used. In the TVRO world, it is called by several names. Some people generically refer to it as the video detector or discriminator, while others use detection circuit or demod, (short for demodulator circuit).

DEMODULATOR CIRCUITS

Regardless of the name and type of circuit, the circuits all do the same thing. Their purpose is to take the carrier wave, which is changing in frequency up to 18 MHz above and 18 MHz below the center frequency, and convert the frequency changes into a video signal. This is accomplished in several ways.

PLL Demodulators

One of the most popular video demodulation methods, used by the first generation TVRO receivers, involved the Phase Lock Loop (PLL) circuit. The most commonly used circuit was built around the NE564 PLL chip from Signetics. One of the first, if not the first, design of a TRO demodulator circuit using a PLL chip was done by Steve Birkill, of England. He first demonstrated its feasability for 4 Ghz TVRO in the summer of 1978. It wasn't long before several TVRO manufacturers were incorporating a similar chip in their receivers.

The PLL chip greatly simplified the demodulation circuit. It reduced the cost and improved the sensitivity in one stroke. The biggest problem with the NE564 IC was (and still is) its maximum operating frequency, which is speced at 50 MHz. Needless to say, some PLL chips worked fine up to 100 MHz, but most would not cut the mustard. The first generation of PLL circuits, which used a 70 MHz input, resulted in an overall poor to OK performance, depending almost entirely on the NE564 high frequency performance. To improve performance and dependability, it was found that if the frequency of the signal could be divided in half, down to 35 MHz, or even by one fourth, to 17.5 MHz, then the operating frequency would be within the PLL's specifications and improved performance would result.

A PLL chip is made up of several discrete parts (see Fig. 10-1). The signal is input through pin 6. It first enters a limiter which produces a constant amplitude output to drive the phase comparator. The phase comparator is a double-balanced mixer which mixes the VCO output (usually pin 9 tied directly to pin 3), with the limiter output. The difference signal is then passed on to an amplifier and also fed back to the VCO. The feedback signal is used to lock the VCO onto the incoming frequency. The amplifier is actually a unity gain transconductance amplifier and comparator. It is used as a post demodulation low-pass filter. There is also a Schmitt trigger included, but it isn't used in TVRO applications.

As the PLL tries to lock onto the incoming signal, the voltage at pin 14 varies in direct proportion to the difference between the input frequency and the VCO's free-running frequency. Pin 14 is thus a voltage corresponding to the frequency changes occurring in the incoming carrier. This voltage is a replica of the original video signal. Pin 12 and pin 13 usually have an adjustable ca-

pacitor to center the VCO's lock-in range to the incoming frequency. On PLL's used for audio detection, the capacitor is usually replaced by a varactor diode. Pins 1, 2, and 10 are usually tied to the VCC line either directly or through small value resistors.

Figure 10-2 is the NE564 circuit from an Amplica R-10 receiver. It is typical of the divide by four usage of the PLL. Its input frequency is 17.5 MHz. It again inputs the divided carrier into pin 6. The video output is pin 14 which goes to a 2N2222 transistor, used as a buffer amp.

Pin 7 is the connection to the bias filter. It is tied to pin 6 via a 1K resistor and to ground via the .01μF capacitor. This same type of input is also on both the 70 MHz and 35 MHz PLL circuits. Pins 3 and 9, the VCO output and phase comparator input, are tied together with a selected-in-tuning (S.I.T.) bypass capacitor to filter out i-f crosstalk. The circuitry between pins 12 and 13 sets the lock-in center frequency. It is what determines whether the chip is to look for signals centered at 70, 35, or 17.5 MHz. Typically the trimmer capacitance is 1 to 10 pF at 70 MHz, 5 to 20 pF at 35 MHz and 15 to 35 pF at 17.5 MHz. Pin 11 is used as a test point for setting the center frequency by use of a frequency counter. Pin

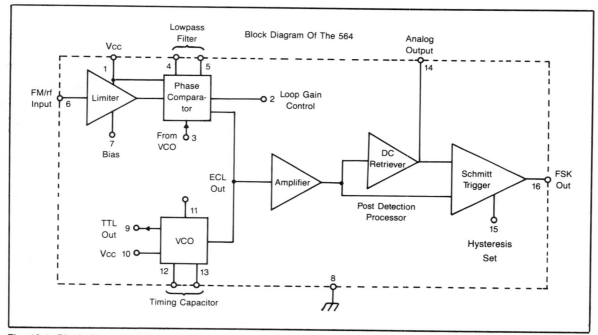

Fig. 10-1. Block diagram of a NE564 Phase Lock Loop (PLL) IC. It consists of an amplifier/limiter, a phase comparator (similar to a DBM) and a VCO. The error voltage between the VCO's frequency and the input frequency is fed back to the VCO and is also amplified. The detected video or audio is output on pin 14.

1 is the VCC (+ 12 Vdc) input. Pins 2 and 10 are also tied to VCC through 8.2K and 240 ohm resistors.

The frequency at which the PLL chip is run has a direct bearing on how hot the chip gets. One of the first problems manufacturers found was that the chips would burn-up at 70 MHz (and even 35 MHz) if there was no heatsink. So most manufacturers added heatsinks. Heatsinks only work correctly if there is a direct thermal contact between surfaces. In some receivers, the heatsinks were stuck on with glue, instead of heatsink compound. This acts like an insulator and thus doesn't adequately transfer heat to the heatsink. If you have to replace a NE564, be sure to scrap off any glue that may be found on the heatsink, and to use heatsink compound between the IC and the heatsink.

You can use used desoldering or wicking braid, filled with solder, for holding the heatsinks onto the IC. Just solder one end to the heatsink and the other end to the ground plane on the board. Use one ground strap on each side of the U-shaped heatsinks. This has the added advantage of grounding the heatsink and thus provides better shielding of the NE564 VCO signal from radiating back out into the circuit.

It was also found that even if the chips were presorted

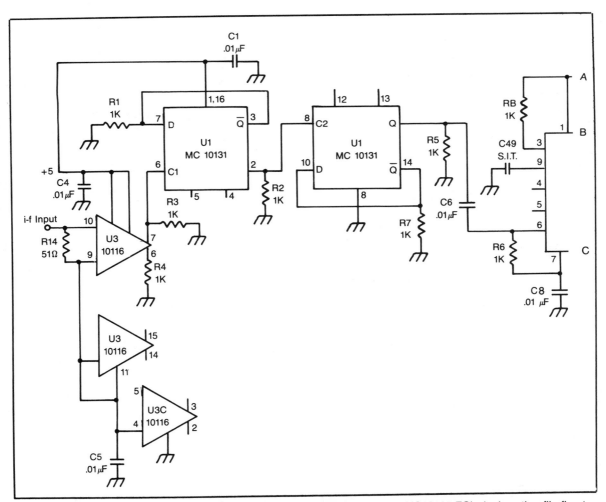

Fig. 10-2. A divide by four PLL circuit from an early Amplica receiver. It uses a MC10131 ECL dual section flip-flop to divide the signal and digitize it so that a clean squarewave is input to the 564 chip. C57 sets the PLL lock-in center frequency at about 17.5 MHz. The PLL frequency is read at E2. Courtesy of Amplica. (Continued on page 118.)

into those that would make it at 70 MHz, there would still be a percentage that would not produce very good video, so IC sockets were added onto the board so that, if need be, the ICs could be easily changed during final test. This carried over to the 35 MHz and to some 17.5 MHz circuits as well.

Figure 10-3 is the demodulation section from a Dexcel DXR 1100 receiver. It uses a 74S74 divide-by-two chip before the NE564 to lower the operating frequency to about 34 MHz. Its circuitry is very similar to that in the Amplica receiver except that it is running at 34 MHz. If a swept waveform is sent into the PLL, then what will

appear at the output is a straight line where the PLL tracked the sweep. This is shown in Fig. 10-4. Here a dual trace scope is used with two probes taking signals out of the receiver. The haystack shape is the output of the i-f amplifier and is taken off of TP1. It is centered at 68 MHz. The line that cross almost diagonally is the lock-in range of the 564 chip. It is taken off of TP2, which is on the output (pin 14) of the 564. The lock-in point is at 21 MHz and the lock-drop point is at 47 MHz, because the frequency is 1/2 of what the i-f amp and filter circuit is running at. The receiver uses 68 MHz to prevent crosstalk between the rf modulator and the i-f strip, which

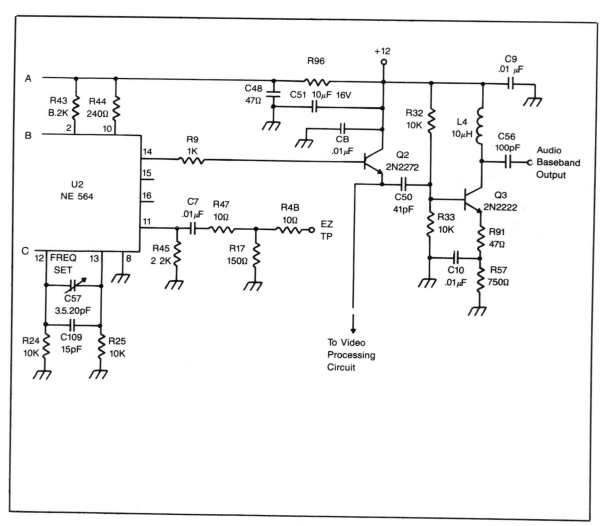

(Continued from page 117.)

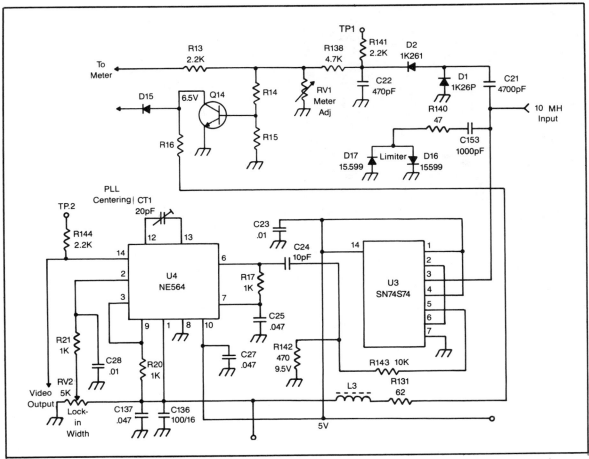

Fig. 10-3. A divide by two PLL circuit from the DXR 1100. A TTL flip-flop (U3) is used to divide the signal's frequency in half before it is sent to the 564 chip. The 74574 also digitizes the signal, which is why the soft limiter (D16 and D17) does not figure into the picture quality. CT1 is the PLL lock-in centering adjustment, while RV2 adjusts the width of the lock-in range. Courtesy of Gould/Dexcel.

may cause herringbones in the rf output signal.

IC Balanced Demodulator

The most popular IC balanced demodulator circuit has at its heart the 1496 balanced modulator/demodulator chip. This chip is manufactured by many companies and thus has many prefixes. The chip is the same whether its called an LM1496, MC1496, NE1496, or NJM1496. The 1496 produces an output voltage that is proportional to the product of the input signal and a carrier signal.

The basic circuit is shown in Fig. 10-5. This is from a KLM Olympiad I. The LM1496 is listed as U1. The 70 MHz input signal is capacitively coupled to pin 1 and

to pin 10. Pin 10 has an adjustable, by trimmer cap C28, phase delayed signal applied to it. Pin 10 is the − carrier input, while pin 8 is the + carrier or return input. These are mixed with the out of phase signal inputs on pins 1 (+ signal input) and 4 (− signal or return input). The result is a balanced baseband video output. The − signal or normal polarity is on pin 12, while the + signal or inverted polarity is on pin 6.

The normal polarity baseband video from Pin 12 is sent to buffer transistor Q6. Pin 6, the inverted polarity baseband output, is sent to the afc circuit. The trim pot, R70, is the afc center adjustment. The trimmer cap, C28, is used to adjust the S-curve for proper symmetry.

This type of circuit is used in all the newer KLM

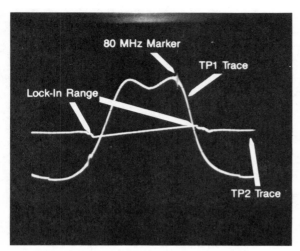

Fig. 10-4. Scope photograph of the signals found when sweeping the DXR 1100 PLL circuit. The i-f filter response from TP1 is shown as well as the lock-in range of the PLL from TP2. The i-f filter is about 26 MHz wide at the −3 dB points.

receivers, in the ICM SR-4600 and in many other receivers. The 1496 is also used in the Release descrambler, to remove the modulating waveform that is added onto the video and audio signals.

Delay Line Discriminator

The delay line discriminator is usually composed of discrete components. In its basic form, the discriminator consists of a delay line and two diodes. The delay line phase shifts the incoming signal which is applied equally between the two diodes. The mixed output is a voltage which is proportional to the frequency shifts applied to the carrier wave. As the frequency goes above the nominal center frequency (70 MHz), the voltage goes positive, as the frequency drops below 70 MHz the output voltage goes negative.

Some delay line discriminators used a balanced bridge arrangement. A typical example of this type of circuit is the R.L. Drake ESR24 (Fig. 10-6). In this circuit, the 70 MHz signal is split between two 47 ohm resistors. One signal line goes to a delay line which delays the signal by 3/4 of a wavelength. The other goes to a transformer. The transformer's secondary is centertapped to ground, which causes two signals, 180 degrees out of phase with each other, to form at each end. The delayed signal is also applied to a transformer with a center tapped secondary. Its signal is also applied to the bridge rectifier, or

double-balanced mixer as it is more appropriately called in this case.

The difference signal is taken from the centertap of the delayed signal's transformer. As the frequency changes, so does the phase of the two signals. These changes are reflected in what is called the S-curve. An S-curve is shown in Fig. 10-7. When the frequency is exactly 70 MHz, then the output will be zero volts. As the frequency rises, the voltage will go positive, and as the frequency goes below 70 MHz, it will go negative. With time, this voltage fluctuation recreates the original video signal. Typically the signal will be about 100 to 200 mV peak-peak. This signal is routed to many different areas in the receiver. It goes to the afc circuit, to the center tune meter, the baseband output, the audio detector circuit and to the video circuitry.

The delay line demodulator from a Conifer RC-2001 is shown in Fig. 10-8. It is the same demod circuit as the Drake ESR24. Notice the coil of coax. It is cut so that it is 3/4 wavelength long from one end to the other. Thus the signals at the other end are phase shifted 270 degrees in relation to the input signal.

Quadrature Detector

The quadrature detector is similar to the Foster-Seely discriminator. It processes the 70 MHz signal by splitting it into two balanced signals whose phase is 90 degrees apart. These are then combined in phase and out of phase to demodulate the baseband video.

The most popular chip that is used is the 1357. It is a quadrature detector designed for the standard FM radio 10.7 MHz i-f. But it will function very nicely at 70 MHz. Again it can be found with many prefixes such as MC1357, LM1357, and NE1357. Figure 10-9 is the quadrature circuit for the Dexcel DXR-900. This receiver has an improved threshold and video performance over that found in the various 70 MHz and below PLL circuits.

The input to the circuit is through pin 4. The signal has been limited and filtered to about 26 MHz at that point. The detector has only three adjustments. The input matching (L4), quadrature lock-in range (L2), and the baseband high-frequency response (CT1). The detected baseband video is output at pin 1. It feeds the audio demodulator, the signal-strength meter and remaining video processing circuitry.

Ratio Detector

The Birdview 20/20 receiver uses a ratio detector.

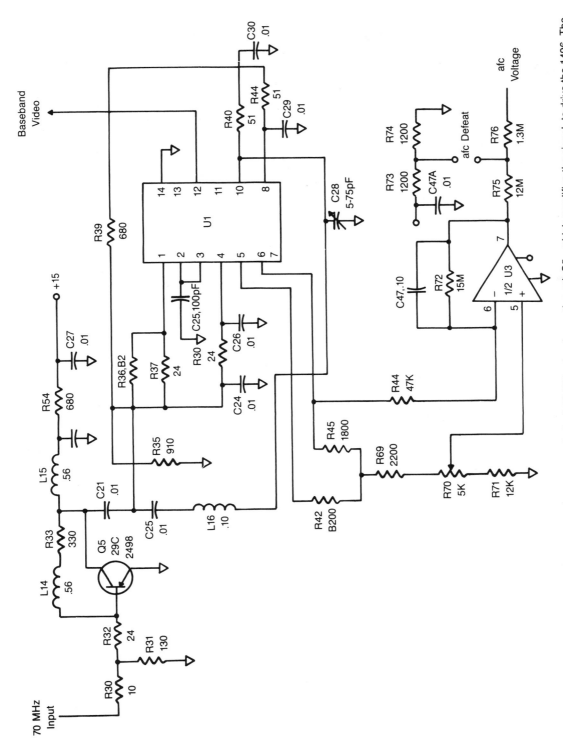

Fig. 10-5. A typical 1496 balanced demodulator circuit. The limited 70 MHz signal goes through Q5, which amplifies the signal to drive the 1496. The output of the 1496 is fed to the video circuitry through pin 12. The balanced signal at pins 5 and 6 is fed to the afc circuit. This circuit is from a KLM Olympiad I receiver. Courtesy of KLM.

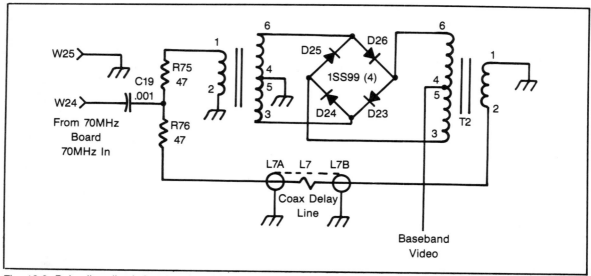

Fig. 10-6. Delay line discriminator as used in all the Drake designed and built receivers. The input is a squarewave of about 5 volts from the limiter circuit. The four diodes and two transformers create a double balanced mixer (DBM) which is used to phase detect the video. Courtesy of R.L. Drake.

The ratio detector circuit is shown in Fig. 10-10. It is like other detectors in that as the signal goes above 70 MHz it causes a positive going voltage to be output to I6, the video amplifier chip. As the signal swings below 70 MHz then that voltage changes to a negative going voltage. This is accomplished by the charges that occur to C45, C46, and C47 as the diodes are turned on and off. If the system is tuned to 70 Mhz and the signal is centered at 70 MHz then the afc output voltage from J4 will be zero.

The 20/20 has a unique video circuit in that it is using a push-pull output (Q7 and Q8) to drive the rf modulator, baseband output, and filtered video output. CR9 is the clamp diode.

Troubleshooting

The most common symptom that accompanies demod circuit failure will be no picture or sound, a picture full of horizontal lines, or a scrambled picture. The screen may also be solid white or black. In almost every case, both picture and sound are equally affected. This is because the sound is detected from the video, so if the video isn't detected properly then surely the sound won't be either. If the sound is coming through, then the problem is most likely in the video processing section, not in the demod.

Tearing in the picture or excessive noise could be due

to a low signal level or a limiter that's not functioning correctly. If a PLL chip is used, then it may be an indication that its lock-in range is too narrow or that its center

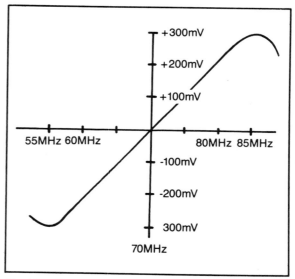

Fig. 10-7. The output voltage from the circuit in Fig. 10-6. As the frequency rises from 55 MHz to 85 MHz the voltage goes from − 300 mV to + 300 mV. Ideally at 70 MHz there would be zero volts output.

Fig. 10-8. A photo of the delay line and i-f strip from a Conifer receiver. It is virtually identical to every Drake manufactured receiver. Courtesy of Conifer Corp.

frequency is misadjusted. It could also mean that the chip's high frequency response is not up to par. NE564 PLL circuits often run at high temperatures and thus their performance can degrade over time. Again not all NE564s can be used at 70 MHz or even at 35 MHz. It's best to install a socket. That way if the chip doesn't make it, you don't have to unsolder it again, which is a sure way to cause lifted pads or cracked circuit traces. Those that don't pass, unless they are completely bad, can be used as audio demodulators since the maximum frequency is under 10 MHz.

The actual signal at the input of most demodulation circuits is a square wave of about 200 to 500 mV. The frequency will vary from about 57 MHz to about 83 MHz. The exact upper and lower limits being set by the i-f filter. If this signal is not a square wave, but is still more like a sine wave, then the limiter is not doing its job. This will cause noisy, streaky pictures because of the phase jitter that will occur in the demodulator, as it tries to follow the frequency changes of the broadcast signal and the phase changes of the remaining carrier wave.

VIDEO CIRCUITS

Once the video has been detected there are still a few changes that must be done to the signal before it can be sent to a video monitor or TV set. The first thing that must be done is to boost up the level of the signals to the 1 volt peak-to-peak that is the standard video level. The next thing that is done is to roll off the high frequency boost that the signals undergo during transmission. This is called video deemphasis. The last thing that must be done to the signal is to rid it of a 30 Hz AM signal that was imposed upon it during uplinking. This signal is called dithering, and it is added onto the video at AT&T's request, to insure that the satellite signals don't interfere with Ma Bell's terrestrial transmissions. It is trapped out by the clamping circuit.

The amount of boost required to raise the video to 1 volt P-P depends upon the method of demodulation used. In a delay line discriminator, the signal level is very low because of the losses in the diodes and coupling circuitry, and thus must be amplified in several stages to

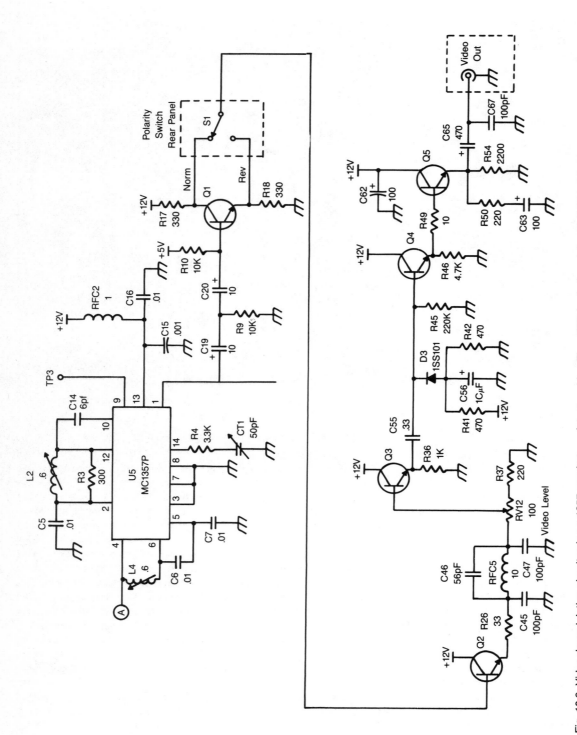

Fig. 10-9. Video demodulation circuit using a 1357 quadrature detector chip. The 1357 is designed for 10.7 MHz i-fs in FM radio, but it works quite nicely at 70 MHz. It puts out a very hot video signal (almost 1 volt p-p) right at pin 1 of the 1357. Courtesy of Gould/Dexcel.

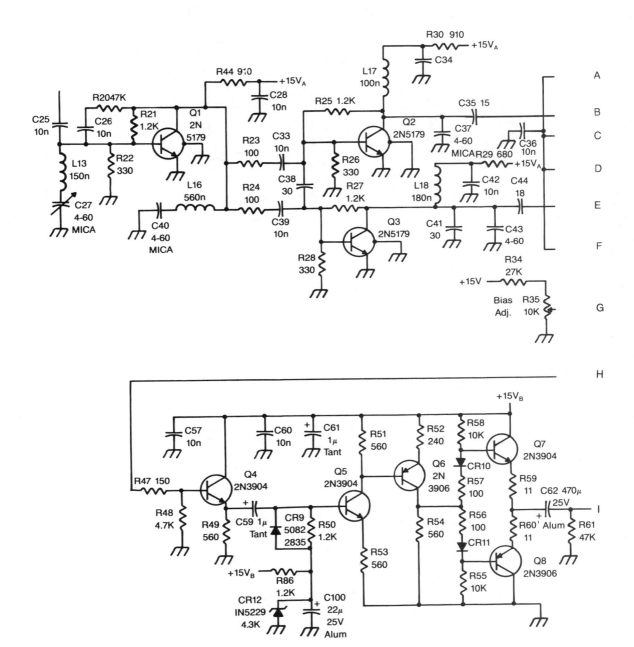

Fig. 10-10. A discrete phase discriminator circuit from a Birdview receiver. It uses the circuitry between Q1 and Q2/Q3 to phase shift the signals so that a response like that shown in Fig. 10-7 is achieved. The Birdview receiver is unique in that it uses a balanced video output amplifier. Courtesy of Birdview. (Continued on page 126.)

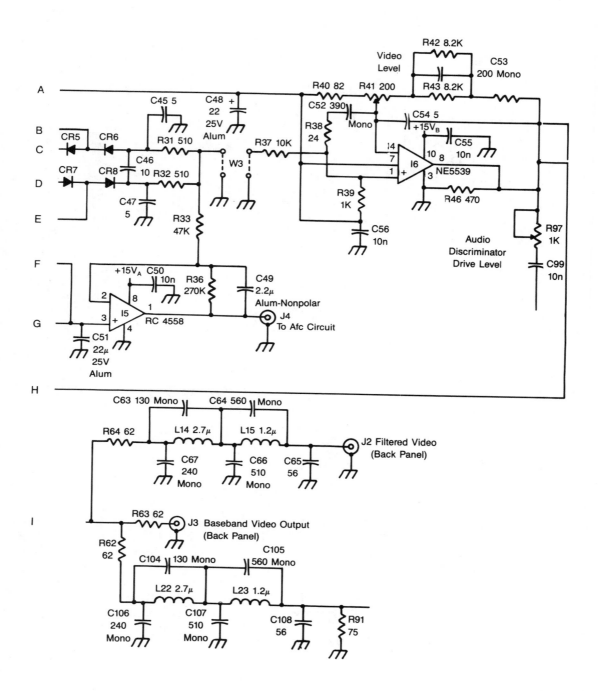

(Continued from page 125.)

126

get the required voltage level. In a quadrature detector like the 1357, the level is already amplified within the chip and is virtually at 1 volt as it comes out of the chip, thus the rest of the video circuit is used for processing rather than amplification.

Usually the first processing circuit that the raw video passes through are two filter networks, the deemphasis filter and the low-pass filter. The deemphasis filter rolls off the higher frequency components in the signal that were boosted during transmission to result in a flat video response. The preemphasis is used to get a better video signal to noise ratio. The second filter is used to attenuate the audio subcarriers located above the highest video frequency, so that they do not interfere with the video. The low-pass filter is designed to roll off the signals above 4.2 MHz. The filter networks consist of several inductors, capacitors, and resistors. A typical filter is shown in Fig. 10-11. It is from the Conifer RC-2001.

Almost all video broadcast today uses the same polarity. A few years ago, there was a method of cheap scrambling that used inverted video. Inverted video is a signal that has the sync pulses going toward the full white level instead of the full black level. To a TV set, trying to lock onto the signals, it is as if there is no horizontal sync. The result is a picture with a wavy black bar (the horizontal sync) down the center, exactly like that seen when the horizontal sync pulses are removed by a suppressed-sync scrambling system.

Once all the satellite receiver manufacturers started putting video polarity switches on their receivers, this method of scrambling was useless. Today the video polarity switch is useless, as there is no programming that uses reversed video on a regular basis. Thus the switch is set to the – position (for negative going sync pulses) and left there. Because the video passes through this switch, there is a possibility for problems occurring with it. The video polarity switch is a good place to use as a check point for video circuits, as it is usually after at least one stage of amplification. If there is no video at all, the polarity switch should be the first suspect.

Most often the video polarity switch will be connected to the output of a 733 or 592 video op-amp chip. The chip outputs a balanced video signal. A balanced signal is actually two signals 180 degrees out of phase from each other. The two signals are output on pins 7 and 8 (the two chips are pin for pin equals). The two outputs go to a single-pole, double-throw slide switch, where one of them is selected to be passed onto the rest of the video circuit. In this type of circuit, the video level is often adjusted by a trim pot connected between pins 3 and 12 or pins 4 and 11 of the IC.

If a 733 or 592 chip is not present in the circuit, then

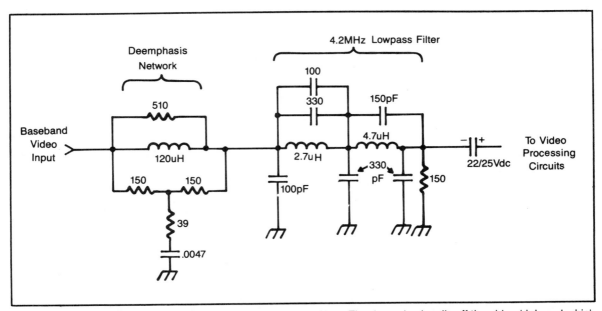

Fig. 10-11. A typical deemphasis network and 4.2 MHz lowpass filter. The deemphasis rolls off the video high end which is boosted during transmission. The 4.2 MHz filter rolls off the audio subcarriers so they do not interfere with the color subcarrier at 3.58 MHz.

often a transistor will be used to invert the input to one of the legs of the switch. In a common emitter configuration, the signal on the collector is 180 degrees out of phase with the signal on the emitter, thus if the collector and emitter are connected to a switch one can choose the video polarity. This type of switching method usually has a buffer stage following it, with the video level adjustment following the buffer stage.

CLAMP CIRCUITS

The next video processing is the clamp circuit. The clamp circuit is used to swamp out the 30 Hz triangle wave that is added onto the video signal during uplinking. This dithering or dispersal waveform spreads the signal more evenly across the transponder so that there are no hot spots occurring in the transponder. This assures that there is no concentrated signal energy occurring in any particular part of the channel. This was set-up originally by AT&T in their deal to share the band with the common carrier satellite communications. This signal supposedly guarantees that the satellite signals will not interfere with AT&T's terrestrial signals. Its too bad we can't get a guarantee from Ma Bell that they won't interfere with us!

After demodulation, this 30 Hz signal causes voltage changes to occur in the video signal. This will appear as rapid brightness changes to the TV set, causing a pulsing of the lighter portions of the picture if it is not swamped out. In most receivers a part called a clamp diode is used to swamp out these signals. The clamp diode is not really a specific part that can be called out by number, as there are many different types of diodes and circuits used to clamp the signal. Figure 10-12 is a photo of an unclamped waveform.

Since there are 30 frames per second in the broadcast video signal and there are 30 Hz (or cycles per second) in the dithering waveform, you might imagine that it is more than mere coincidence that both have 30 occurrences per second. Indeed the two are synced together. Scrambler circuits use this to their advantage, which is why most descrambler boxes require that the video be unclamped. They use the 30 Hz waveform to help sync their circuitry to the broadcast signal so that their decoding circuitry will be looking in the right places for the decoding information. Figure 10-13 is a technical bulletin from Dexcel on unclamping their receivers for using decoders. In most cases, it is best to contact the manufacturer for specific instructions on unclamping the video in their receivers.

If the clamp diode fails, then the picture will take on a pulsing quality to it. In most cases, the diode will have opened if this happens. Many kinds of diodes are used: the standard signal diode (1N4148), Schottky diodes (5280-2800), and hot carrier diodes (1SS101). One cannot substitute a signal diode for a Schottky, although you may be able to substitute the other way if the circuit voltages are right. After the clamp circuit is a buffer and further video amplification. Typically, the last two stages consist of 2N2222A transistors or a FET and a transistor. They are used for voltage gain and to insure that there is a low impedance output. In almost every receiver, the video output impedance is 75 ohms. This means that it must be terminated by a 75 ohm load to get a 1 volt signal out. In most cases, the video input to the rf modulator is also taken directly from the final buffer stage.

VIDEO SIGNAL

A video signal contains all the information necessary for the TV or video monitor to display moving pictures. It contains a synchronizing signal to match the television to the broadcasting station's timing pulses, it contains luminance or picture information, and it contains color information.

In the United States, we use a television broadcasting standard derived by the National Television Standards Committee. Thus we use what is known as the NTSC

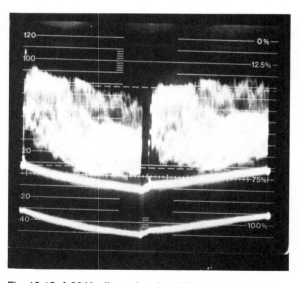

Fig. 10-12. A 30 Hz dispersion signal that is added onto the video signals to prevent interference with terrestrial transmissions. It is clamped or trapped out by the clamping circuit in the receiver. If the clamp is turned off, then the picture will pulsate at a 30 Hz rate.

Gould Inc., Dexcel Division
2580 Junction Avenue
San Jose, California 95134
Telephone (408) 943-9055
TWX 910-338-0180

GOULD

Electronics

TECHNICAL BULLETIN

To: Innovision Distributers   From: Customer Service   Date: 11/19/84

Model #: A11   Bulletin #: I-001

Subject: Using A Decoder With Gould Receivers

This procedure explains how to use a Gould Innovision Satellite Receiver with a decoder.

Gould/Dexcel recommends the unit first be unplugged from its ac source before beginning any internal switching or modification. All models except for the DCR-4000 require that the top cover be removed.

Observe the following steps:

1. To remove the cover, take out four Phillips head screws, two from each side and slide the cover off toward the back of the receiver.

2. DXR-900: Remove clamp diode D3 and replace it with a 47K, 1/4W resistor.

3. DXR-1100: Remove clamp diode D4 and replace it with a 47K, 1/4 W resisitor.

4. DXR-1200: Move internal switch SW2 to the OFF position.

5. DXR-1300: Move internal switch SW101 to the OFF position.

6. DCR-4000: Set the CLAMP SWITCH on the front panel to "0", clamp off position.

On receivers that require internal switching or modification, all part numbers are identified on the PCB and added resistors may be soldered on the top side of the PCB due to plated through holes.

A common experience using Orion decoders indicates that the video level affects the decoder almost as much as unclamping the video does. Gould/Dexcel has found that that a video output level of 1.4V P-P going into the Orion worked the best. A level above this caused the video to wash out in the center of the screen. The best level varies depending on the receiver and decoder.

All audio is digitally encoded onto the video signal in place of the horizontal sync pulses. Since all Gould Innovision receivers have a flat response to 4.0 MHz, there should be no problem in decoding the audio.

Any modifications attempted in the field may void the warranty if done incorrectly. All Gould/Dexcel specifications are subject to change without prior notice.

Fig. 10-13. Some processing equipment, like descramblers, require unclamped video. This Technical Bulletin lists the procedure for the various Dexcel receivers. Courtesy of Gould/Dexcel.

system. This system is also used in Japan, Canada, and in part of South America. PAL and SECAM are the two other widely used broadcast standard throughout the world. Both are noncompatible with NTSC broadcasts.

Because the U.S. is the world leader in TV broadcasting, the NTSC system is the oldest standard. Unfortunately, this also means that it's not the best system. It was developed specifically for black and white broadcasts, and it was not until the late 1950's that standards for adding color during broadcasting were developed.

Because there were so many black and white TV's in service, the color signals had to be added onto the existing black and white signals to insure that black and white receivers would still be able to receive the new color broadcasts, albeit in black and white. This is why the luminance (black and white) information and the chrominance (color) information overlap, yet are two separate signals.

Synchronization

To portray moving pictures, whether by film or by video, requires a series of still pictures. If there are at least 30 pictures or frames per second, then there will appear to be a smoothly moving image. Less than 30 frames per second and there will be a noticeable lack of smoothness as well as a flicker to the image. In the NTSC system, the frame rate is 30 per second.

To insure that flicker will not be a problem, it was decided that there would be 60 fields per second. As can be surmised, that means there are two fields per frame. But what constitutes a field and a frame? A field consists of 262 and 1/2 scan lines. By scanning the screen twice per frame, flicker is reduced. This process is called interlaced scanning. By this it is meant that the scan lines from field one fall in between the scan lines from field two. To accomplish this requires that the video display be in exact step with the broadcasting signal. Thus we have two types of synchronization pulses for the picture sweep.

The *horizontal sync* triggers the horizontal oscillator, located in the television or monitor. The oscillator signal in turn returns the electron beam from the right side of the screen to the left side of the screen, without leaving a trace across the screen. Thus is derived the name, horizontal blanking interval. This is the time it takes for the electron beam to be moved back to the other side of the screen where it will start another trace across the screen for the next line. The horizontal rate is equal to the number of scan lines used in a frame, which is 525, times the frame rate, which is 30. Thus the horizontal

sync pulses occur 15,750 times per second.

The *vertical sync* triggers the vertical oscillator (also in the TV or monitor) to return the electron beam from the bottom of the screen to the top of the screen, also without leaving a retrace line. The vertical sync has a frequency equal to the field rate or 60 times per second. The vertical sync is much longer than the horizontal sync, and because of this is used to carry several pieces of information. VITS, VIRS, captioned titles, videotext and digital data are some of the things sent along with the video during the vertical interval.

Another synchronizing signal is the *color burst*. The color burst is located during the horizontal sync pulse and consists of eight (or more) cycles of a 3.58 MHz signal. This signal is compared in the video section of the TV or monitor to the color oscillator's output, which is also running at 3.58 MHz. The phasing of the two signals must be exact or the colors will be incorrect. Adjusting the tint control on the TV basically is adjusting the phasing of the burst to the TV's color oscillator.

MEASURING THE VIDEO

Video is commonly measured by a system designed by the IRE or Institute of Radio Engineers. In this system, a 1 volt peak-to-peak video signal is divided into 140 IRE units. The 140 IRE units are broken down into 40 units of horizontal blanking, and 100 units of picture information above the horizontal blanking level.

In a properly set-up receiver, the video output should be like that shown in Fig. 10-14. The horizontal sync extends from -40 IRE to 0 IRE. The color burst extends equally above and below the 0 reference (blanking) line $+20$ and -20 IRE. The highest video signal is at 100 IRE. This signal is equal to one volt peak-to-peak into a 75 ohm load.

Video Frequency Response

The crispness of the picture is directly related to the video frequency response. It can be compared to definition in audio reproduction, where if the highs are not present then the sound will be muddy or muffled. If the video high frequencies are rolled off or attenuated, then the picture becomes softer, or even blurred if extreme rolloff occurs. The color depth will also be affected by high frequency rolloff. The frequency response of most TV's and monitors can be adjusted by using the sharpness control. By rolling off the high end, you can cover up noisy chrominance and make sparklies less noticeable.

In almost every satellite receiver, there are video fre-

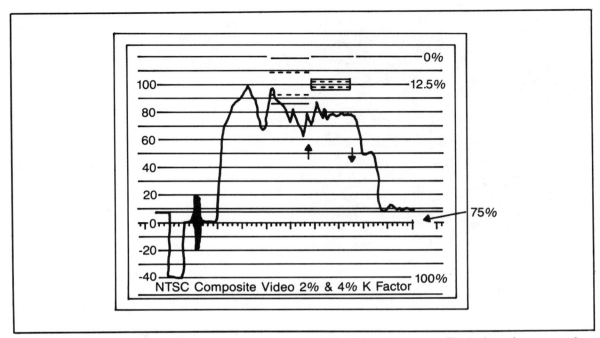

Fig. 10-14. Drawing of a typical waveform monitor display of one line of a video signal. The horizontal sync goes from +7.5 IRE to −40 IRE. The color burst stretches from +20 to −20 IRE. The video signal goes from +7.5 IRE to 100 IRE.

quency response adjustments that are used to compensate for circuit tolerences. These should only be adjusted by using a waveform monitor.

Video Level

Not all transponders put out a video signal that is exactly 1 volt peak-to-peak. Some transponders broadcast slightly lower levels because of the number of audio and digital subcarriers that they broadcast along with the video. Some seem to broadcast a slightly higher level, to possibly increase the signal-to-noise ratio slightly.

For satellite receiver set-up, a waveform monitor is used to set the video output level as well as the frequency response level. Differential gain, chrominance-to-luminance delay and other measurements which a waveform monitor can perform are usually not used during troubleshooting of TVRO receivers.

To properly view the video output of a receiver, a waveform monitor and a video monitor should be looped through and terminated by a 75 ohm load. Check the video monitor's owners manual to verify whether it is internally terminated or not.

Almost every receiver must be terminated to give the proper video level. One exception to the rule is the

Cosmos II by Northwest Sat Labs, which is already terminated internally through a 75 Ohm resistor. If it is connected up to a terminated load, the video level will be about .5 volt. To compensate, the 75 ohm resistor that is in series with the output must be shorted out.

INTERPRETING VITS

Testing video circuits usually involves looking at the VITS or the Vertical Interval Test Signals. These are test signals that are broadcast along with the video during the vertical blanking period. The vertical blanking period is the horizontal black bar that can be seen in between frames if the picture is rolled down by adjusting the vertical hold control. If you look closely at it you can see the VITS. They are the colored bars at the lower edge of the vertical sync bar. Of course, looking at the VITS in this manner won't really help you much. Usually one would use a waveform monitor to actually look at the test signals.

Before you can make use of these signals in troubleshooting a TVRO receiver, you must have two things, a waveform monitor and a knowledge of what constitutes a video signal. Waveform monitors are not inexpensive pieces of test gear, but they are crucial if you are to properly repair and set-up the video section of a satellite re-

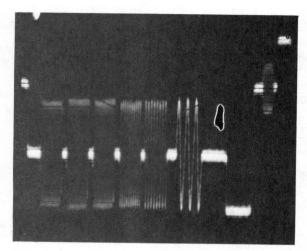

Fig. 10-15. A multiburst waveform. It is sent during the vertical blanking interval. It is usually sent on line 17. It may be at 100 or 70 IRE. It consists of 6 bursts of frequencies from 500 kHz to 4.0 MHz.

ceiver. The knowledge of what constitutes a correct video signal can be gained by reading the following paragraphs.

The quality of a receiver's video can be found by looking at the video waveform on the waveform monitor. The frequency response and amplitude are the main things that need to be checked and adjusted in a satellite receiver. Such other important signal qualities as differen-

tial gain, differential phase, and line time distortion can be measured, but in most receivers, they cannot be changed without redesigning the receiver. Severe distortion will be indicative of component failure, for in most cases receivers still have passable video even at their worst.

There are basically three different VITS that are used to service satellite receivers. These are the multiburst, the composite test signal and the sine-squared pulse. They are shown in Fig. 10-15 through 10-17. The composite and the multiburst are the best to use for setting the video level. Multiburst is usually found on line 17 of Field 1, while the composite signal is often found on line 18 of either Field 1 or 2.

Since satellite signals are not mandated to include test signals by the FCC, some of them do not have them (most noticably C-span). Also since satellite transmissions are not truly NTSC compatible signals, they do not have to broadcast exactly 100 IRE of video. If you look at some of the transponders on F3-R, you'll find a fairly wide difference between video levels. Typically MTV offers the most consistent video level.

The video level should be set by using the IRE setting on the waveform monitor. This setting rolls off the higher frequencies, which results in a cleaner line with which to set the level. The sync pulse should be set to −40 IRE and the white flag to 100 IRE for the composite signal or to 70 IRE for the multiburst.

Note: Some multiburst signals have the white flag

Fig. 10-16. A composite test signal, that is also part of the VITS. It consists of a white flag (100 IRE signal), a 2-T signal, and a stairstep waveform.

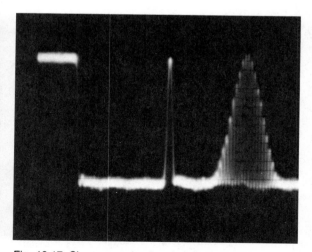

Fig. 10-17. Sine-squared pulse signal that can be used to determine phase and gain between the luminance or detail information and the chrominance or color information. It is also part of the VITS.

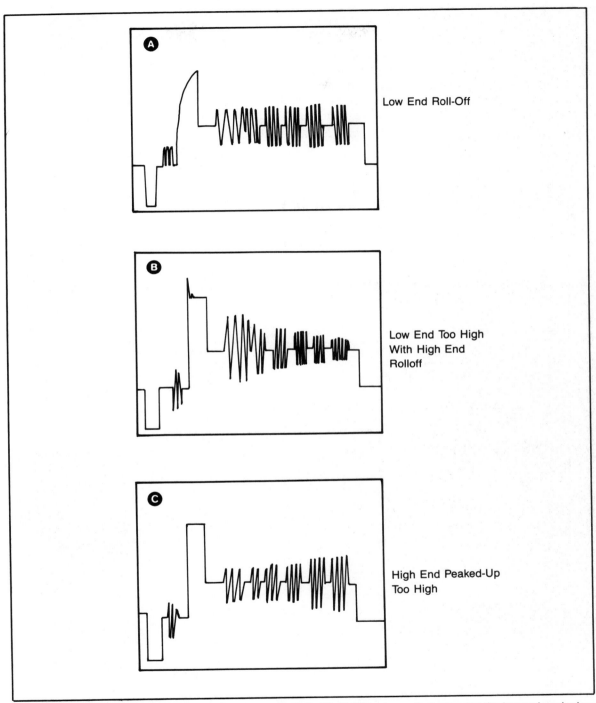

Low End Roll-Off

Low End Too High
With High End
Rolloff

High End Peaked-Up
Too High

Fig. 10-18. (A). Multiburst waveform showing a low end rolloff. (B). Multiburst waveform showing the low end peaked up combined with a high frequency rolloff. (C). Multiburst signal showing a peaked high end response.

at 100 IRE, and the multibursts at 70 IRE.

The multiburst is used to set the frequency response of the system. It consists of a white flag and six bursts of sinewave signals at 500 KHz, 1 MHz, 2 MHz, 3 MHz, 3.58 MHz, and 4.2 MHz. Ideally all of the signals will be exactly the same in amplitude. Realistically there will be some rolloff of the higher frequencies in almost every receiver. The leading edge of the white flag can be used to set the low frequencies. If the top edge is a clean square then the low end is OK, if there is a tilt to it, then the low end is difficult. If there is a spike or overshoot on the leading edge, then the low end is too high in amplitude and is ringing. Figures 10-18A, 18B, and 18C show the different frequency problems that can occur. Compare these to Fig. 10-15.

Typically there are three adjustments in a satellite

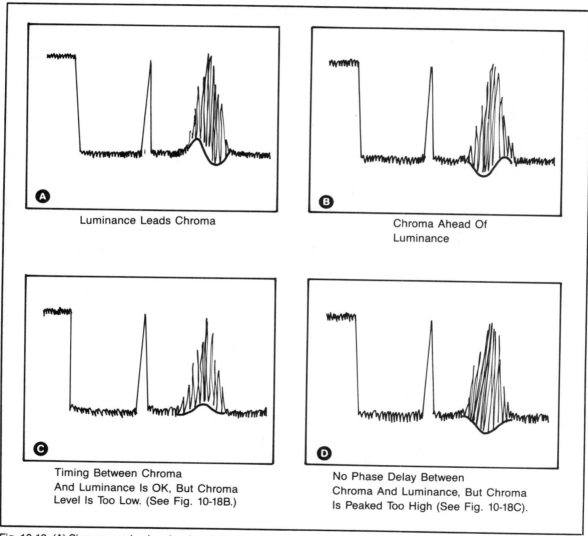

A Luminance Leads Chroma

B Chroma Ahead Of Luminance

C Timing Between Chroma And Luminance Is OK, But Chroma Level Is Too Low. (See Fig. 10-18B.)

D No Phase Delay Between Chroma And Luminance, But Chroma Is Peaked Too High (See Fig. 10-18C).

Fig. 10-19. (A) Sine-squared pulse showing chrominance delay compared to the luminance. (B). Sine-squared pulse showing the chrominance leading the luminance. (C). Sine-squared pulse showing no timing problems between the chroma and luminance, but there is a chroma level problem. In this case, the chroma is too low. (D). Sine-squared pulse showing a boosted chroma response. Again the timing between chroma and luminance is OK, but the luminance gain is less than the chroma gain.

receiver for video. These are the video level, which should be set to 1 volt peak-to-peak; the high frequency response, which is set to give the flattest response in the multiburst waveform; and the low frequency response, which is set viewing the leading edge of the white flag. Occasionally the frequency adjustments will be called out as luminance and chroma adjustments, but even if they are, they are still just low (luminance) and high (chroma) frequency adjustments.

The sine-squared pulse is the waveform that Earth Terminal used in their ads for a long time. It is a special pulse signal that can tell a lot about the receiver's differential gain and phase. It is used to qualitatively judge receivers on a side by side technical comparison. It will show any chroma or high frequency problems that the receiver has. If there is a high end rolloff, then the chroma will be to low. This will result in the bottom of the waveform being bowed up. If the high end is peaked up too high, then the pulse will be enlarged at the bottom. If there is chroma delay in relation to the luminance signal, then there will be an S-shaped curve at the bottom of the pulse that corresponds to whether it is ahead or behind the luminance. Figures 10-19A through 10-19D show the four problems that can arise. Compare these with the waveform shown in Fig. 10-17.

Audio Processing

T HE AUDIO DEMODULATION CIRCUIT IS USED TO SEP-
arate the audio information from its carrier. The
audio carrier is actually a subcarrier that is impressed onto
the video signal. The audio subcarriers are above the
video, in the 4.5 to 10.0 MHz range. In normal practice,
no audio carriers are found below 5.0 MHz or above 8.0
MHz.

There is much more than just the picture audio that
is carried in the audio subcarrier band. There are independent FM radio stations, national radio network feeds,
regional radio network feeds, teletex, stock exchange
reports and a host of other digital and analog information. Figure 11-1 is an an analyzer photo taken at baseband video of TR3 on G-1 (WGN) showing the great
number of audio subcarriers present. The frequency
range, from left to right, is 4.3 to 9.3 MHz with each
horizontal division being 500 KHz. The center line is 6.8
MHz.

As can be seen, WGN broadcasts over 20 different
subcarriers within this band. The separate carrier in the
middle, located at 6.8 MHz and isolated from those on
either side of it, is the WGN program audio. Normally
the program audio carrier has a much wider deviation as
well as a hotter signal. In WGN's case it is broadcast using almost narrowband deviation, with a signal level that's
not much higher than the other subcarriers. This can
cause problems in many home receivers, since they may
only be set up for standard wideband reception.

AUDIO SUBCARRIER SPECIFICATIONS

The audio that is broadcast via satellite can be subdivided
between those being broadcast in a wideband mode and
those using narrowband. *Wideband,* which is sometimes
referred to as normal deviation, is used for the video program's audio. *Narrowband* is used for FM radio transmissions. Four narrowband channels can be sent in the same
space used by one wideband subcarrier.

Wideband and narrowband refer to the deviation or
width of the audio channel as it is broadcast. Wideband
transmissions are normally ± 200 KHz (maximum being
± 237 KHz) in deviation and have a frequency range of
50 Hz to 15 KHz, with a signal to noise ratio of 70 dB.
Narrowband deviation can be ± 25 KHz with a frequency
range of 50 Hz to 7.5 KHz, or ± 50 KHz with a frequency
range of 50 Hz to 15 KHz. Their signal to noise ratio ap-
proaches 65 dB to 70 dB when the signals are companded.

There are no FCC mandated standards and thus not
all audio subcarriers are broadcast using deviations of exactly ± 200 KHz or ± 50 KHz. In addition, some narrow

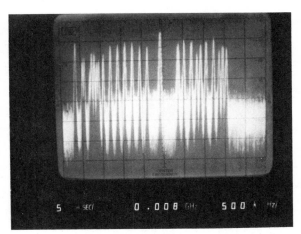

Fig. 11-1. A Spectrum Analyzer photo of all the audio subcarriers found on G-1, TR3 (WGN). It was taken at the baseband video output. The center frequency is 6.8 MHz with a resolution of 500 kHz/div.

deviation subcarriers combine two audio channels, if 7.5 KHz is their highest frequency, into the space used by one channel with a 15 KHz frequency response. Table 11-1 is a listing of some of the subcarrier services on WGN. Look back at Fig. 11-1 and compare their listed frequencies to their analyzer photograph position. This should graphically show why a receiver, set in the wideband position picks up a lot of noise and distortion as it is tuned across WGN.

Deviation variations, close spacing of subcarriers, and the use of companding is why some channels cannot be cleanly received even with receivers which feature a narrowband/wideband selection switch. If the customer is interested in receiving the FM radio subcarriers, make sure the receiver can handle the channels that are desired, as the deviation bandwidth is sometimes not adjustable from that selected during circuit design.

AUDIO CIRCUITS

The input signal to the audio demodulation, or audio demod, circuit must be baseband video because that is where the audio subcarriers are located. Therefore it must come after the video has been detected in its demod circuit. To illustrate the signal flow in the audio circuit, a Luxor 9550 block diagram is shown in Fig. 11-2.

In the 9550, the detected video out is split and is run through two filters. These filters are used to separate the baseband signals into those above 4.5 MHz and those below 4.5 MHz. The top line is the video circuit, and the low-pass filter is needed to prevent any audio IM (Inter-

Modulation) distortion in the video. Likewise a high-pass filter, that rolls off all the signals below 5 MHz, is used to prevent video interference from the chroma components of the video signal. This type of circuit is used in all receivers.

After the HPF (high-pass filter), the signals are split so that two independently tuned audio channels can be detected simultaneously. This allows stereo reception. Since both channels are identical, only one will be covered.

The Luxor circuit uses a balanced demodulator IC that operates at an i-f of 10.7 MHz. If 10.7 MHz sounds familiar, it should be. It's the i-f used in almost every FM radio made today. By using a standard i-f like 10.7 MHz, all the special ICs and filters made for FM radios can be used. This simplifies the design and manufacturing as well as the troubleshooting. The less parts one has to check the better!

To derive a 10.7 MHz i-f from the 5.0 MHz to 8.0 MHz subcarriers requires that the signals be mixed with a tuneable carrier that is 10.7 MHz higher than the desired audio subcarrier frequency. A tuning voltage is applied to a VCO (voltage controlled oscillator) that in turn puts out a sine wave that varies from 15.7 to 18.7 MHz. When it is mixed with the baseband signals, the desired channel is output at 10.7 MHz. In the Luxor receiver, this is done by a monolithic IC (SO42P).

A ceramic FM radio-type BPF stops all spurious signals from entering the following detector IC, except for those centered around 10.7 MHz, which are passed on through. There are actually two 10.7 MHz ceramic filters. One is used for wideband reception and a second is added for narrowband reception.

The BPF output is coupled into the demodulator, which is a TBA 120T IC. The block diagram is shown in Fig. 11-3. The 10.7 MHz signal enters the IC at pin 14 and is amplified by an 8-stage differential op-amp, which is set up as a limiter. The op-amp has two outputs, 180 degrees out of phase from each other. These are available at pins 6 and 10, and are applied to a phase shift network to derive two signals that are 90 degrees out of phase from the first two signals. There are now two balanced signals 90 degrees in timing difference from each other. This is called being *in quadrature*. See Fig. 11-4.

These two signals are mixed together in the balanced mixer. The resulting signal is a rapidly changing dc voltage which is proportional to the frequency changes of the carrier. Thus the output is a reproduction of the audio signal that was transmitted. It is output from pin 8 of the

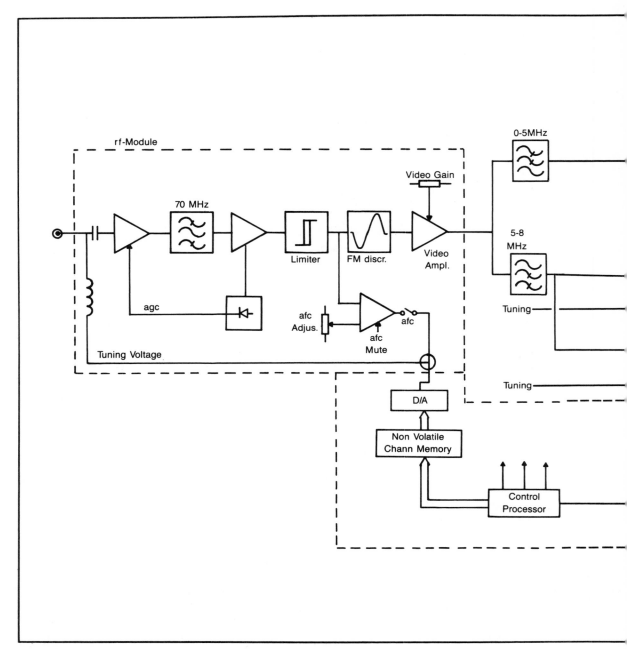

Fig. 11-2. Block diagram of the Luxor 9550 receiver. Courtesy Luxor North America.

IC. Pins 4 and 5 are used for volume control signals that affect the output circuit. One signal mutes the volume when changing channels, and the other is sued to compensate for the difference in volume when detecting narrowband vs. wideband carriers.

The two channels are not run into the stereo processor section. Channels 1 and 2 go to both a solid state switch (CD4066) and separate dual op-amps (1458). The

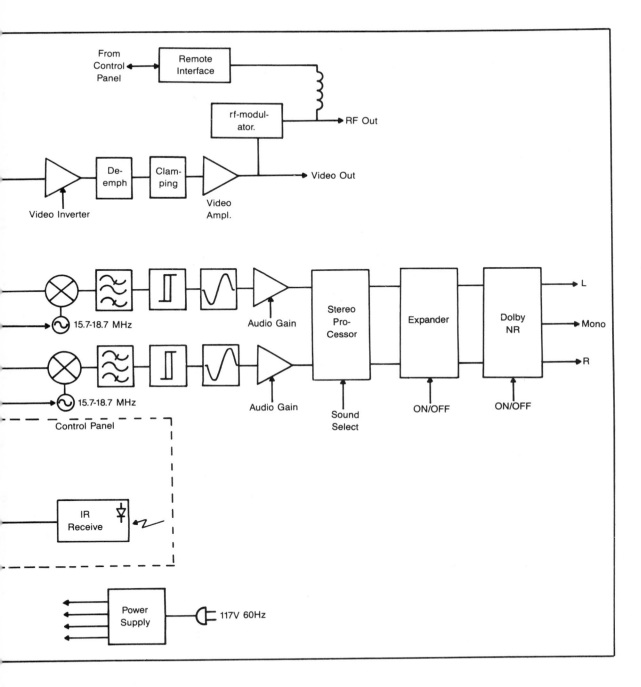

op-amps are tied together in such a manner that they form a network that algebraically adds and subtracts the signals together. The result is separate left and right channels if the input signals were from a matrix stereo source. This

is covered in more detail in the Matrix Decoder Section.

The solid-state switch routes the audio signals to different circuits depending on the chosen audio format. If it is in the matrix mode, the audio is obtained from the

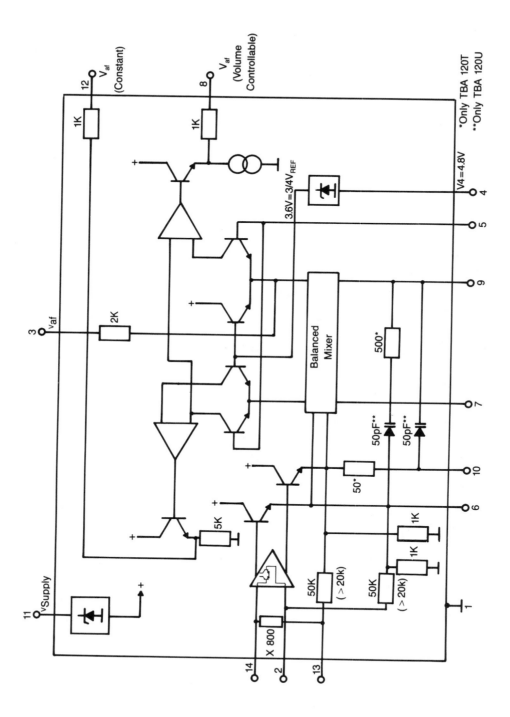

Fig. 11-3. Block diagram of a TBA 120T i-f amplifier and detector. It is used in the Luxor 9530, 9540 and 9550.

140

two 1458 chips mentioned in the last paragraph. If mono 1 is selected, then channel 1's output is selected while 2's output is turned off. If mono 2 is selected, then channel 2's output is selected instead. If the format is direct (discrete) stereo then the two outputs from the 1458 IC's are shut off and the separate channel 1 and 2 outputs are sent to the next 4066 switch.

A second switch is used to select whether to send the two channels to the compander IC (NE571), which is used in an expansion mode to put back the original dynamics of the music when compression was used during broadcasting. Almost all narrowband subcarriers are broadcast using this method. The audio goes through one final IC before reaching the back panel RCA jacks. It is processed by a Dolby noise reduction chip (LM1112). The Dolby chip passes the audio without change unless the Dolby NR switch is in the On position. Dolby is a second form of noise reduction that is being used on MTV and TMC as well as some narrowband carriers. It is also found in many of the newer receivers. Both of these circuits are covered in more detail later in this chapter.

In some receivers, the audio is tuned and detected by using an NE564 Phase Lock Loop IC. This same IC is used in many receivers as the video detector and its operation is basically the same as that, except that it is also doing the tuning.

Figure 11-5 is the audio circuit schematic from a Gould/Dexcel DXR 900. It uses a varactor diode to tune the 564 across the audio subcarrier band. The baseband video signal is input on pin 6. It is filtered by R63, RFC6, C73 and C70 which form a high pass filter with a cutoff frequency of 5 MHz. Pin 2 controls the bandwidth that the PLL chip tracks, which in this case is strictly for wideband reception.

Figure 11-6 is a block diagram of the PLL chip showing the main components. There is a Limiter that prevents AM noise from affecting the detector. A Phase Comparator which compares the incoming signals to a reference frequency output from a VCO. The VCO which is driven by the phase comparator's output and tuned by a varactor diode. And an amplifier which amplifies the dc difference signal to form the audio.

If there is a difference between the incoming signal and the VCO output signal, there is an error voltage that is developed in the phase comparator. This is fed back to the VCO and also appears at pin 14 as a rapidly changing dc voltage. As the tuning voltage is changed by the front panel audio tune control, the capacitance of the varactor diode, D1 is changed. This changes the output frequency of the VCO, which in turn changes the phase comparator's output which causes the VCO to follow the frequency changes in the signal.

The rapidly changing dc voltage is a reproduction of the audio signal that was broadcast. Since the audio is emphasized or boosted at the high end during broadcasting, it must be in turn, deemphasized in the receiver.

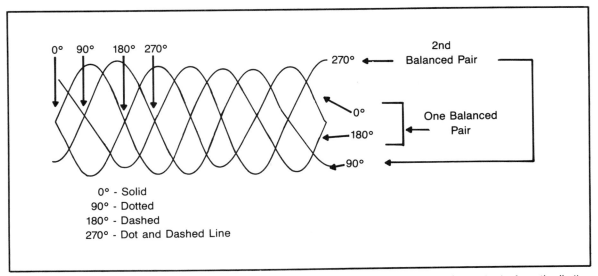

Fig. 11-4. Sine waves in quadrature. The 0 degree and 180 degree signals are the + and − outputs from the limiter, while the 90 degree and 270 degree signals are the result of the phase shifting of the 0 and 180 signals 1/4 wavelength. They are used to demodulate the audio.

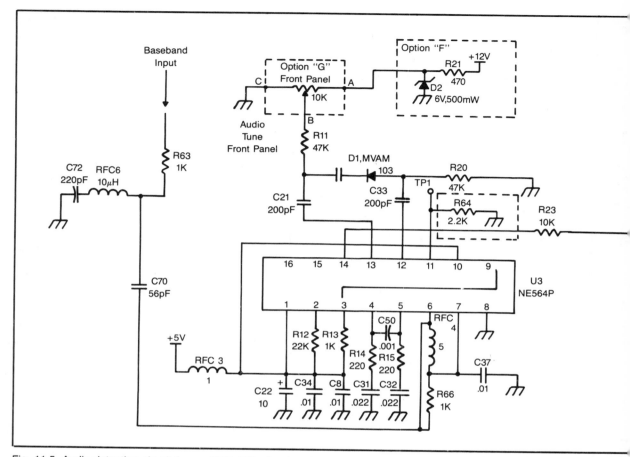

Fig. 11-5. Audio detection circuit from a DXR 900, showing the use of a NE564 PLL chip as a combination subcarrier selector and demodulator. Courtesy of Gould/Dexcel.

Table 11-1. A Listing of Some Audio Programs Broadcast on WGN.

#	Frequency	B.W.	Use
1	5.40 MHz	15 kHz	Moody Broadcasting—Left
2	5.58 MHz	15 kHz	SMN Pop Adult—Left
3	5.76 MHz	15 kHz	SMN Pop Adult—Right
4	5.94 MHz	15 kHz	SMN Country—Left
5	6.12 MHz	15 kHz	SMN Country—Right
6	6.30 MHz	15 kHz	WFMT—Left
7	6.48 MHz	15 kHz	WFMT—Right
8	6.80 MHz	15 kHz	WGN TV Audio
9	7.38 MHz	15 kHz	Bonneville Beautiful Music—Left
10	7.56 MHz	15 kHz	Bonneville Beautiful Music—Right
11	7.695 MHz	7.5 kHz	Seeburg
12	7.785 MHz	7.5 kHz	Spare
13	7.92 MHz	15 kHz	Moody Broadcasting—Right
14	8.055 MHz	7.5 kHz	SMN Stardust—Left
15	8.145 MHz	7.5 KHz	SMN Stardust—Right

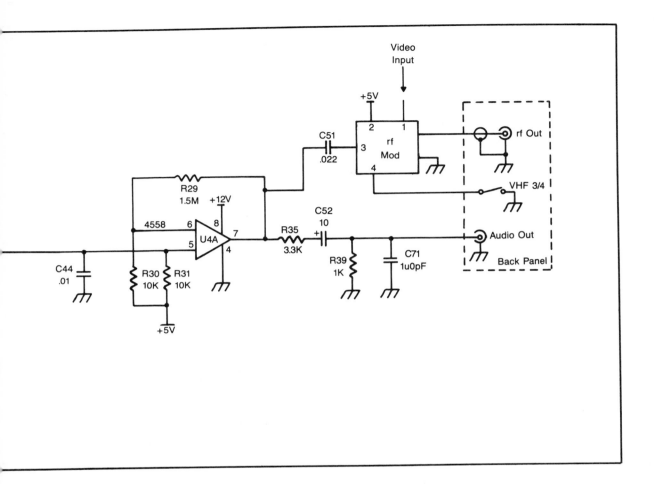

This is accomplished by R23, C43, and C44. The audio is then amplified to drive the rf modulator and the audio output jack.

STEREO BROADCASTING METHODS

There are six types of audio being broadcast via satellite. The most common is a single channel wideband mono subcarrier located at 6.8 MHz. The second most common is discrete stereo, both in wideband and in a narrowband format. The next most common method is a single carrier with narrowband deviation. It is often used in a half channel format with a top frequency of only 7.5 KHz. The other two methods are matrix stereo and multiplex stereo. Table 11-2 lists the six types of transmissions. A satellite program guide such as Orbit Magazine or OnSat should be consulted to determine what type of audio is being broadcast and on what frequencies the subcarrier(s) may be found.

Both discrete and matrix stereo require two separate tuning sections. This is why there are two audio tuning controls on most stereo receivers. The only type of stereo that can be tuned in using only one control is multiplex stereo. Some of the earlier satellite receivers offered built-in multiplex stereo, as it was assumed that it would be the predominant method of stereo broadcasting. As it has developed, discrete stereo has taken the lead with matrix and multiplex being about equal in use.

Discrete Stereo

Discrete stereo is one form of stereo audio that is broadcast using two subcarriers. One subcarrier contains

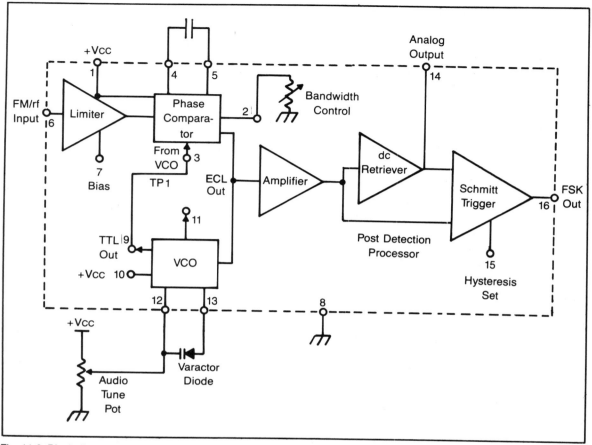

Fig. 11-6. Block diagram of the NE564 chip as used for audio subcarrier tuning and detection. The 564 consists of a limiter, phase comparator, error amplifier, and VCO.

Table 11-2. The Six Types of Unencoded Analog Audio Currently Broadcast by Satellite.

Mono, Single Carrier. 6.8 MHz. Most common TV sound frequency. Normal deviation.

Single Carrier, Narrow Deviation. Used for special audio programs such as Seeburg Muzak.

Discrete Stereo. Dual carriers. Separate left and right channels with normal deviation. Used for stereo TV sound.

Wegener 1600 System, Dual Carriers. One for the left channel, one for the right channel. Narrow deviation. Used for FM radio subcarriers.

Matrix Stereo, Warner Amex System, Dual Carriers. L + R for mono reception, with L − R on a second subcarrier. Normal deviation.

Multiplex Stereo, Leaming System. Similar to standard FM stereo. Single carrier of L + R with a double-sideband suppressed-carrier of 38 kHz containing the L − R information. Normal deviation.

the left channel information, and the other subcarrier contains the right channel information. Usually the lower frequency channel is the left channel.

In most cases, a discrete stereo broadcast is not compatable with mono receivers. They would only receive the left or right channel, although on some transponders a third channel is also broadcast, which is a monophonic program signal, or combination of the two discrete channels. In this instance, the mono signal would be used to feed the rf modulator.

Discrete stereo is broadcast using both narrowband and wideband deviations. In most cases, the wideband deviation is used for the video program's audio, while narrowband is used for FM radio broadcasts. Most narrowband channels are broadcast next to each other. As an example, the Rock-A-Robics channel (TR3 on G-1) is broadcast on 7.38 MHz and 7.56 MHz. The left channel is on 7.38 MHz and the right is on 7.56 MHz.

Matrix Stereo

Matrix stereo also uses two subcarriers. One carries the L + R (left plus right) signal, while the second carries the L − R (left minus right signal). To decode the two channels into stereo requires a matrix decoder. This system uses the L + R or mono signal to feed the rf modulator. It is also compatable with mono receivers, whereas the discrete method is only compatable if there is a mono carrier in addition to the two stereo carriers.

Table 11-3 is the matrix algebra. The matrix decoder algebraically sums and subtracts the two subcarriers from each other, with the output being the stereo left and stereo right channels. A typical simple decoder circuit is shown in Fig. 11-7, which is the matrix decoder section of a Gould/Dexcel DXR 1200.

The circuit consists of two 4558 op-amps. To correctly decode matrix stereo the L + R signal must enter through C66, and the L − R signal must be input to R104. Reversing this will cause the right channel to be out-of-phase with the left channel.

With the stereo/normal switch (S6) in the stereo position, the output is obtained from the second section of each op-amp. Thus the two signals must go through the resistor network which comprises the matrix circuit. U10b is set up as a differential amp, while U11b is set up as a summing amp. Resistors R89, R90, R86 and R108 (all 24K ohm) are cross-connected so that the two signals are mixed equally.

Pin 5, the noninverting input of U11b, is the summing point for the two channels, thus the output of the op-amp is the Left channel, since the Right channel is cancelled out when they were summed together. Getting the right channel is a bit more involved. The L + R carrier is applied to the noninverting input (Pin 5 of U10), and the L − R signal is connected to the inverting input (pin 6 of U10). The result is the difference between the two signals, or the Right channel. If switch S6 is put into the normal position, then the upper channel is output directly to the right RCA jack, while the lower channel,

Table 11-3. Matrix Decoder Algebra.

Correct: (L + R) + (L − R) = 2L Left

A = main (L + R)
B = sub (L − R)   (L + R) − (L − R) = 2R Right

Incorrect: (L + R) + (L + R) = 2L + 2R Left

A + B tuned to
same carrier   (L + R) − (L + R) = 0 Right

Incorrect: (L − R) + (L − R) = 2L + 2R (Out of phase) Left

A + B tuned
to same carrier   (L − R) − (L − R) = 0 Right

Incorrect: (L − R) + (L + R) = 2L Left

A = sub (L − R)
B = main (L + R)   (L − R) − (L + R) = 2R (Out of phase)

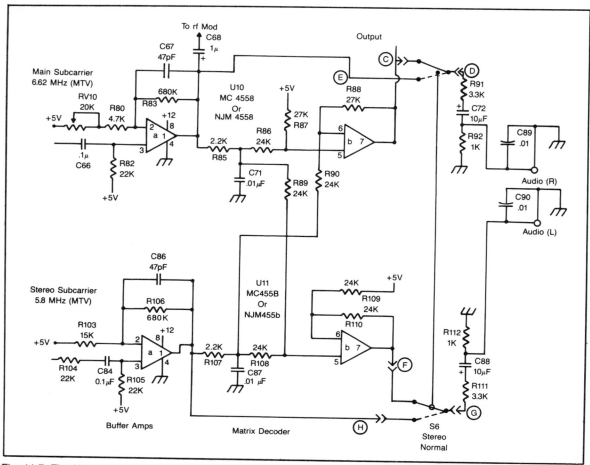

Fig. 11-7. The Matrix decoding circuit from a DXR 1200 receiver. It uses a resistor matrix and an op-amp summer and differential amp to separate the left and right information from a matrix stereo broadcast. Courtesy of Gould/Dexcel.

is output to the left RCA jack, bypassing the matrix.

The only adjustment required is the matrix set-up control (RV10). It is used to balance the level of the L + R signal to the L – R signal. It is adjusted by first tuning both channels to the same carrier, and then turning on the matrix circuit. This will cause the left channel to almost double in volume, but it will cause the right channel to disappear. Set RV10 to null the right channel so that the least amount of sound possible is present at the right channel output.

Multiplex Stereo

The multiplex system is similar to that used to broadcast standard over-the-air FM radio. It consists of one subcarrier that is the L + R channel and a double-

sideband suppressed-carrier modulated at 38 KHz with a 19 KHz pilot signal that is the L – R channel. The main L + R channel is used to feed the audio to the rf modulator. This system is also compatible with mono receivers in that both channels can be received.

The main difference between standard FM-multiplex and satellite multiplex is in the deviation and modulation processes. The total deviation is much greater in the satellite system and is a function of the degree of stereo separation. This type of deviation is called adaptive deviation.

Audio Companding

Companding is an audio processing that takes the signal's dynamic range and compresses it before uplink and then re-expands it after demodulation. This expanded signal restores the original dynamics of the signal and thus overcomes the lower signal-to-noise ratio inherent in narrowband channels. Figure 11-18 is the expander circuit from a Luxor 9550 receiver. It is one of the only receivers which has a built-in expander circuit.

The Luxor 9550's uses an NE571 IC to accomplish its audio expanding. This IC is dual channel dynamic range compression or expansion chip. In this case, it is used to expand the signal. When the front panel EXPansion ON switch is in the on position, the left and right expanded audio channels that are connected to pins 1 and 8 of the CD4066 switch (IA17) are switched to pins 2 and 9 which in turn go to pin 5 of the two Dolby IC chips (LM1112). The signals pass through the Dolby chips without modification if the Dolby switch is off. If it is on, then the signals are further processed by the Dolby noise reduction circuit.

Expansion is controlled by signal level, thus a small change in input signal level is expanded to present a greater change in output level. It is a fixed compression/expansion ratio of 2:1. For example, if the signal level changed by 100 mV at the compressor's input (at the uplink),only a 50 mV change would occur at its output. This compressed signal is then transmitted and received where the expander takes the same 50 mV signal change and expands it back to a 100 mV change.

If the expansion circuit is used on signals that were not compressed during broadcast, it will cause the audio to have a pumping quality to it as the circuit tries to expand the level according to what it thinks are compressed signals.

Dolby Noise Reduction

There are several types of Dolby noise reduction circuits in use. These are labelled Dolby A, Dolby B, and Dolby C. Almost every cassette deck made today features one of these types of noise reduction. The satellite transmissions use Dolby B.

A Dolby circuit divides the audio-frequency spectrum into several different bands of frequencies. Each band gets a different amount of boost or preemphasis. This boost increases the signal (in effect expanding it). Thus when the signal is reduced or compressed to the original levels during playback, the noise added during transmission is also reduced.

Dolby processing takes advantage of the fact that low frequencies take up most of the power transmitted in a signal and yet are fairly immune from noise because noise is more noticeable and predominant at higher frequencies. Thus if the low frequencies are left alone, or even backed off slightly, the higher frequencies can be selectively boosted to optimize the deviation and power available.

The Dolby system does this by dividing the audio spectrum into several bands. It then takes those bands and applies varying amounts of level boost, depending on the dynamics and frequency. On the receiving end, it is done in reverse. The frequencies are equalized back to their original relationships, and the dynamic range is continuously adjusted according to the signal's level and frequency. This makes the Dolby system very sensitive to signal level, as that is its reference. If it is incorrectly set-up, then the circuit will compress too much, or not enough, distorting the harmonic structure of the sound. In extreme cases this can cause a shrillness to the sound or cause it to be muffled. It may also cause noticeable signal level changes, similar to that caused by the expander circuit operating on a noncompressed signal.

AUDIO BROADCASTING SYSTEMS

There are four main types of audio broadcasting systems being used today. Even though these are usually referred to as if they were generic (i.e. matrix, discrete, or multiplex), each type actually represents a complete transmit and receive system, which is usually designed and sold by one company and which is not truly compatible with equipment made by other companies. Since there are few real standards for audio broadcasting by satellite and since every company's got their own way of doing things, it's amazing that home receivers can pick up as many different audio broadcasts as well as they do.

Wegener 1600 System

The Wegener system is the most popular FM radio subcarrier broadcasting scheme. It is used to great extent on the subcarriers on WGN and WTBS (TR-3 and TR-8 on G-1). Such FM radio channels as the Moody Broadcasting Service, the Cable Jazz Network, Satellite Jazz Channel (KKGO), WFMT, and all the Satellite Cable Audio Networks (SCAN) use this system.

The Wegener system can be broadcast in discrete stereo or in mono. The channels are broadcast using a companded signal and a narrowband deviation of ±50 kHz. The biggest advantage to this system is the number of channels that can be broadcast in a limited space. The main user of this type of broadcasting is the FM radio services.

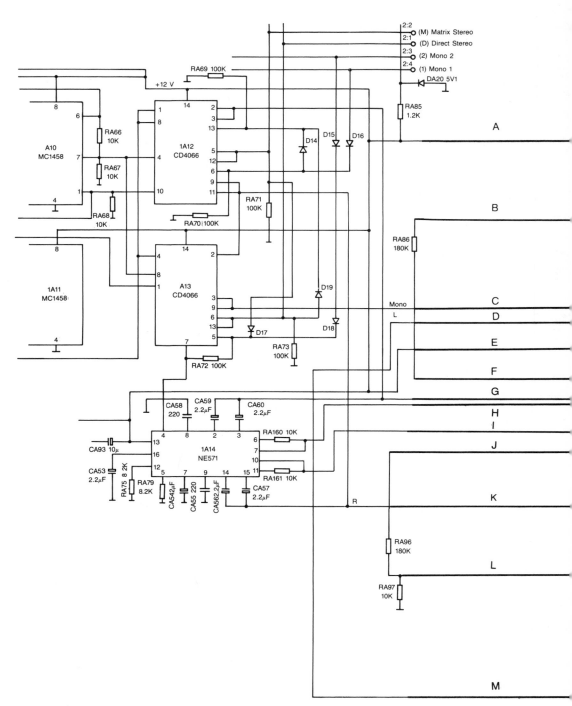

Fig. 11-8. Audio processing circuit from a Luxor 9550, showing the expander IC, the two Dolby chips and the 4066 switches that are used to select the audio mode. Courtesy of Luxor North America.

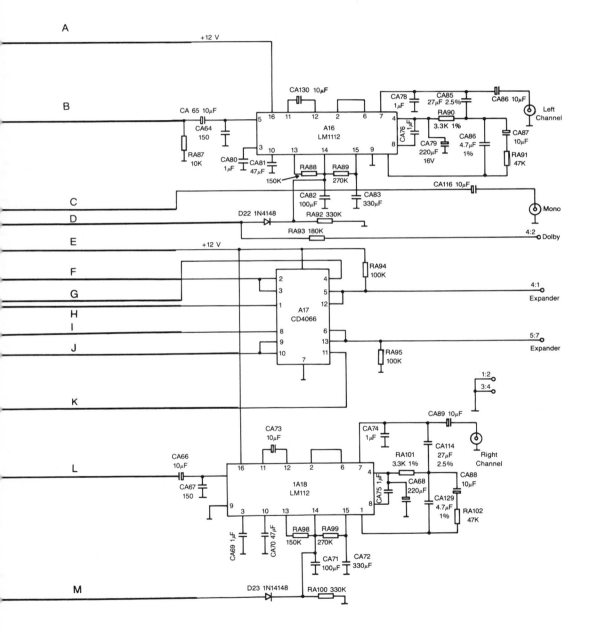

A

+12 V

B

CA130 10μF

CA 65 10μF
CA64
150
RA87
10K

A16
LM1112

CA78
1μF
CA76
1μF

CA85
27μF 2.5%
CA86 10μF

RA90
3.3K 1%

CA86
4.7μF
1%

CA79
220μF
16V

Left
Channel

CA87
10μF
RA91
47K

CA80
1μF
CA81
47μF

RA88
150K
RA89
270K

CA82
100μF
CA83
330μF

CA116 10μF

Mono

C

D

D22 1N4148

RA92 330K

RA93 180K

4:2 Dolby

E

+12 V

F

RA94
100K

G

A17
CD4066

4:1
Expander

H

I

J

RA95
100K

5:7
Expander

K

1:2

3:4

CA89 10μF

CA66
10μF
CA67
150

1A18
LM112

CA73
10μF

CA74
1μF
CA75 1μF

RA101
3.3K 1%

CA68
220μF

CA114
27μF
2.5%

CA129
4.7μF
1%

Right
Channel

CA88
10μF
RA102
47K

CA69 1μF
CA70 47μF

RA98
150K
RA99
270K

CA71
100μF
CA72
330μF

L

M

D23 1N14148

RA100 330K

149

Warner-Amex Stereo Transmission System

The Warner system is used on MTV, the Movie Channel and the Disney Channel. It is a wideband matrix stereo broadcast. The Disney and Movie Channel frequencies are 6.8 MHz for the L + R subcarrier and 5.8 MHz for the L − R subcarrier. MTV uses the same L − R subcarrier frequency but it uses 6.62 MHz for the L + R subcarrier.

A matrix broadcast can be identified by one loud subcarrier and one blank or very low subcarrier, which in the case of MTV, will consist mainly of reverberation and echo. There are also supersonic control tones for local ad insertion by cable companies. These can bleed through during demod and can cause a harshness to the sound, especially in the case of MTV, when the VeeJays are talking.

Times-Mirror Stereo Transmission System

This system is wideband discrete. The left channel is on one subcarrier, while the right channel is on a separate subcarrier. There is usually no noise reduction or companding applied to the signals. Examples of this type of system are CMTV (Country Music TV), the Nashville Network, and ESPN.

There is no special processor involved in receiving these signals. All that is required is two tunable audio demodulators.

Leaming Stereo Transmission System

The Leaming system transmits its signals using a multiplex format. It is the oldest system that is currently in use, having been originally designed for cable transmission over 15 years ago. It has been used for almost 6 years on satellite transmissions.

TROUBLESHOOTING GUIDE

Some of the most commonly encountered problems with the audio processing circuits are given here. In each case, I present things to check and suggest cures.

One Channel Missing when using Matrix Stereo. Probably the most common stereo problem is the, "I pushed in the Matrix button and I lost one channel" problem. Almost 100% of the time, it is due to an incorrect setting of the front panel tuning controls. If both controls are tuned to the same subcarrier, it will remove one channel. This was shown algebraically in Table 11-3. if both inputs to the matrix decoder are L + R or if they are L − R, the answer is the same. The right channel's

output is cancelled out. The cure is to set the A (or main) subcarrier to the L + R signal, either 6.8 MHz or 6.62 MHz for MTV, and set the B (or sub) control to the L − R subcarrier, which is usually 5.8 MHz.

The right channel will also disappear if the program is broadcast in discrete stereo and matrix stereo is selected.

Popping Noises in the Audio. Audio noise can usually be isolated into one of four areas in the TVRO system. The first area is low or noisy signal strength coming into the receiver. The second is misalignment or drift in the video i-f, video demod, or audio demod circuit. The third is a cold solder joint or malfunctioning part, such as a noisy resistor, transistor, IC, or diode. The fourth is drift or component change in the rf modulator.

Popping or static noises are the audio equivalent of video sparklies. Their presence may indicate drop-outs during video or audio demodulation. In most cases, the audio is fairly immune to video sparklies, as the picture may be barely watchable and the audio can be crystal clear.

The first area can be discounted if the video is clean. If there are mostly sparklie-free channels throughout the belt, then the signal up to and including the video demod is OK. In most receivers, there is a hard limiter in front of the demod circuit. Thus the demodulated signal coming from the video demod circuit will be a fairly constant level. It is at this point that the audio subcarriers are split from the video.

If the S-meter level is good, always check the receiver switch and control settings. This check includes the wideband/narrowband switch setting, the selected audio mode, and the setting of the audio channel-tuning controls. Make sure that the program that the receiver is tuned to is broadcasting the same kind of audio that the receiver is set up for. Finally, try fine tuning the TV or switch the modulator and the TV to the other output channel.

If the video is noisy and the S-meter is very low (less than a 1 or 2 on a 10 scale), then there could be problems with the video i-f , the down converter, the LNA, dish aiming, or any of the cabling between the dish and receiver. The video may be watchable, but the sound may be filled with pops, clicks, and static noises that cannot be fine tuned away.

Misaligned i-f Strip. A misaligned 70 MHz i-f strip can cause the picture and sound to not line up. In other words, the best sound does not result in the best picture. If the audio clears up, but the picture gets worse as the video fine tune control is adjusted, it almost always is an indication of i-f misalignment.

Another problem is misalignment and drift in the audio demod circuit. In extreme cases, there is popping and static noises on every channel. If the audio demod or i-f is slightly out, tuning in the narrowband subcarriers will be hard if not possible and cleaning up the noise in the program audio on WGN and WTBS will be very difficult. A full audio alignment is called for in this case.

Thermal, Noisy, or Intermittent Component. The second problem, a noisy component, can be the hardest to track down, especially if it is intermittent or thermal. The use of an oscilloscope will help somewhat but perserverance is what is really called for. The best place to start is at the demod output, if the sound is clean at that point, both the demod alignments and any components in the demod can be cleared.

Active components such as op-amps and transistors are first on the suspect list. Here the various dc voltages should be checked for abnormalities. Short of replacement, the only way to determine if the noise is coming from an IC is to check the input signal and then check the output signal. Noise appearing on the output and not on the input could indicate a bad IC. Remember some ICs have tremendous signal-voltage gains so that must be considered. The noise may have been there on the input, but have been discounted.

Any suspected feedback or biasing resistors could be bypassed with a .01 μF cap to see if there is a drop in the noise output. The most likely resistors to develop noise across them in transistor circuits are the base biasing resistors and by unbypassed emitter resistors.

If the noise is only apparent after a coupling capacitor, then replace the coupling cap. Exact replacement values are usually not critical in coupling capacitors, except those used in tuned circuits.

Using freeze spray and a soldering pencil to cool and heat components will often cause the intermittent component to start acting up. Running the receiver at 125 Vac will also help to cause on-the-edge components to start acting up, or to fail completely.

RF Modulator Noise. The last area that can be the source of noise is the rf modulator. The modulator can have the same kinds of transistor, IC, resistor and capacitor noise problems. A quick check of the modulator is to send the audio directly to a stereo system, if there is no noise problem in the stereo then the rf modulator is the best suspect.

The only area where the modulator will almost always be at fault is if the noise is a hum or buzz, that changes with the video. If lettering on the screen or an outdoor scene causes a loud buzzing in the TV but not in the stereo system, then the video level is too high or the rf modulator's audio frequency is too low (i.e. not exactly at 4.5 MHz).

RF Modulators

SINCE SATELLITE SIGNALS ARE NOT COMPATIBLE with standard TV sets, there must be some device which translates the satellite signals into something that regular TV sets can receive. This device is called a radio frequency modulator, or in its everyday nomenclature; the rf modulator, rf mod, or the modulator.

On satellite receivers with built-in modulators, the rf modulator is usually contained in a metal box located somewhere in the satellite receiver with several wires coming out of it via feed-thrus. The box is used to shield the modulator from the rest of the circuitry to prevent video interference.

The modulator combines the video and audio into one signal that can be used by a standard TV set. Its output is 75 ohms, unbalanced. RG-59 cable is usually used to connect the rf output to the TV set. If the television has a type-F connector, the RG-59 cable from the rf output of the receiver can be directly connected to it. If the TV only has two screw terminals that read "antenna in" or "300 ohm input," then a balun (a 75 ohm to 300 ohm transformer) will be required to impedance match the modulator to the television.

On the rf modulator itself, or sometimes on the back panel of the receiver, you will find a switch to select which channel is output to the TV. Most modulators used in home video equipment are designed to switch between channel 3 or 4, although channel 2 or 3 modulators are occasionally found. These channels (2, 3, 4) have become the defacto standard modulator frequencies even though they are not ideal choices. Unfortunately the three VHF channels are all adjacent channels, which means there may be possible co-channel interference. They are also susceptible to interference from auto ignitions and CB and amateur radio harmonic transmissions. The 70 MHz i-f frequency of the satellite receiver itself also falls into the pass band of channel 4.

TV SIGNAL STANDARDS

In the U.S. and Canada, television is broadcast using the NTSC (National Television Standards Committee) standards. NTSC TV signals are broadcast using AM for the video and a 4.5 MHz FM subcarrier for the audio. These are combined at the transmitter and sent out as one signal.

AM is a process whereby a fixed frequency carrier is modulated or changed in amplitude. Unfortunately this type of broadcasting system is sensitive to noise spikes from such sources as lightning and car ignitions. A noise spike will create a pop or crackle in the audio or a black or white streak or spot in a video signal. Since wideband

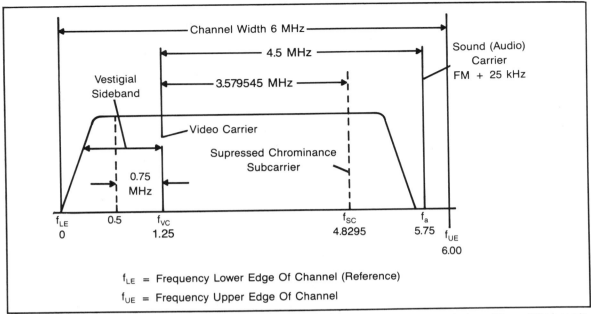

f_LE = Frequency Lower Edge Of Channel (Reference)

f_UE = Frequency Upper Edge Of Channel

Fig. 12-1. Frequency chart of the NTSC broadcasting system using vestigial sideband transmission. It is an AM transmission, except for the FM audio subcarrier which is 4.5 MHz above the channels center frequency. There is also a 3.58 MHz suppressed chrominance carrier for syncing the TV with the transmission.

FM systems were not perfected back when TV standards were established, AM was the only way to go, so we're stuck with it for terrestrial TV broadcasting.

Standard TV is broadcast using a bandwidth of 6 MHz to broadcast the video and audio. Figure 12-1 shows the channel frequency utilization for a NTSC transmission using vestigial-sideband transmission. The video carrier is 1.25 MHz above the lower edge of the channel, while the audio carrier is .25 MHz below the upper edge of the channel or 4.5 MHz above the video carrier. There is also a color or chrominance subcarrier 3.58 MHz above the video carrier.

Compare Fig. 12-1 to the frequency plot for a satellite broadcast channel in Fig. 12-2. Here the video fre-

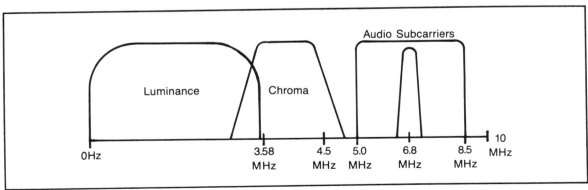

Fig. 12-2. Frequency chart for a satellite TV broadcast. Since both the audio and the video are FM, the nominal center frequency would have symmetrical signals stretching ± 10 MHz from it. The video information still takes up 4.2 MHz, and the audio is above the video and stretches from about 5.0 to 8.5 MHz. Typically, the program audio is broadcast at 6.8 MHz.

153

Color Television Systems in Use or Planned

COUNTRY	COLOR SYSTEM	COUNTRY	COLOR SYSTEM	COUNTRY	COLOR SYSTEM	COUNTRY	COLOR SYSTEM
Afghanistan	PAL	Dominican Republic	NTSC	Kuwait	PAL	Saudi Arabia	SECAM & PAL
Albania	SECAM	Ecuador	NTSC	Lebanon	SECAM	Senegal	N.A.
Algeria	PAL	El Salvador	NTSC	Liberia	PAL	Sierra Leone	PAL
Angola	N.A.	Equatorial Guinea	N.A.	Libyan Arab Republic	SECAM	Singapore	PAL
Antigua	NTSC	Ethiopia	N.A.	Luxembourg	SECAM & PAL	South Africa	PAL
Arab Republic of Egypt	SECAM	Finland	PAL			Spain	PAL
Argentina	PAL	France	SECAM	Madagascar	N.A.	Sri Lanka	PAL
Australia	PAL	French Guiana	SECAM	Malaysia	PAL	Sudan	N.A.
Austria	PAL	French Polynesia	SECAM	Maldives	PAL	Surinam	NTSC
Bahamas	NTSC	Gabon	SECAM	Malta	N.A.	Swaziland	PAL
Bahrain	PAL	German Democratic Republic	SECAM III	Martinique	SECAM	Sweden	PAL
Bangladesh	PAL			Mauritius	SECAM	Switzerland	PAL
Barbados	NTSC	Federal Republic of Germany	PAL	Mexico	NTSC	Syria	SECAM
Belgium	PAL			Monaco	SECAM & PAL	Tanzania	PAL
Benin	SECAM	Ghana	PAL			Thailand	PAL
Bermuda	NTSC	Gibraltar	PAL	Mongolia	N.A.	Togo	SECAM
Bolivia	NTSC	Greece	SECAM	Morocco	SECAM	Trinidad & Tobago	NTSC
Brazil	PAL	Guadeloupe	SECAM	Netherlands	PAL	Tunisia	SECAM
British Virgin Islands	NTSC	Guatemala	NTSC	Netherlands Antilles	NTSC	Turkey	PAL
Brunei	PAL	Guinea	N.A.	New Caledonia	SECAM	Uganda	N.A.
Bulgaria	SECAM	Haiti	SECAM	New Zealand	PAL	United Arab Emirates	PAL
Burma	N.A.	Honduras	N.A.	Nicaragua	NTSC	United Kingdom	PAL
Cambodia	N.A.	Hong Kong	PAL	Niger	SECAM	United States	NTSC
Canada	NTSC	Hungary	SECAM	Nigeria	PAL	Upper Volta	N.A.
Central African Republic	N.A.	Iceland	PAL	Northern Mariana Islands	NTSC	Uruguay	PAL
Chad	N.A.	India	N.A.	Norway	PAL	USSR	SECAM
Chile	NTSC	Indonesia	PAL	Oman	PAL	Venezuela	NTSC
China	PAL	Iran	SECAM	Pakistan	PAL	Vietnam	N.A.
Taiwan Province	NTSC	Iraq	SECAM	Panama	NTSC	Yemen Arab Republic	N.A.
Colombia	NTSC	Ireland	PAL	Paraguay	PAL	People's Democratic Republic of Yemen	N.A.
Republic of the Congo	N.A.	Israel	PAL	Peru	NTSC		
Costa Rica	NTSC	Italy	PAL	Philippines	NTSC	Yugoslavia	PAL
Cuba	NTSC or SECAM	Ivory Coast	SECAM	Poland	SECAM	Zaire Republic	SECAM
		Jamaica	N.A.	Portugal	PAL	Zambia	PAL
Cyprus	PAL	Japan	NTSC	Qatar	PAL	Zimbabwe	N.A.
Czechoslovakia	SECAM	Jordan	PAL	Reunion	SECAM		
Denmark	PAL	Kenya	N.A.	Romania	SECAM		
Djibouti	SECAM	Korea	NTSC	St. Kitts	NTSC		
		People's Democratic Republic of Korea	N.A.	St. Pierre & Miquelon	SECAM		
				Samoa (American)	NTSC		

Fig. 12-3. Country to standard that is used. The big three are NTSC, PAL and SECAM.

quency modulates a carrier that is the nominal center frequency of the channel. There is usually an audio subcarrier located at 6.8 MHz. The subcarrier frequency can range from 5.0 MHz to 8.5 MHz, and can consist of multiple subcarriers for audio or data. This subcarrier shows up as two sidebands located on either side of the center frequency. The video frequency extends from about 30 Hz to 4.2 MHz and is standard NTSC video.

There are other standards in use throughout the world for broadcasting over-the-air TV. There is also the PAL and SECAM systems. Many countries actually use their own variation of each system. Thus, there is no universal rf modulator. What will work in the U.S. will not work properly with the TV sets in France or Germany. To figure out what country uses what system, refer to the country-to-system cross-reference in Fig. 12-3.

TYPES

Rf modulators are used not only in satellite receivers but also in home computers, VCRs, and video games. Mainly because of their wide-spread usage in inexpensive home video game controllers, there have been several ICs developed specifically for this purpose.

Modulators can be classified into two categories, crystal controlled or LC tuned. Crystal controlled modulators use a quartz crystal as the tuning device, thus frequency stability over time and temperature is accurately maintained. An LC (inductor/capacitor) tuned modulator is less expensive than a crystal modulator, but its long-term frequency stability may not be as good (especially in inexpensive LC modulators).

LC Modulator

The LM1889 IC modulator is used in the SatTec R-5000 receiver, among others. The circuit consists of the IC (in an 18-pin DIP package) with an external 4.5 MHz tunable tank circuit for the audio subcarrier and a 54 MHz to 88 MHz adjustable tank circuit for the video carrier. It is factory tuned to output channel 4 but may be changed in the field to any channel from 2 through 6, by changing one capacitor value and by adjusting one coil. Figure 12-4 is the schematic for an 1889 modulator.

The LM1889 consists of two separate channel

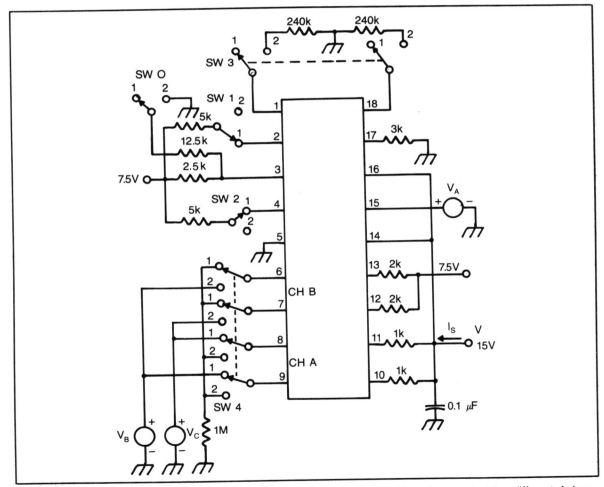

Fig. 12-4. LM1889 rf modulator IC. It is virtually a miniature all-in-one TV transmitter. It can output two different rf channels (typically 3 and 4). It is used in the Sat-Tec R-5000 receiver.

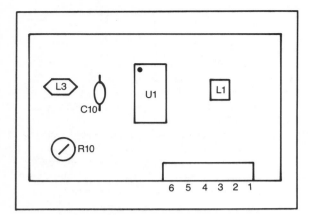

Fig. 12-5. Drawing of the JM-7 rf modulator board from a Sat-Tec 5000 showing the adjustment locations for changing the output channel.

oscillator circuits that can be switch selected to easily choose between two channels. The SatTec design only uses one of these oscillators and thus can only switch channels by adjusting a coil setting or capacitor value. The channel tank circuit is connected to pins 6 and 7 of the IC.

C10 is the capacitor value that must be changed. The smaller the capacitance the higher the modulated channel. It is usually shipped with a 30 pF capacitor installed. This allows the coil (L3) to be tuned to either channel 3 or channel 4. If C10's value is changed to about 15 pF (most easily done by paralleling another 30 pF capacitor across C10), then L3 will tune between channels 5 and 6. To tune down to channel 2, C10 should be changed to about 40 pF in value. The exact capacitance is not critical as the fine tuning is done by adjusting coil L3. Unfortunately there is no overlay on the rf modulator board so none of the parts are identified by number, but Fig. 12-5 is a drawing of the board with locations of the essential parts called out.

Before adjusting the coil be sure to turn off the afc control on the TV as it will attempt to track your adjustments. Alternately a frequency counter could be used to measure the exact carrier frequency. In some cases, the 4.5 MHz oscillator must be disabled to get an accurate reading. The unmodulated picture carrier should be 55.25 MHz for channel 2, 61.25 MHz for channel 3, 67.25 MHz for channel 4, 77.25 MHz for channel 5 and 83.25 MHz for channel 6. A spectrum analyzer could also be used to check for tuning of both the picture carrier as well as the sound subcarrier.

The 4.5 MHz audio subcarrier does not change,

regardless of which picture channel is selected. The only time the 4.5 MHz coil (L1, which is in the metal can) would need adjusting is if there was chroma buzz in the audio. Chroma buzz could be due to excessive video level or drift of the 4.5 MHz oscillator. In either case, a slight adjustment of the coil (usually less than one turn) will clean up the audio. Adjust L1 for the most noise-free sound. If the sound is still buzzy after adjusting the coil, try lowering the video level (R10) slightly. The video is probably overdriving the modulator.

A similar circuit is used in all the receivers that use the LM1889 modulator, and the above procedures are applicable to them all.

Crystal Controlled Modulator

An example of a crystal controlled modulator is the one shown in Fig. 12-6. It is used in Drake, Winegard, and Conifer receivers. It uses a 1374 IC as the modulator. The channel is selected by changing between the two crystals in the tank circuit between pins 6 and 7 of the IC. Video fine tuning is done by spreading or compressing the two .29 μH coils. The audio is coupled onto the video by the 43 pF capacitor connected to pin 3 and to pin 1 through the 2.2 K resistor. The adjustable coil (6.4 μH to 15 μH) sets the audio oscillator to 4.5 MHz. The capacitor/resistor network on the audio input is for preemphasis. It rolls off the low audio frequencies, thereby seemingly boosting the high end.

The 38 μH coil and 18 pF capacitor on the video input form a low pass filter to prevent any audio subcarriers from mixing with the video signals. The three .15 μH coils and the 43 and 82 pF capacitors form a low-pass filter for channel 4 and below. The output is double side-band, instead of vestigial sideband, which has less distortion but which will cause interference on the next lower channel due to the presense of modulation information which is normally suppressed being present.

All rf modulators sold in the United States must be approved by the FCC for use. But not all of the receiver manufacturers have applied for FCC approval of the receiver/modulator combination. Those that have been FCC approved have a FCC number on their serial number tag. Even though most modulators are interchangeable on FCC approved receivers, if it becomes necessary to replace the modulator, only an exact replacement part can be used. This is to keep the receiver in compliance with FCC regulations.

Interference

In the major markets in the United States, both

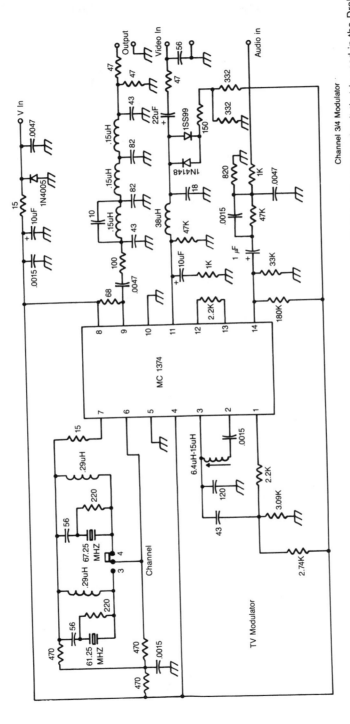

Fig. 12-6. The rf modulator schematic from a Conifer receiver. It is built around the 1374 modulator IC. This is the same circuit that is used in the Drake receiver. Courtesy of Conifer Corp.

157

channels 2 and 4 are usually broadcast, while in smaller markets channel 3 is broadcast. Because of the frequency allocations and method used to broadcast, channels 2 and 3 cannot be broadcast together over-the-air without causing interference to one another. This is the same for any other adjacent channel pair. Figure 12-7 is an allocation chart for the frequencies from 54 MHz to 216 MHz. This covers the low and high VHF bands. The only nonadjacent channels are 4 and 5, and 6 and 7.

If the satellite receiver location is close to a transmitter broadcasting on an adjacent channel, there may be interference in the picture depending on the distance to the transmitter and the rf shielding in the TV itself. If the transmitter is on channel 2, interference may be seen on channels 2 and 3. If the transmitter is on channel 3, interference could show up on channels 2, 3, or 4. If the

transmitter is on channel 4, interference may be on channels 3 or 4.

As a quick check, see if the TV will receive a picture from the local transmitter without having an antenna hooked up. If a picture can be seen then there probably will be interference on that channel and possibly problems on the adjacent channel as well. In most cases, herringbones, thin wavy white or dark lines running diagonally through the picture, of varying strength would be the interference. Although in severe cases a second picture (from the local transmitter) can actually be seen floating through the background.

AUXILIARY COMPONENTS

In many cases, the satellite receiver will be hooked up

Channel No.	Bandwidth	Video Carrier	Color Carrier	Audio Carrier
2	54.0-60.0	55.25	58.83	59.75
3	60.0-66.0	61.25	64.83	65.75
4	66.0-72.0	67.25	70.83	71.75
5	76.0-82.0	77.25	80.83	81.75
6	82.0-88.0	83.25	86.83	87.75
FM	88.0-108.0	-----	-----	-----
A	120.0-126.0	121.25	124.83	125.75
B	126.0-132.0	127.25	130.83	131.75
C	132.0-138.0	133.25	136.83	137.75
D	138.0-144.0	139.25	142.83	143.75
E	144.0-150.0	145.25	148.83	149.75
F	150.0-156.0	151.25	154.83	155.75
G	156.0-162.0	157.25	160.83	161.75
H	162.0-168.0	163.25	166.83	167.75
I	168.0-174.0	169.25	172.83	173.75
7	174.0-180.0	175.25	178.83	179.75
8	180.0-186.0	181.25	184.83	185.75
9	186.0-192.0	187.25	190.83	191.75
10	192.0-198.0	193.25	196.83	197.75
11	198.0-204.0	199.25	202.83	203.75
12	204.0-210.0	205.25	208.83	209.75
13	210.0-216.0	211.25	214.83	215.75

Fig. 12-7. Listing of the rf channel number to the video carrier, audio subcarrier and color subcarrier frequencies, as used by the NTSC broadcasting system.

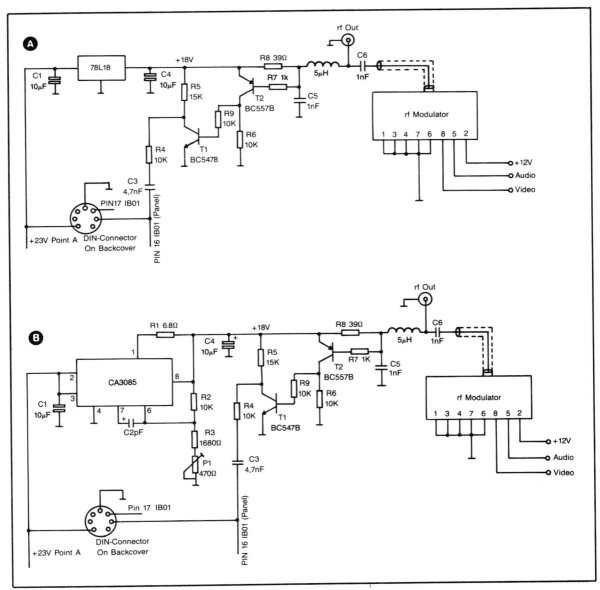

Fig. 12-8. Schematic of the interface and power supply circuitry for the 9536. It is located in the 9550 and 9540 receivers. Courtesy of Luxor North America.

to more than one TV. If the receiver has a UHF remote or a wired remote like Maspro's, then there is no problem changing channels from another room in the house, but if the remote is of the infrared variety then an external system must be used to send the signals back to the receiver. Since we already have a coax cable running from the receiver to the TV set, it logically follows that we

should use that same coax to carry our remote control signals back to the receiver. Luxor has designed this type of system into their receivers so that if a Luxor 9536 remote sensor is used, an infrared remote can control the receiver from any location in the house.

The remote sensor works by translating the infrared pulses into a series of voltage pulses that are carried on the coax cable back to the receiver. At the receiver, these pulses are amplified and sent into the decoder control circuit. Figure 12-8 shows the two different receiving circuits that have been used in the remote-controllable Luxor receivers. The only difference is in the +18 Vdc regulator. It was changed from a CA3085 to a 7818. The 3085 could not handle short circuits on the +18 volt line, whereas the 7818 will merely shut down or current limit if a short occurs. The +18 Vdc is used to power the 9536 sensor.

Figure 12-9 is a block diagram of how the 9536 is used. The biggest precaution is that there is dc on the coax cable. Normally there is not dc on the cable, thus any splitters, A/B switches, etc. that are put into the

system must be capable of passing dc to the 9536. As seen in Fig. 12-10, there is no dc that is passed onto the TV set because of blocking capacitors C2 and C3. The TEA1009 is a detector and decoder chip that outputs a series of pulses according to the incoming code that it received from the remote control's infrared signal.

A remote sensor made by Video Link, called the Xtra-Link, is very much like the Luxor 9536, but it was designed as an add-on for infrared remote controlled VCR's and video disc players. It can also be used with just about any infrared remote controlled satellite receiver as well. The Xtra-Link sets on top of the remote TV and is operated by a wall mounted transformer.

It detects the infrared signals and again translates

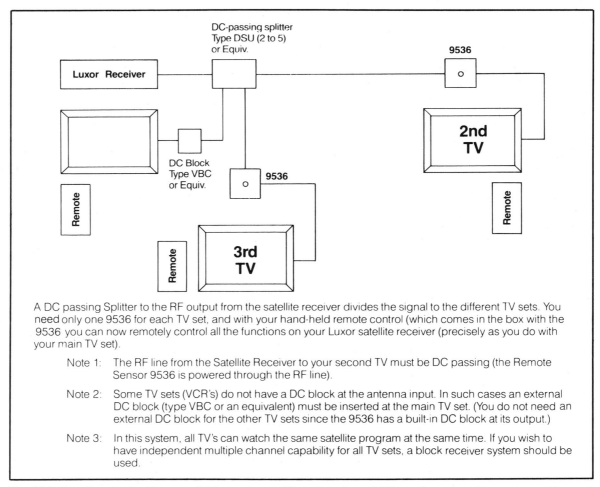

A DC passing Splitter to the RF output from the satellite receiver divides the signal to the different TV sets. You need only one 9536 for each TV set, and with your hand-held remote control (which comes in the box with the 9536 you can now remotely control all the functions on your Luxor satellite receiver (precisely as you do with your main TV set).

Note 1: The RF line from the Satellite Receiver to your second TV must be DC passing (the Remote Sensor 9536 is powered through the RF line).

Note 2: Some TV sets (VCR's) do not have a DC block at the antenna input. In such cases an external DC block (type VBC or an equivalent) must be inserted at the main TV set. (You do not need an external DC block for the other TV sets since the 9536 has a built-in DC block at its output.)

Note 3: In this system, all TV's can watch the same satellite program at the same time. If you wish to have independent multiple channel capability for all TV sets, a block receiver system should be used.

Fig. 12-9. Directions for hooking-up a 9536 remote eye. Since there is +18 Vdc on the coax cable, care must be used to prevent regulator damage due to a shorted cable. Courtesy of Luxor North America.

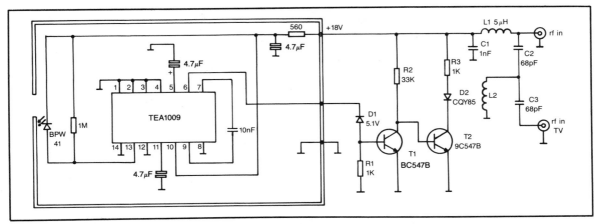

Fig. 12-10. 9536 schematic. Courtesy of Luxor North America.

them into a series of voltage pulses. It varies from the Luxor model in that the Xtra-Link's series of pulses are used to turn on an infrared LED that is placed in front of the receiver's infrared sensor. This way no modifications have to be done to the receiver. The splitter that comes with the Xtra-Link is used to block the dc signals from getting into the receiver or the TV. The LED can be placed up to 2 feet away from the window, so that a motor drive, a VCR and a satellite receiver could all use the same LED if they were positioned properly. The splitter/dc block does have an output for a second LED if necessary.

If there are two sources that need to be hooked into a TV set, then an A/B switch box can be used. Many receivers supply these as part of their package. Their main use is to hook a 300 ohm antenna lead-in cable and a video game or other rf input to the same TV. By switching to input B, the local TV antenna is hooked to the TV. If it is set to A then the satellite receiver is hooked to the TV. It can be used as a 75 ohm to 300 ohm balun by just leaving it set to the A position.

Using VCR's with the satellite receiver can be done in three ways. The best way is to use the video and audio outputs from the receiver. These are directly connected to the audio and video inputs on the VCR. If the VCR does not have audio and video inputs or if an easier switching method is desired, then the A/B switch could be used to switch between the satellite receiver and the VHF antenna. For full flexibility, a third method could be used that would allow the VCR to record off air or satellite signals while the TV set is hooked up to either source. This method requires two A/B switches: one to choose between the VCR's rf output and the direct source

and the other to choose between the satellite receiver and the VHF antenna (the direct source). A splitter would be used on the antenna line to couple the antenna to both the VCR and the direct source switch. The three methods are shown in Fig. 12-11, Fig. 12-12 and Fig. 12-13.

If more than two inputs are to be switched to one TV, then an rf switcher or video selector is needed. Drake has a video selector which can select between five inputs and three outputs. The outputs are designed for two TV sets and one VCR. It is shown in Fig. 12-14.

TROUBLESHOOTING

The modulators used in home satellite receivers do not, by any means, transmit broadcast quality audio or video. To achieve high-quality signals requires an external cable-type modulator. An external modulator may also be required if there is more than 250 feet of cable or if several TVs are to be driven by the modulator. In a Private Cable system, external modulators must be used to mix several different channels together for combining onto one cable. These external modulators could be used on adjacent channels if they have the proper bandpass filtering.

Herringbones can be caused by an outside TV station's signals or by a defective modulator in the receiver. If the isolation between the modulator and receiver is lost through bad grounds, loose covers, or poor power supply bypassing, the signal can be fed back into the video circuits and cause wavy lines. Excessive video or audio levels into the modulator can also cause herringbone patterns as well as a buzzing sound. Component failures in the modulator will cause similar symptoms. To improve power supply bypassing, try tack soldering a .0015 to

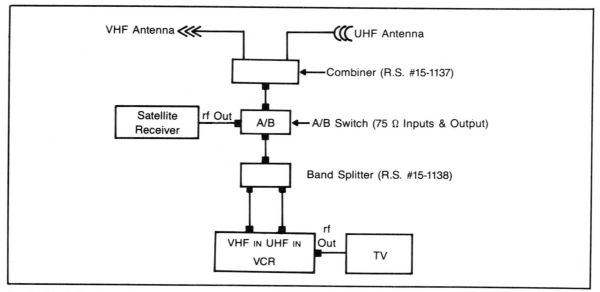

Fig. 12-11. Using the modulator in a VCR to drive a TV set.

.0047 μF capacitor from the dc line to ground. Use short leads and put it as close as possible to the modulator.

LC tuned modulators will often need to be tuned to the TV set. If the video carrier is far enough off, then it will result in herringbones or worse. One popular LC modulator is made by Astec. It has two coils that are reached through holes in the top cover. The coil closet to the four leads that hook-up the modulator tunes the video channel centering. The other coil tunes the 4.5 MHz audio subcarrier.

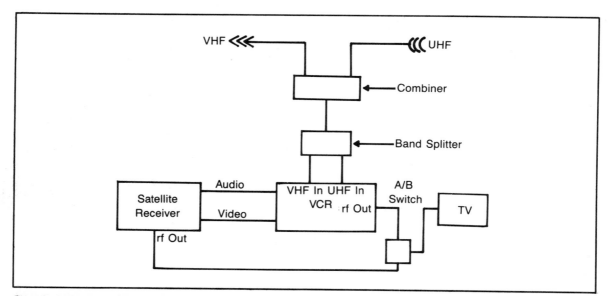

Fig. 12-12. Using an A/B switch to choose between the rf output from the VCR and the rf output from the satellite receiver. This allows recording an off-air channel while watching a satellite channel at the same time. It cannot record a satellite channel and watch an off-air channel though.

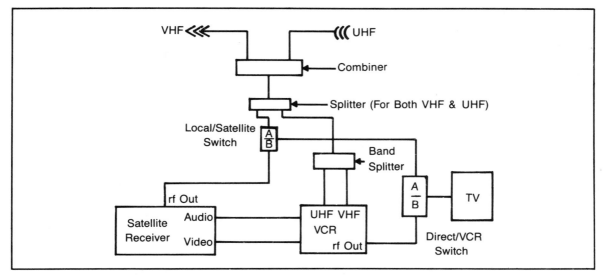

Fig. 12-13. A fully switchable set-up for recording and independently watching either satellite transmissions or off-air signals. It uses two A/B switches to choose between direct and VCR, with the direct input coming from either the satellite receiver or the off-air antenna.

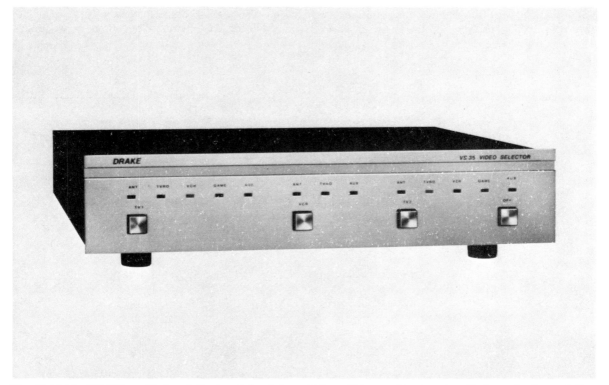

Fig. 12-14. The VS-35 from Drake. A five input rf switcher which could be used to record and independently watch five sources. Courtesy of R.L. Drake.

163

If the TV set that is to be used is a synthesized model, then it will be set exactly on the FCC approved frequencies for all the channels. However, if an LC modulator is used, its output may not be even close to the FCC frequency depending on which channel the manufacturer set it up on and which channel is to be used in the field. This is because LC modulators, on the whole, do not change channels exactly to frequency. If the modulator is set up on channel 3 and then set to channel 4 in the field, then the video tuning will be off slightly. This is no problem for most sets, but synthesized sets may not be able to lock onto the signal. Usually the synthesized sets have a cable/TV switch or a narrow/wide switch located near the adjustment controls. The set should be put into the cable or wideband mode. Often this will enable the TV to lock onto the modulator's output.

If the picture or sound is still not right, then the audio and video coils will need to be adjusted for the best picture and sound. Do this by using an insulated coil tool, which is like a small plastic flatblade screwdriver. They are available at most electronics parts shops. Never turn the coils more than one full turn in either direction. If you've turned it one turn and the picture is still not there, then the TV is probably set on the wrong channel or

something else is wrong in the video circuit. The slugs are very delicate, being a powdered ferrite composition, and will be broken if forced into the bottom of the coil.

If the 4.5 MHz audio carrier is not at 4.5 MHz, then the sound will be noisy, low in volume or even nonexistent. If it is tuned below 4.5 MHz, then buzzing sounds will be heard in the audio, especially when lettering or saturated colors are on the screen. This same kind of symptom will be experienced if the video level is set too high or if the high frequencies are boosted.

In summary, if a video or audio problem is occurring in the rf output, always check the video and audio outputs to see if the problem is present there also. If the sound is clean and if the video looks OK, then chances are the modulator is at fault. Before swapping out the modulator, try resoldering the connections to it, especially the ground joints. Try reseating the metal covers, as well as any plugs that are used. If there is an RCA jack on the modulator, try plugging the rf line into an F female-to-RCA male adapter, and then plug that into the modulator directly. If that works, then the coax cable between the modulator and the back panel is shorted or open. Check the F connector on the back panel. You might find that it is connected by only one or two strands.

Miscellaneous
Receiver Circuits

T HIS CHAPTER COVERS THE REMAINING CIRCUITS in a satellite receiver. Such things as indicating circuits, tuning circuits, and remote controls are discussed.

INDICATOR CIRCUITS

There are a variety of indicators that could be used in a satellite receiver. LEDs in various forms and combinations in anything from power on to channel number. Signal strength can be indicated by LEDs or by a meter. Incandescent lamps can also be used.

Meter Circuits

Probably the first indicator circuit that comes to mind is the signal-strength meter. The signal-strength meter is used to peak up the dish, the channel tuning, and the polarization. It is usually a relative meter, although in the Earth Terminal's receiver, it is accurately calibrated in dB. In some receivers, the meter is calibrated to the i-f strip. If the signal is the right level, the meter will read between half and full scale. In other receivers, the meter has an adjustment so that it can be set anywhere in its range. A typical signal-strength meter circuit is shown in Fig. 13-1.

The signal-strength meter is tapped off of the i-f strip after it has been filtered and before it enters the limiter. The 70 MHz signal is usually amplified by a transistor and then detected by a diode. This detected signal is applied to the meter. Meters are rated in microamps, with typical meters showing a full scale indication when their rated current, 50 to about 200 microamps, is run through them. Usually a low meter reading is indicative of DC, LNA, cable, or dish problems. Although it could also be i-f strip or meter driver problems.

A second type of meter is the center-tune meter. It is typically hooked into the afc feedback loop, and is used to detect an unbalanced condition in the afc loop. If the channel is properly center tuned, then the afc voltage would be at its nominal point, which is often a positive 1 or 2 volts. The center tune meter would be centered up at this point. If the channel was to drift off above its proper centering, then the afc voltage would also change to a more positive voltage. This would affect the tuning voltage so that the down-converter's LO frequency would be adjusted automatically to rebalance the afc circuit. If the channel was to go below its nominal center frequency, then the circuit would be unbalanced and the afc would be more negative, which would add or subtract

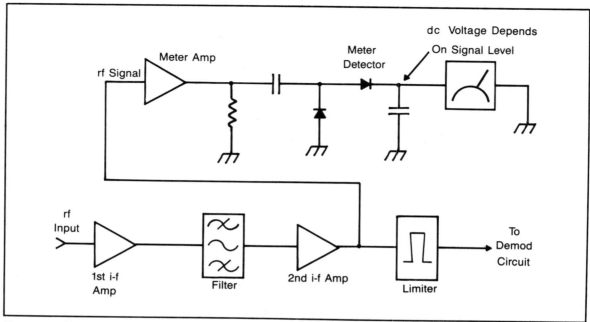

Fig. 13-1. In a typical signal meter circuit, the input is usually tapped off of the i-f signal just before the limiter.

depending on the circuit, with the tuning voltage raising the channel back to the afc's balance point. Figure 13-2 shows a typical center-tune meter circuit.

LED Circuits

An LED or Light Emitting Diode, does just what the name says. It emits light when a current is passed through it. LED's are used for such display functions as showing which audio channel has been selected, relative dish position, relative signal strength, and, for synthesized receivers, receiver lock-in indication.

Regardless of what is being displayed, all LED circuits consist of a voltage of the proper polarity and a current limiting resistor in series with the LED. There is usually a control circuit that turns the voltage off and on. Normally this control circuit is an npn transistor, although it can be a IC comparator, IC display driver, or a mechanical or solid state switch.

Figure 13-3 shows a basic LED circuit consisting of + 12 Vdc going through a 470 ohm resistor and the LED. Notice that the LED symbol is like a diode's with a circle around it. That means that the LED, which has two leads, is polarized. It MUST be connected correctly or it will be damaged. LEDs have a much lower reverse voltage tolerance than most diodes, and they are real easy

to destroy if they are hooked up backwards or if the current limiting resistor shorts, is too low in value, or is forgotten.

LED's can be run on just about any voltage from 1 Vdc on up. The main determining factor is the current limiting resistor. Its value must increase if the voltage is increased. As an example, at + 15 Vdc it is normally 680 ohms, at + 12 Vdc it is 470 ohms, at + 5 Vdc it is 330 ohms and so on. As seen in Fig. 13-3, the positioning of the resistor and the LED, in reference to the voltage and ground is noncritical.

Figure 13-4 is the LED display for a KLM Sky Eye IV. Here the LEDs are used to display signal strength. The ICs, U9, U10, and U11 are LM358 low-power dual op-amps.

One section of U11 is run as a voltage follower. It receives an input dc voltage into pin 3 that is proportional to the incoming 70 MHz signal. The signal is rectified by D1 and D2 and the overall dc level is adjusted by R29. U11 supplies voltage to all the other ICs' inverting inputs.

The other ICs (both sections of U9 and U10 and the remaining section of U11) are configured as summing amps. The output of each IC is equal to the + input voltage minus the − input voltage. The + input is set so that the voltage is highest on U9 pin 5, and lowest on U11 pin 5.

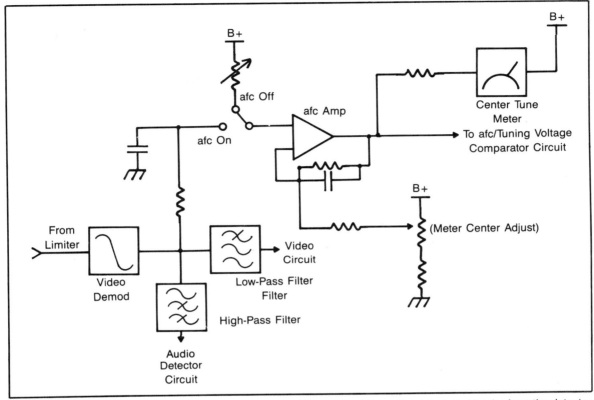

Fig. 13-2. Typical center-tune meter circuit. The input signal is tapped off from the detected video (or from the detector itself) and is then amplified to provide an afc correction voltage. This voltage also drives the center-tune meter.

The output voltage of each IC is high enough to keep the LEDs off unless there is a voltage coming from U11 pin 1. Once the − input exceeds the + input the output of the IC goes to ground. The LED is then turned on. As the − input goes below the fixed + input voltage, the LED is gradually turned off.

D12 will therefore turn on first followed by D11, D10, D9, and finally D8. Even though there are only five levels, this type of display can easily be used to peak up the dish or the polarization.

LED Readouts

Several LEDs when properly combined into one housing, can form an alphanumeric (letter or number) character. This readout character, when combined with other readout characters, can then be used to denote channel number, audio subcarrier frequency, the dish location count, or the selected satellite's name and number.

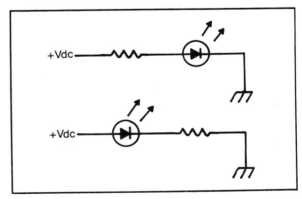

Fig. 13-3. An LED uses a resistor to limit the current that it can draw. If the resistor shorts or is not the correct value, then the LED can burn out. It uses a diode symbol, because it is a two-connector semiconductor device that emits light when a current is passed through it in the proper direction. An alphanumeric LED display is just several LEDs that have their cathodes or anodes connected together.

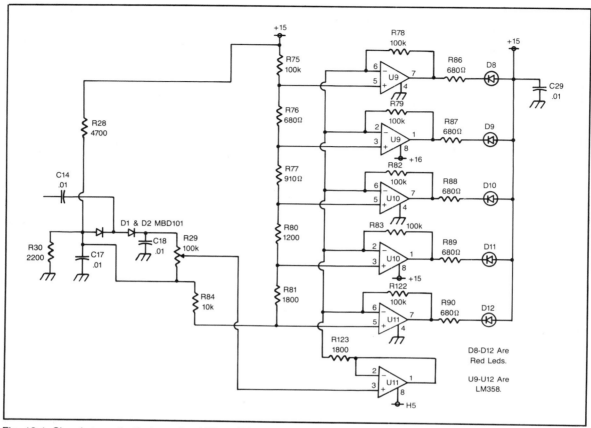

Fig. 13-4. Signal-strength display using LEDs. D8 through D12 turn on as the input voltage to pin 3 of U11 rises. The level is adjusted by R29. It is from a KLM receiver. Courtesy of KLM.

Figure 13-5 is a typical two section 7-segment display used to display channel number. The small letters "a" through "g", label the seven segments in the least significant digit or LSD. The capital letters label the sections in the most significant digit of MSD. If a number one is displayed, then only segments b and c are turned on. For a number 8, all segments are turned on. This type of display is best suited for numbers, although limited alphabetical characters (such as A, C, E, F, H, J, L, P, S, and U) can also be displayed.

An alphanumeric display works in much the same manner as the solitary LED. Each section is turned on or off by applying a voltage and a ground. LED displays can be divided up into two general types: common cathode and common anode.

Common Cathode. In the common-cathode display, the cathodes of each individual LED segment are tied together. This point is usually run directly to ground,

although there are circuits that have a common negative voltage and each segment is turned on by grounding the input.

This below ground method of driving a display is often used when there are positive and negative voltages available. Typically the negative supply has plenty of current to spare, whereas the positive supply is usually at its limit. LED displays draw a fair amount of current, so by taking it from the negative supply the total current used by the receiver is more equally divided between the positive and negative voltage regulators.

Two popular receivers that use a below ground display are the Drake ESR-24 and the Winegard SC-7035. Figure 13-6 is the display and latch/driver IC from the Winegard SC-7035. It is located on a separate board which connects to the main board via P1.

The two channel number readouts are MAN 74As and the ICs (U1 and U2) are 4511s. A 4511 is a 7-segment latch and display driver CMOS IC. Its truth table is shown in Fig. 13-7 along with the various outputs that will occur. The 4511 takes a BCD (Binary Coded Decimal) input and outputs the proper series of voltages to turn on the display so that the correct number is displayed. The 4511, or any 7-segment latch and display chip, takes a valid BCD input (0-9) on pins 1, 2, 6, and 7 and converts it into a driving signal that puts out the voltage found on pin 16. If all inputs are low, then the output voltages are held at the voltage on pin 8. In most circuits, pin 8 of a 4511 is ground and pin 16 is positive, but here pin 8 is a negative 15 volts and pin 16 is ground. As long as 16 is more positive than pin 8 (or less negative as in this case),

then the circuit will function properly. Notice that it will take two BCD words to get a full display.

A combination of grounds (which in this circuit is high) on pins 7 through 10 of P1 will cause various MSD sections to light, while a combination of grounds of pins 1 through 4 of P1 will cause various LSD sections to light. A point to remember, especially in negative voltage circuits, is that a logic high is equal to the voltage on pin 16 of the 4511, while a logic low is equal to the voltage on pin 8 of the 4511.

A similar 4511 circuit is shown in Fig. 13-8. It is the channel number display section of the Gould/Dexcel DXR 1300. It uses a 4511 latch and driver for one digit (the

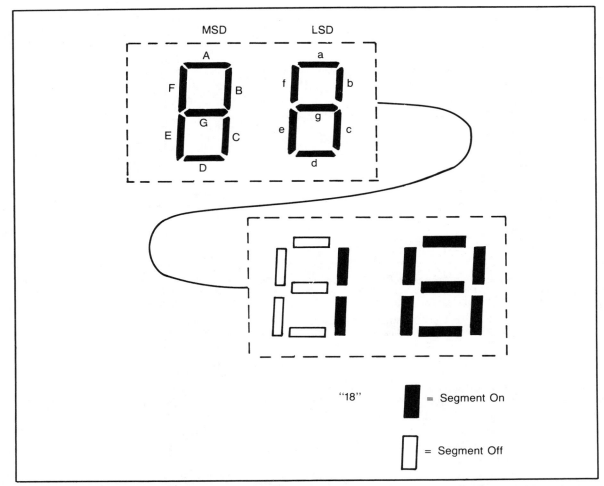

Fig. 13-5. A dual seven-segment LED readout for displaying two numbers. The segments are usually labeled a-g for the LSD (Least Significant Digit) and A-G for the MSD (Most Significant Digit), although this is not universally used. If segments B, C, and a through g are turned on, then the number 18 is displayed.

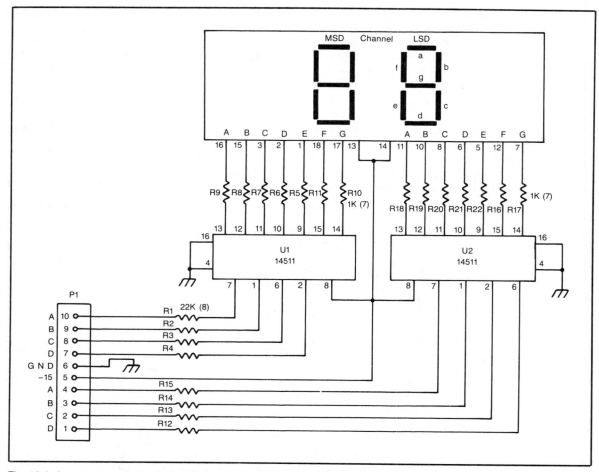

Fig. 13-6. A common cathode display. Here the driving source voltage is supplied by the 4511s. Since the readout's common point (pins 13 and 14) is at −15 Vdc instead of ground (which it usually is at) it is harder to follow, but the circuit acts the same. As each input goes high (ground in this circuit), it turns on the corresponding segment. Courtesy Winegard.

LSD), and a transistor driver for the MSD. In this circuit the 4511 is used with a positive voltage.

The MSD doesn't need to display anything higher than a 2, since the highest channel number is 24. Thus a dedicated matrix driver circuit using two transistors is all that is necessary. In this circuit, if Q304 and Q305's bases are both low, then there will be no voltage on the 560 ohm resistors and thus no readout.

Putting a logic high on Q304 will turn on sections "B" and "C" only, resulting in a "1." A logic high on Q305 will turn on all sections except "C" and "F," resulting in a "2." Since section "B" is common, two diodes (D317) are used to block the voltage from the undesired sections.

Common Anode. In a common-anode circuit, positive voltage is applied to the common terminal and the segments are turned on by selectively grounding the inputs.

An example of a common-anode readout circuit is shown in Fig. 13-9. Here both pnp and npn transistors are used to turn-on the display. A positive 15 volts is applied to the common anode pin of D1 or D2 whenever the lines MSD or LSD go low (remember a pnp turns on with a low). This turns on the voltage going to pin 14. When outputs "A" through "G" go high, the corresponding npn transistors are turned on, pulling the collectors to ground, which turns on the individual segments that are connected to the transistors through the 910 ohm resistors. Notice that two readouts are driven by the same circuit. To accomplish this the MSD and LSD inputs are toggled on and off at a high speed. At the same time, inputs A, B, C, and D are also toggled between the two digits to be displayed so that during one cycle the MS digit is triggered and during the next cycle the LS digit is triggered. This is done at such a high speed that the switching is not even noticeable to the eye.

Incandescent Lamps

Meter lamps are probably one of the most often changed components, and yet in many receivers the lamps are soldered in place or need some oddball voltage to operate on. This should not stop you from replacing them, since in 90% of the cases, using a generic replacement lamp will work fine if a voltage dropping resistor is used along with the replacement lamp. Since most lamps are not critical circuit components and are basically only there to illuminate the meter, indicate that it is on or indicate some change of state, buying one type of bulb, and then putting in a selected dropping resistor, means that almost every circuit can use the same bulb.

I use a 12 volt bulb that draws about 25 mA (similar to Radio Shack #272-1141) or for low voltage circuits a 6 volt bulb that also draws 25 mA (#272-1140). I use one or the other for every meter or indicator that uses a soldered-in lamp. In most cases, I don't bother to open up the meter, because it is easier to simply glue the other lamp to the bottom or the side of the meter housing. To the customer, it looks virtually identical to the original illumination. For some receivers, like the Sat-Tec and the STS, the bulb mounts behind the meter, making it easier to replace.

I use a test lamp which has a 5K trim pot in series with the meter. I have a pair of mini test clips on the ends. After attaching the lamp and pot to the meter's lamp terminals, I adjust the pot until the lamp glows with the correct brightness. Ohmming out the pot determines the dropping resistor's value, which is put in series with the lamp across the meter' terminals. Typically the resistor value is 47 to 470 ohms (1/4 watt) in value, depending on the voltage in the circuit.

Of course, Radio Shack is not the place to buy lamps,

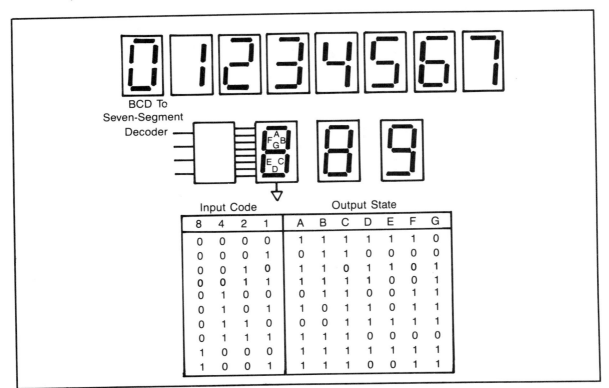

Input Code				Output State						
8	4	2	1	A	B	C	D	E	F	G
0	0	0	0	1	1	1	1	1	1	0
0	0	0	1	0	1	1	0	0	0	0
0	0	1	0	1	1	0	1	1	0	1
0	0	1	1	1	1	1	1	0	0	1
0	1	0	0	0	1	1	0	0	1	1
0	1	0	1	1	0	1	1	0	1	1
0	1	1	0	0	0	1	1	1	1	1
0	1	1	1	1	1	1	0	0	0	0
1	0	0	0	1	1	1	1	1	1	1
1	0	0	1	1	1	1	0	0	1	1

Fig. 13-7. The truth table for the circuit shown in Fig. 13-6. The input is a BCD or Binary Coded Decimal, which consists of 4 lines weighted 1, 2, 4 and 8. The 4511 decodes the four lines and latches the outputs high or low according to the table.

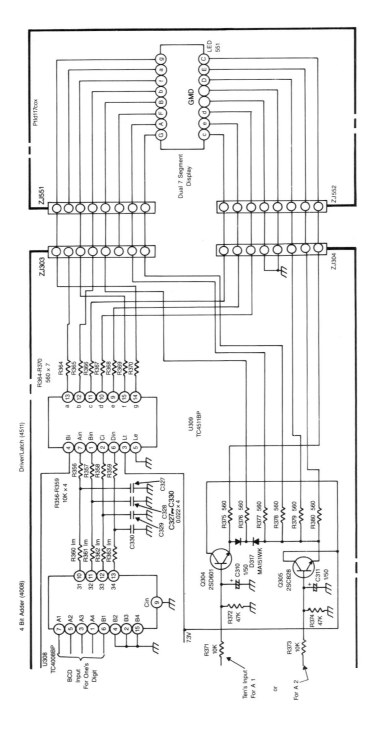

Fig. 13-8. LED display from a DXR 1300 receiver. It also uses the 4511 driver/decoder chip. Since the tens digit will always be blank, a 1 or a 2, then a simple transistor driver is all that is necessary. When Q304 is turned on, then a 1 is displayed (segments B and C). If Q305 is turned on, then a 2 is displayed (segments A, B, G, E, D). Courtesy of Gould/Dexcel.

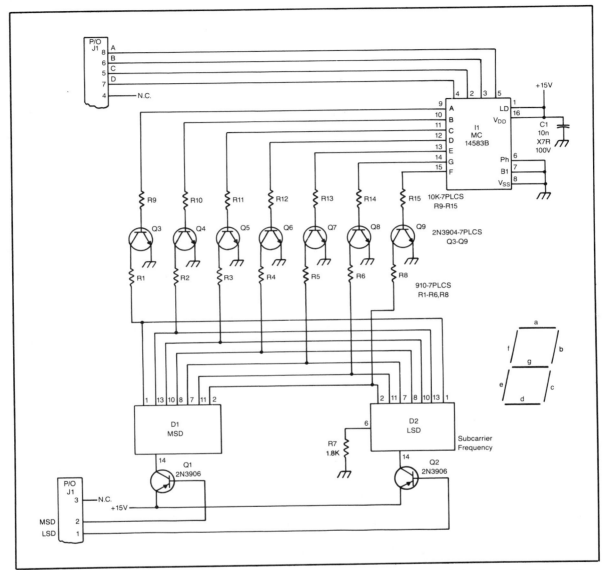

Fig. 13-9. A common anode display. In this type of display, the individual segments are pulled to ground and the positive voltage is common to all. Courtesy of Birdview.

except in emergencies. Buying spare parts is always cheaper if you shop through the mail order places. By using one lamp, you can buy in hundred lot quantities and save a few cents on each one, which more than makes up for the added resistor. The time saved in not having to search for a particular voltage lamp is worth more than the money saved by buying in bulk. Plus the resistor limits the current, so if there wasn't any current limiting before,

there is now. This guarantees that the receiver won't be back in for a burned out lamp before your warranty period is up.

About the only other lamp that must be stocked are the ones for the screw-in lamp fixtures, like those found in some meters. They are available both mail order and at Radio Shack, and, if you want to pay extra for the exact replacement part, from the manufacturer.

TUNING CIRCUITS

Satellite receivers were originally designed with a pot to tune in the channel with. This is the simplest way to tune something, as all that is required is one pot for both fine tuning and channel selection. Simple is a relative statement of course, since a little old lady in Nebraska may not think it's so easy when she's trying to find her favorite program using a pot that is only marked in relative channels.

That is why detent tuning was developed. In a detent system, each channel has a specific location. In the early detent models, this meant that there was a 12 position switch, with a fine tune control that chose between the odd and even channels. This was OK if the system did not have any drift. Unfortunately almost every system except the synthesized ones have some drift. After a while, the channels either don't line up with the odd and even marks, or the channel selector is one or more channels off.

Some people think that if all the channels are off by one number, that by loosening the knob and moving the pointer to the right number it will take care of the error. Boy are they wrong! To set up a system that has drifted requires the receiver and the down converter to be rematched together. This is required for almost every satellite receiver that uses detent or random access tuning (12 push buttons). There are some of the newer satellite receivers that use what is called synthesized tuning. In their systems, the temperature drift is concealed because of the synthesizer circuit, which is set for a certain frequency for a certain channel. It will lock onto that frequency no matter what by adjusting itself accordingly.

But for the rest of the receivers, a rematching is occasionally necessary. Luxor even goes so far as to suggest that you rematch your receiver four times a year, in the spring, summer, fall, and winter. Try telling that to that lady in Nebraska. "It's easy, just get a screwdriver and take the top cover off. See those little black things there. Well, turn them until your channels come in . . . .''

In most receivers, rematching consists of setting the start voltage and end voltage that goes to the channel select resistor string. Typically this is done on channels 1 and 23 or 1 and 24. In many of the detent, up-down, and random access tuning receivers, there will also be internal channel fine tune pots for presetting the tuning on all 24 channels or for preset tuning 12 pairs of channels.

To change channels usually involves changing a voltage that is sent to the DC. This voltage may be positive or negative. Typically it is between 4 and 15 volts, although there are models that have gone as high as 20 Vdc

for tuning in the top channel. Most receivers use a positive voltage. The most noticeable exception is Drake's DC which uses a negative voltage.

Some receivers use a microprocessor to do the tuning, but in many cases, it is not a true synthesized receiver as some would suspect. In many microprocessor designs, the microprocessor merely selects which resistors (or trim pots) in the voltage divider will be used to select the channel. Sometimes a digital-to-analog converter IC will be used. The D/A chip takes a digital word of usually 8-bits and outputs a dc voltage corresponding to the weight of the word. If the word is all ones or all the inputs are high, then the maximum voltage is used. If all the inputs were low, then the lowest voltage would be output. By biasing the dc output voltage, the minimum and maximum voltages could be positive or negative and could be 4 to 8 volts, 10 to 20 or any other combination desired. This type of tuning is used by Houston Tracker in their Super Plus board. A block diagram of a microprocessor and a D/A circuit is shown in Fig. 13-10.

The most abused part on a satellite receiver will be the channel selector pot or switch. The video fine tune control or the PR II switch probably runs a close second in failure rate. In most receivers that use a pot, it will be a 10 K linear model. The fine-tune control will usually be the same or similar. The PR II switch is a momentary single-pole, double-throw type with a center return. Radio Shack has two types that can be used (#275-637 and #275-709).

In many cases, switches are unique to one model. Some may be very hard to come by, especially on those receivers made in Taiwan or Korea. They must have people that just dream up weird switch combinations all day. Why else would you have a double-pole, double-throw switch mounted on a stereo pot in a push-pull configuration that is soldered into the board, and yet only use one pole of the switch? Needless to say, it's not a switch that will be found at Radio Shack.

Many receivers use detent tuning. Normally this is accomplished by using a rotary switch which has a small ballbearing to get the detent feel. If this type of switch is turned too fast, the ballbearing can be jarred loose. Instead of replacing the switch, go to your local bicycle repair shop with one of the ballbearings (from a good switch) and buy a handful for 50 cents. It's a lot cheaper than buying the switch from the receiver manufacturer for $8, which means you have to charge your customer $10 to $15 for it. Instead, you can charge him a bit extra for labor for switch repair.

Imported receivers, especially from Taiwan, use

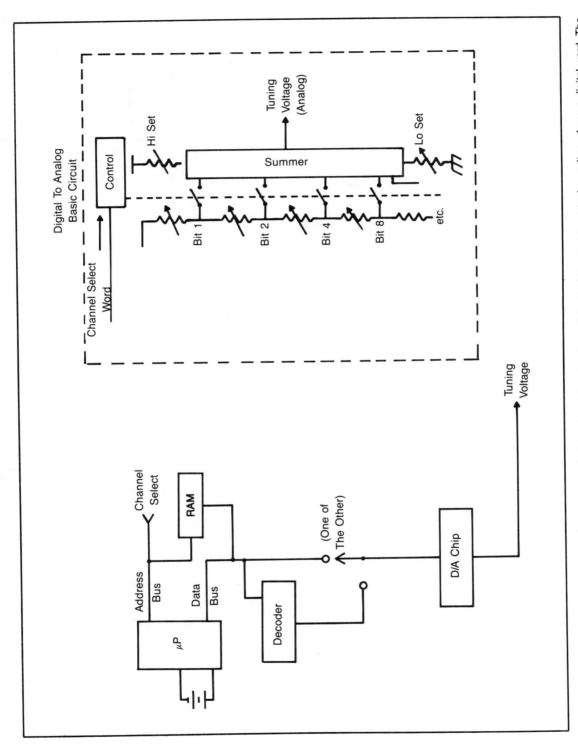

Fig. 13-10. Block diagram of a microprocessor controlled digital-to-analog interface. It is used to generate a tuning voltage from a digital word. The Super Plus from Houston Tracker uses this technique. It can differentiate between 256 individual voltage steps by using 8 bits of information.

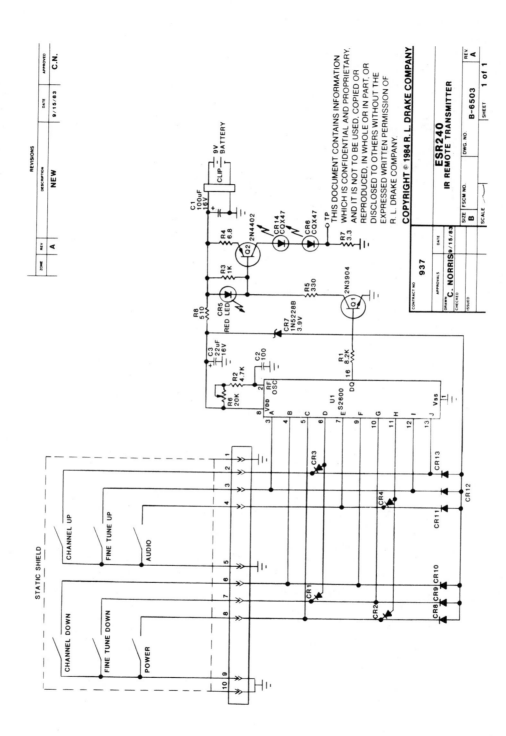

Fig. 13-11. Schematic of the Drake ESR 240 remote control. It is slightly more complicated in that it also allows control of the video fine tuning. The 240A remote allows full control of the actuator as well. Courtesy of R.L. Drake.

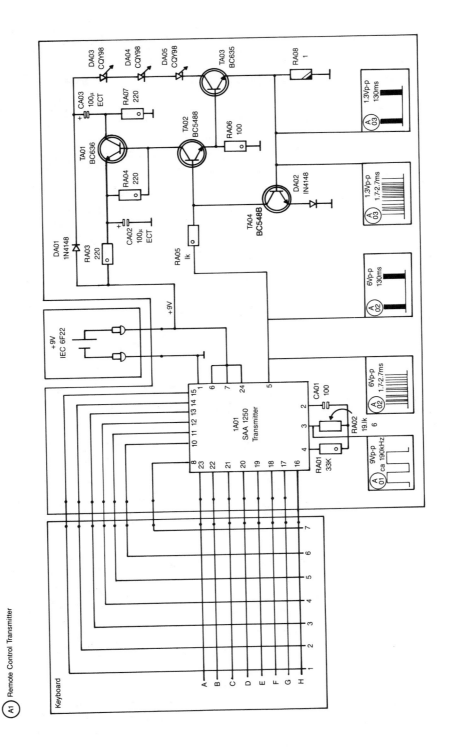

Fig. 13-12. Remote control using a 1986 IC. Most of the remote controls are based on a 455 kHz clock that is divided down to the required frequency. It has random access channel selection, a format switch, and video fine tuning. This remote uses two infrared LEDs and one red LED as operator feedback.

177

cheap pots. In most cases of intermittent action or an open pot, if a pair of needle nose pliers are used to squeeze the rivet that fastens the terminal to the carbon, the pot can be saved. I always try it first. My success rate is probably more than 75%. It's a lot easier and faster than cutting out the pot and replacing it, especially if it is a value that you don't have! This same thing holds true for slide switches. They can become open because of oxidation or because of the back not being tight enough to keep good contact with the movable part of the switch. By squeezing the switch slightly, it will often improve the contact so that it will function correctly.

REMOTE CONTROLS

Infrared remote controls work by sending a series of pulses of infrared light from an infrared LED. These pulses usually are sent using PCM or Pulse Code Modulation at a rate of about 90 to 100 KHz. A typical remote control merely changes the channel up and down and turns the receiver on and off.

Drake's remote goes a step farther. It also includes fine tuning and an audio mute. It is shown schematically in Fig. 13-11.

Going a few steps farther, we come to the all-in-one remote controls. The best example of this type is the Luxor. This remote is intimidating at first, but after using it awhile, you get used to its idiosyncrasies. You can change satellites, video channels, and audio channels. You can also fine tune the polarity, and the two audio channels. It's only missing a few buttons that are found on other remotes. It doesn't have an audio mute or video fine tuning, which when using the older DC's is sorely missed. The schematic is shown in Fig. 13-12. It uses the SAA1250 transmitter chip from IT&T and three LED driver transistors.

Antenna Positioning Systems

BEFORE 1983 THERE WERE VERY FEW SYSTEMS that were sold with a motor drive included as part of the package. Polar mount dishes featured a hand crank to manually move the dish through the Clarke Belt. The older az-el mounts had two pivot points which had to be adjusted if the dish was moved to another satellite.

This usually meant that a TV and receiver would be carried out to the dish and that the channels would be scanned while moving the dish. Or it meant that two people were required to move the dish, one person who watched the TV inside the house, and the other, outside, who moved the dish by hand. Usually a lot of guessing went into which channels were active and even as to what satellite the dish was pointing at. Putting hash marks on the jack's shaft was the most sophisticated feedback system going.

When there were only a few satellites carrying TV transmissions and no regular satellite TV guides, this means of moving the dish was not too bad if it wasn't freezing or raining outside. But today, with twelve satellites that have regular video programming, motor drives have gotten to be part of the standard package.

FUNDAMENTALS

There are three basic things needed for an APS or Antenna Positioning System. One is the motor drive or actuator, which is fastened onto the dish and which does the manual labor. Second is the actuator controller, which sits on top of the TV or the satellite receiver. When someone pushes some buttons or turns a knob or flips a switch, it supplies power to the motor, which turns a jackscrew, which moves the dish across the belt.

At this point, you have no idea where the dish is, except by figuring out what program the receiver is tuned to and looking up which satellite and transponder it is broadcast on. Thus the necessary third ingredient in a complete system is location feedback. In some systems, this is used by the controller to automatically position the dish, in other systems it is strictly used as a visual indicator of dish position, so that the operator will know when to release the button or switch that will in turn stop the dish. Typically this visual indicator is a series of LED's, a three digit LED readout, a fluorescent display, or an analog meter.

Every basic Antenna Positioning System has these three elements. But several other items are needed before an APS becomes truly user friendly:

☐ It should know where each satellite is and remember that location, even with extended power failures.

□ It should be able to list the satellites in a common form (i.e. F3R, G1, T301) for easy selection.

□ It should be able to control polarization, because of the format change, or at least have a control signal to interface to the receiver with.

□ It should be easily reprogrammed in case of position or memory loss.

□ It should be fully remote controlled.

With a microprocessor chip, all of these items are easily implemented if the software design is good and if the motor drive itself is properly matched to the dish type and is properly installed.

Linear Actuators

A linear actuator is the most widely used TVRO antenna mover used today. It consists of a motor, a set of reduction gears and a jack which is run in and out using either a reciprocating ball or an acme screw. Figure 14-1 shows the components that make up a reciprocating ball actuator, while Fig. 14-2 shows the acme screw. The basic difference is in the way the inner tube is hooked to the jackscrew.

There are only a couple combined controller and actuator manufacturers. Superwinch, Inc., who entered the TVRO actuator field after 12 years of manufacturing electric winches and hoists is one of them. For the most part, linear actuators are bought by the controller manufacturer from a separate transmission gear manufacturer. Probably the biggest supplier is the Saginaw Steering Gear Division of GM. Other manufacturers include Warner Electric, Duff-Norton, Wesbar, and General Transmission. A fairly recent development is imported actuators, some of which are cosmetic knock-offs of the major brands. Be aware of these imitations, for a dollar saved today will most certainly be lost on service calls later.

Common practice calls for the motor to be mounted onto the polar mount with the jackscrew end mounted to the dish. Which side the jackscrew mounts on is determined by your geographic location. Usually those positions east of the Rocky Mountains mount the jack on the right or western side of the dish and those west of the Rockies mount it on the opposite side.

Since the linear actuator is exposed to the elements, a weathercover for the jackscrew is essential to prevent excessive wear due to water or dirt accumulation on the shaft. Even though most actuators have an O-ring between the fixed and movable shafts, extra protection beforehand is always better than having to do a warranty service call later. Especially if your location freezes dur-

ing the winter, which is when the first call would surely come in.

Other Actuator Types

There are three other types of actuators used today. These are the direct-gear drive, the chain drive and the az-e1 drive. The first two are built by the dish manufacturer for their particular dish. Some examples of these can be found in the product line from Birdview, KLM, and Paraclipse. The az-e1 drive is built by JRC and marketed by Nissho Iwai in the U.S.

Birdview's APS is a direct-gear drive which consists

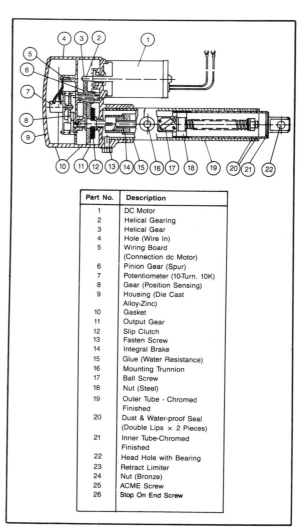

Part No.	Description
1	DC Motor
2	Helical Gearing
3	Helical Gear
4	Hole (Wire In)
5	Wiring Board (Connection dc Motor)
6	Pinion Gear (Spur)
7	Potentiometer (10-Turn. 10K)
8	Gear (Position Sensing)
9	Housing (Die Cast Alloy-Zinc)
10	Gasket
11	Output Gear
12	Slip Clutch
13	Fasten Screw
14	Integral Brake
15	Glue (Water Resistance)
16	Mounting Trunnion
17	Ball Screw
18	Nut (Steel)
19	Outer Tube - Chromed Finished
20	Dust & Water-proof Seal (Double Lips × 2 Pieces)
21	Inner Tube-Chromed Finished
22	Head Hole with Bearing
23	Retract Limiter
24	Nut (Bronze)
25	ACME Screw
26	Stop On End Screw

Fig. 14-1. Typical Ball screw actuator.

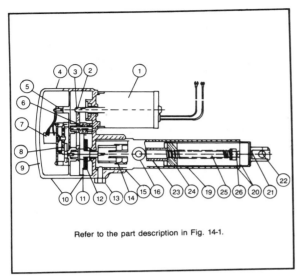

Refer to the part description in Fig. 14-1.

Fig. 14-2. Typical Acme screw actuator.

of a motor and threaded screw which are mounted to the polar mount. These drive a half-moon shaped gear, which is attached to the dish. It is a compact, fully enclosed APS which also features an automatic stop to prevent jamming in case an obstruction is encountered by the dish. The only drawback to the design is that it has a fixed declination of 5.5 degrees. This is fine for the middle of the country but will not properly track the belt at the upper or lower latitudes. Figure 14-3 shows the Birdview APS.

The chain drive uses the same type of motor as the direct drive and is also attached to the polar mount. But it drives a sprocket gear that pulls a chain that is attached to the east and west sides of the dish and thus moves the dish across the belt. It is sometimes called a horizon-to-horizon drive as the dish can go almost 180 degrees.

The last type is really a unique and rarely used design. It is the azimuth-elevation drive. This type of drive uses two separate movements for each satellite change, thus two motors are needed. It is the most accurate of the motor drives but also the most expensive. At this time, there is only one controller designed like this, the Eagles Prey robotic arm from JRC. It is programmed with the latitude and longitude of the dish site, and the microprocessor program takes it from there. Its only drawback at this time is the price, which is still about twice as expensive as the top of the line linear actuator.

MOTOR CONTROLLERS

The first antenna movers were fairly crude. Most used

dc motors with a momentary DPDT (double-pole, double-throw) switch put in series with the motor supply wires. Most designs did not include feedback. The operator merely watched the dish to make sure it didn't hit the ground, or some other object, and that it was stopped before it got to the end of the actuator stroke.

Figure 14-4 is a basic schematic of this type. All that is required is a transformer that puts out the correct current and voltage for the motor being used (usually 3 to 6 amps at 12, 24, 36, or 90 volts), a bridge rectifier to supply both a plus and minus voltage, and a momentary DPDT switch wired in an X pattern.

The switch applies dc voltage to the motor in two polarities. Putting a positive voltage on the red wire moves the screwjack out, while putting a negative voltage on the red wire moves the screwjack in. Since there is no feedback, caution must be used when using such a simple controller, especially if the dish is out of sight from the controller's location.

This simple circuit is useful for checking motor drives, as it gives a go/no go test to the motor. If a multitapped secondary transformer is used, with a four position switch, then most motors could be checked with only one test unit. Caution is called for, as applying a voltage beyond what the motor is rated for could cause damage to the motor or to the gears that it drives.

Figure 14-5 is the KLM Polar-Trak controller schematic. It is a typical example of a basic motor-control circuit. It consists of a 35 volt centertapped transformer and bridge rectifier which drives the motor through two switches, a DPDT momentary switch and a SPST fast-slow switch. The motor can be driven by two voltages, ± 36 Vdc, or ± 18 Vdc. The higher the voltage the faster the motor turns. So far, it is basically the same circuit as the motor test circuit shown in Fig. 14-4, but two additional circuits are also included.

One is the visual feedback of dish location, and the other is a polarizer control circuit. The visual feedback consists of a meter movement that is set to act as a voltmeter. The Pot is a ten turn potentiometer, located in the motor drive housing, that is rotated by the motor. When the screwjack is all the way out, the meter is set to read full-scale by adjusting the 5,000 ohm trim pot. As the motor turns, the voltage reaching the meter increases or decreases depending on direction, thus a crude antenna position can be determined by the meter reading.

The polarizer circuit is for a Polarotor 2 or equivalent polarization controller that uses a dc motor to turn the waveguide probe. Again the Polarity switch is a momentary DPDT. Holding it one way turns the motor clockwise. Holding it the opposite direction turns the motor counterclockwise. The polarity trim control is used to set pin 2 of the LM317 regulator IC to 10 volts.

FEEDBACK CIRCUITS

There are three types of motor position feedback systems used in APSs. All are located within the motor housing and use the motor's rotation to supply the information used to determine the antenna's position.

The most reliable feedback system is the ten-turn potentiometer, if it is a sealed mil-spec potentiometer which is connected directly to the driving gears by a spur gear. It is usually hooked up as a voltage divider. The feedback consists of a varying voltage that is continuously changing as the motor is turning. This voltage is used to drive a circuit much like a voltmeter. This system has the advantage of being immune to lightning damage, power outage, interfering noise pulses and moisture.

Other types of feedback use pulse counting to deter-

Fig. 14-3. The Birdview antenna mount and motorized gear drive. Courtesy Birdview.

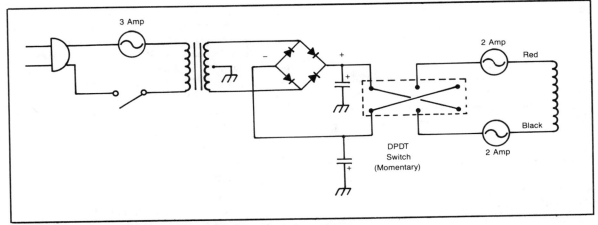

Fig. 14-4. Circuit diagram for a test controller for checking motor drives.

mine the motor's position. One method uses reed switch closures. This type has a round plate with one or more magnets fastened to it that is turned by the motor. Located next to this assembly is a normally open reed switch. When a magnet goes by, the switch closes. This method will produce one pulse per magnet/per rotation. Usually there are four magnets so one rotation of the motor will send out four pulses.

The third type of feedback utilizes something called the Hall Effect. A Hall-Effect sensor is a solid-state circuit that detects the presence of a magnetic field. It can be thought of as a solid-state reed switch. It is driven in the same fashion. It does require a bias voltage to operate and so there are three wires required compared to only two for a reed switch. There is a +5 Vdc line and the two switch contacts. Ground is supplied through the case.

The last type of feedback is optical. There are two methods used. One is a light and detector that are inter-

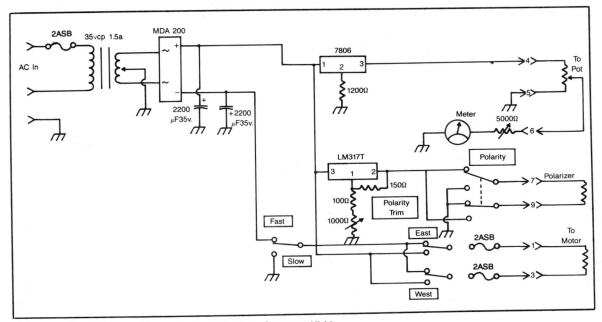

Fig. 14-5. The KLM Polar-Trak actuator controller. Courtesy KLM.

rupted by a rotating shutter, the other is a light that reflects off a glass plate to drive a detector. A glass plate can be used in both versions. In version one, it would have clear and opaque sections screened on it. This would allow pulses of light to hit the detector, which would drive the feedback circuit. The second version uses a chromed glass plate, again with strips of black or opaque paint. The chrome would reflect the LED light into the detector to cause a pulse. These also require a bias voltage to turn on the LED and to supply the detector (which is a phototransistor) with a pull-up voltage.

Any of the pulse counting schemes can be effected by noise spikes from lightning, lawn mowers, drills and other motors (including the motor drive itself), static electricity, power-line spikes or surges, and even spurious radio transmissions from CB radios, etc. This causes false pulses to be sent to the controller which counts them as true pulses. This will result in mispositioning of the antenna. If pulse counting feedback is used, shielded cable must be used on the sensor wires.

The light activated systems are less immune from noise spikes than either the reed switch or Hall Effect sensor. In addition, the Hall Effect sensor is easily damaged by lightning and static discharge. Thus it is doubly important to shield the cables and ground the dish when these types are used.

Another important factor in positional feedback systems is the ability to detect if any of the motor interface wires are open or shorted. This can lead to serious consequences if the feedback does not change as the antenna is moved. The actuator can be driven to its end points and be damaged or the dish could be run into the ground or other object.

To protect the actuator from damage by either travelling too far out or by forcing the inner tube into its stop, there must be some kind of switch or circuit to cut the voltage to the motor before any damage occurs. There should be two switches, one to set the outward limit, the other to set the inward limit.

The Skywalker APS places two switches in series with the motor. One switch is set so that it opens up when the motor has reached into outward end travel limit. The other opens when the actuator tube has pulled inward to its limit. Typically the first switch is a mercury reed switch, when is also referred to as a tilt switch. It is fastened to the dish, and its angle is set so that driving the dish to a point just beyond the farthest satellite causes the switch to close. This type of switch must be surely fastened or else the end point will change as the dish is rotated. This can then lead to the actuator tube travelling too far out and causing the damage that the switch

was supposed to prevent.

A mechanical pressure switch is used on the inner limit. It is activated by the movement of the limit stop plate, which is mounted on the movable inner shaft. As the inner shaft is pulled into the main or outer housing, the stop plate will hit the pressure switch which is mounted to the end of the outer housing. This opens up the voltage to the motor. If the stop plate is not secured properly or if the switch is not secured properly then this can allow the motor to pull the shaft all the way into the housing and up against the stop, possibly damaging the actuator.

Another method of setting the upper and lower limits is to use the feedback from a ten-turn pot. This is probably the most popular method. This type of end limit protection is set by two trim pots, usually marked high set and low set. Two voltages are set by these trim pots that correspond to the end limits. These voltages are supplied to comparator circuits. When the positional feedback reaches either voltage, one of the comparator circuits will go high. This change in state is used to turn off an SCR (Silicon Controlled Rectifier) or similar high current switch which controls the voltage to the motor. If the end limits are set incorrectly, then the motor will drive the actuator tube too far out or in and cause damage.

Some actuators have built-in limit switches. These are usually set so that the power is interrupted to the motor 1/2 inch before the actual end limit is reached. Additional limits can usually be set within the controller if they are needed.

An example of a controller that uses positional feedback with a comparator circuit is the Drake APS 24. It is a typical commercial version of the simple controller which consists of two pushbuttons which are held down to drive the antenna either east or west. Inside the controller there are two relays which are configured in a DPDT arrangement. Only one can be activated at a time, unless there is a component failure. If they were both turned on then the + and − voltages from the bridge rectifier would short together causing the fuse to blow.

The controller's schematic is shown in Fig. 14-6. It uses a 10K pot to supply positional feedback. The pot has + 15 Vdc applied to one end terminal while the bottom and centertap terminals are sent to a voltage divider circuit. To choose between an 18 inch or 24 inch stroke actuator requires that a switch be set. This switch determines the values used in the voltage divider. If this switch is incorrectly set, then the actuator could be damaged (if a 24 inch throw actuator is set for 18 inches) or may not travel through its full range (if an 18 inch actuator is used and the switch is set for a 24 inch actuator). The dividend

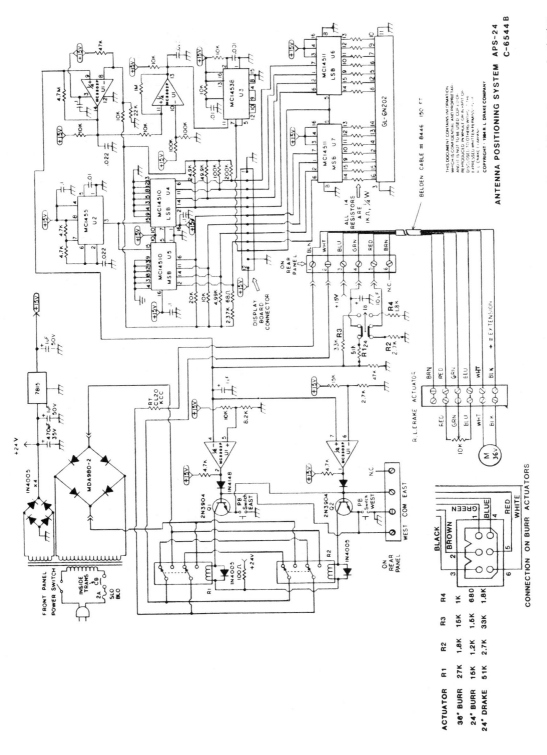

Fig. 14-6. The schematic for the APS-24. It uses 3302 comparators to determine whether the actuator has reached its end limits, to prevent actuator damage. It is dependent upon the installer to properly set them up though. Courtesy R.L. Drake.

ACTUATOR	R1	R2	R3	R4
36" BURR	27K	1.8K	15K	1K
24" BURR	15K	1.2K	1.5K	680
24" DRAKE	51K	2.7K	33K	1.8K

185

voltage is then routed to two MC3302 dual op-amp ICs, which are set up as comparators. One set of comparators is used to determine the actuator end limits. This prevents the actuator from being driven beyond its tube ends, if the correct actuator is used with the unit. There are several resistors that must be changed to match the throw of the actuators originally used with the APS 24. If the actuator type is changed, these must be checked for proper end limit points.

The other two comparators are used to compare the positional feedback voltage with a BCD (Binary Coded Decimal) code. This is in turn used to drive the display which is incremented or decremented according to whether this comparison voltage is higher or lower than the BCD voltage. The APS 24 uses several ICs that are listed in Chapter 17: the BCD to 7-segment latch/decoder/driver ICs (4511s), the BCD up/down counter chips (4510s), the one-shot monostable which drives the 4511 (4538), and the comparators (3302).

SOPHISTICATED CONTROLLERS

Today, of course, it's a whole different ballgame. There are microprocessor-controlled motor drives that tell which satellite the dish is on, that automatically adjust the polarization for the Westar/Satcom reversal, that can lock-out certain satellites from view, and that give error codes or messages when a command is given improperly. Some even have an option to supply a tuning voltage to the down converter. One example of a microprocessor controlled APS is the Luxor 9534.

Unfortunately, this microprocessor control has a few drawbacks. It costs more money to manufacture, it is a hundred or maybe even a thousand times more complicated to operate and repair, and it is much more susceptible to lightning damage and power-line spikes.

But with proper installation procedures, the second generation of microprocessor controlled antenna positioning systems is proving to be very reliable, electronically. The biggest problems still lie with the motor drive and feedback systems. As a general rule, operator error is the main cause of difficulty. If that can be ruled out, then suspect cable/connector problems, an outside component (motor drive), and the controller last.

APS TROUBLE AREAS

The number one problem throughout the industry is weather. If it was always sunny and warm, with 20% humidity and never any wind, a motor drive's life would be easy. Temperature and moisture are the two main problems, but a more subtle one, that is often overlooked,

is the wind. Even slight breezes put hundreds of pounds of extra torque onto the drive, which in a polar mount, is the only thing that is holding the dish steady in an azimuth direction. Appendix B has a chart listing the wind speed versus dish size and the loading that is encountered. While reading the chart, remember that it is the actuator that is carrying the bulk of the loading.

Some of the most common problems that occur with APS's are: putting on an actuator that is either too small or too large for the dish, not aligning the drive shaft so that there is no side torque, not installing spike protectors on the ac power line, not using shielded feedback wires, improper mounting of the drive, forgetting to drill the required drain holes, and not using rubber boots to protect the shaft and motor from water contamination.

As can be seen, most of the problems relating to motor drives is really in the drive and dish rather than the controlling electronics. Of course there are exceptions to every rule, and there are some very poor controller designs on the market. Luckily the second, and upcoming third generation, is learning from their forefathers mistakes.

TROUBLESHOOTING GUIDE

The following troubleshooting guide can be used to isolate and repair most common APS problems. If a simple controller like that shown in Fig. 14-4 is built and used to drive the motor directly, then it will be readily apparent which component, either the motor or controller, has failed. If the motor works, then the cable, controller, or feedback circuit is at fault. If the motor does not work, then internal limit switches may be open, the brushes may be worn, or a motor winding may be open. If the fuse on the test controller blows, then the motor is shorted or the actuator tube is binding.

Fuse on Controller Blows

1. Motor is overloaded. Check for possible obstruction, shaft binding, build-up of ice or snow, and loose hardware.

2. The inner tube needs to be regreased. Tube is binding requiring too much current.

3. There is an internal short in the controller circuit. If fuse blows with all interconnecting cables removed, then controller is bad. Check transformer, bridge rectifiers and SCRs (see Chapters 8 and 16).

Satellite Positioning Readout Is Incorrect

1. Play in the antenna or drive. Sometimes the ac-

tuator can slip in the clamp that mounts it to the polar mount, this will cause all the satellites to be off.

2. Positional feedback is wrong. Check for loose pot if that type is used. Check that the rotating magnet assembly in the motor drive is not loose if a Hall or reed switch is used. Check feedback wiring connections.

3. Unshielded wires are used on the feedback lines allowing false pulses to trigger the counter circuit. Replace with shielded wires. Be sure to ground the shield to the motor chassis.

4. End limits changing. If the drive counts from an end limit and if it is a mechanical switch, then it might be moving when the end limit is reached.

5. If controller has a memory circuit, check the back-up battery.

No Motor Movement

1. If the motor shuts down suddenly after several minutes of movement, then there may be an integral transformer overload protector built into the controller. Usually this will reset itself after 15 minutes when the transformer cools down.

2. One of the limit switches is open. This can be due to a faulty switch or broken wire.

3. One of the motor wires is disconnected.

4. The controller senses that a wire is shorted or open and is shut down. Usually there will be some kind of readout indicator in a controller that can do this.

5. Controller is in the parental lock position.

Motor Moves in only One Direction

1. End limit is improperly set.

2. Controller failure. One of the relays or driver circuitry has opened up.

3. The push button or switch on the controller is defective.

Motor Is Sluggish

1. If it is sluggish in the retracting direction, then there may be a build-up of ice, dirt, or other foreign material on the shaft. Clean the shaft and use an accordian cover to prevent any future occurrences.

2. The wire size is too small for the cable run. Using the test controller at the dish should clear up the problem. If it does then the wire size should be increased.

3. If it occurs at the ends of the travel only, check the pivot angle. It may be too great.

Dish Will Not Track the Entire Belt

1. Improper setup of limit switches or actuator mounting during initial installation.

2. Dish is hitting an obstruction.

3. Dish is actually moving across the belt, but there is an obstruction between the dish and the satellite blocking the signals.

4. Dish mount is not set up correctly to track the arc. This will usually result in only one or two satellites being received properly, the rest will be missing or have excess sparklies.

Motor Oscillates

1. Feedback or gain adjustment not properly set.

2. Motor torque too much for weight of dish.

Motor Moves Out but Won't Return

1. If the dish is moved too far, it can become parallel with the actuator. This will cause the actuator to bind.

2. End limit switch is bad or set incorrectly.

No Indication that the Controller Is On

1. Plug a lamp or other electrical device into the same outlet to make sure that ac is present.

2. Check the controller's back panel fuse. If blown replace with same rated fuse. Disconnect motor drive before turning unit back on. If fuse goes again there is an internal controller short.

3. If the external fuse is OK, check for internal fuses. If blown replace with ones of same rating.

4. Check for key switch being in the proper position on units that have a dish lock that uses a key.

TROUBLESHOOTING TIPS FOR SPECIFIC APSs

Each system has its own quirks and nuances that must be addressed. Some things that will wipe out one system will not even faze another one. The most common problems are those which were outlined in the last section. Such things as wire size and cable run, proper actuator attachment and weatherproofing are common to everyone. Unique problems like certain circuits blowing and failure modes are generally confined to a certain class of actuators.

Microprocessor driven actuators, as well as receivers, must be protected from line spikes by a good line protector. These are the same type of protectors that are used with computers. Be sure that the model you buy can supply the necessary current for the actuator used. If these items are not used, then your memory will be erratic, the actuator could take off by itself or the interfacing chips could be destroyed.

The Luxor 9534 and the STS MBS-AA

Even though this section is mainly addressed to the Luxor 9534, the two controllers are virtually identical in circuit design and functioning. Therefore almost all of the hints and tips will apply to both models. The biggest difference in the two models is in the interface protection circuitry, where STS has an edge. It appears to be less sensitive to lightning strikes, ignition noise, and other forms of interference. Another difference is that the STS has a Ni-Cad battery vs. the standard nonrechargeable alkaline type used in the Luxor.

Always stack the 9534 on top of the 9550. Likewise for the STS. Doing it the other way can lead to overheating and/or memory problems.

The Polarotor I is used with the 9534. Disconnect it from the 9550 if so attached. The red wire from the PR I goes to +5 V terminal. The white wire goes to the PULSE, and the black wire goes to GND (ground). Be sure to use shielded wire such as Belden 8451. The STS is identical.

The actuator should be installed on the right side of the dish if you are east of the Rockies and should be installed on the left side if you are located west of the Rockies. This affects which Molex pin the red and black motor wires hook to. If the arm is mounted on the right side for the Eastern U.S., then pin 1 is connected to the red lead and pin 4 connects to the black lead. This is the wiring as it comes from the factory.

Reverse pins 1 and 4 if the actuator is to be mounted to the left (eastern) side of the dish. Setting the actuator on the left side also affects the end-limit settings. Thus it is necessary to set the lower limit when the actuator is fully extended and the upper limit when it is retracted. The controller always begins its counting from the western end of the belt. This applies to both systems.

Pins 3 and 6 are the feedback wires from the reed switch. They are not polarized. Be sure that they are shielded as the pulses that they carry control the actuator positioning. Be careful of shorting the positional feedback wires to the +36 Vdc motor wires. This will just about guarantee the interface components and/or resistors will be fried.

The F or Function button is used to do five things:

F, 1 sets the lower limit; F, 2 sets the upper limit; F, 3 locks the drive so that it can't be accidentally moved (good thing to do if you are working on the dish); F, 4 unlocks the drive (make sure you take the remote with you when you work on the dish); and F, 5 clears the memory of all contents (oops). To activate the Function button, press the SATELLITE button first.

There are 30 satellite positions that can be remembered. Before programming can be done though, the upper and lower limits must be set. To set them, manually move the dish slightly beyond the westernmost satellite, regardless of the side the actuator is mounted on. Press SATELLITE, F, 1, STORE to set the western limit. Move the dish to the opposite end of the belt, just beyond the eastern-most satellite. Press SATELLITE, F, 2, STORE to set the eastern limit. The upper and lower limits are now set. If the limits are incorrectly set, the dish may not move. If that is the case, press SATELLITE, F, 5 to erase the memory and start all over again.

To enter satellite information, press SATELLITE, the name of the satellite to be used (SATCOM, WESTAR, COMSTAR or OTHER) and the satellite number (1 through 6). Move the dish to the satellite selected. Tune in an odd or even channel. Press the CW and CCW buttons for the best polarization. Select the opposite polarization and do the same thing. Press STORE to keep the satellite position and polarization skew and format in memory. The satellite location is always referenced from the western limit. Thus the numbers go higher as the dish is moved east. Be sure to make a list of your "other" satellites to reference back to.

When entering normal polarity satellite information, be sure to first check that the POL. INV. (Polarization Inversion) LED is off. To program Satcom 3, press SATELLITE, SATCOM, and 3. Find the satellite using the WEST or EAST buttons. Press ODD and either the CCW or CW buttons to fine tune the polarization. Press EVEN and do the same thing. Once they are both tuned in press STORE.

If the satellite uses reverse polarity, then press the INVert button on the front panel (make sure the POL. INV. LED comes on) before adjusting the polarity with the CW and CCW buttons.

If a satellite position is misprogrammed, merely go through the programming steps and correct the information, following up with a STORE command, which will write over the old information.

To program the channels and the audio format, push the CHANNEL button and the channel to be programmed. If the correct channel is not received, then press and hold either the up or down button on the front panel (the two buttons with the arrows above them) until the proper channel comes in. It is easiest to start with channel 1, since then the down button is pressed until there is only video noise received indicating the receiver is tuned below channel 1. Press the up button until channel 1 comes in.

Once the video is tuned, the audio format should be set. Usually MONO 1 is set to 6.8 MHz and MONO 2 is set to 6.2 MHz. Set the Matrix stereo for MTV and try to balance it so that The Movie Channel will also come in when set for Matrix stereo. Once the channel's audio is set, press STORE.

Since the Luxor is a microprocessor controlled actuator, it is very susceptible to lightning damage. At least one warranty center has known when lightning storms went through a section of the country by the number of 9534s that were returned the following week.

To prevent lightning damage, several actions should be undertaken. First ground the dish; this is the cause of more problems then anything else. Second make sure the feedback wires are shielded. If possible shield the motor wires also. Make sure the receiver and actuator are both plugged into a spike or surge protector. Install varistors on the motor and feedback lines. Varistors are like fast acting zener diodes. They react to over-voltage spikes by shunting them to ground. Hook the varistors between the terminals on the Molex connector and the chassis ground. Use varistors rated at 36 Vdc for the motor wires. The smallest voltage varistor possible should

be used for the feedback wires. See Fig. 14-7 for connection.

If there is a noisy picture with the 9534 hooked into the system, then motor noise or Polarotor noise is probably the source of your problem. Motor noise can be verified by disconnecting the Molex connector and seeing if there is an improvement in the picture. If there is, then install the following components at the Molex connector, following Fig. 14-8. Put a 0.1 μF cap between the two sensor leads, put a 0.1 μF cap from each motor wire to ground and put a 1 ohm, 10 watt resistor in series with each motor lead. Be sure the resistors are isolated from surrounding components as they will get hot when the motor is driven across the belt. Figure 14-8 is a drawing of their placement inside the 9534.

If there was no change after disconnecting the motor plug, then disconnect the polarization device. If there is a change for the better, then the wires need to be shielded. If they already are, then install a 1,000 μF electrolytic capacitor at the PDs splice. The + lead should go to the red wire and the − lead should go to the black wire. Be sure to insulate the capacitor from the weather.

If there is an intermittent operation of the actuator or receiver, check the interface cable and motor cable to ensure that they are properly seated in their connectors.

If the dish does not go back to the same position each time, there may be a problem in the reed switch or there may be a problem in the controller. This can be checked by artificially sending pulses to the counter. To do this,

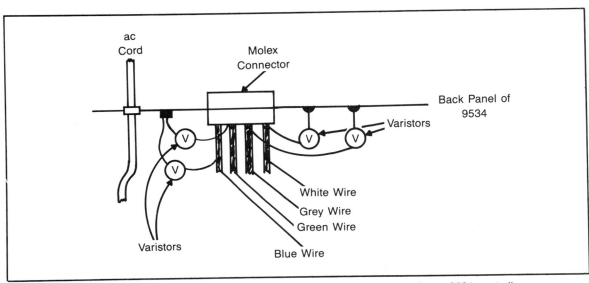

Fig. 14-7. Drawing showing the positioning of the varistors that can be added to a Luxor 9534 controller.

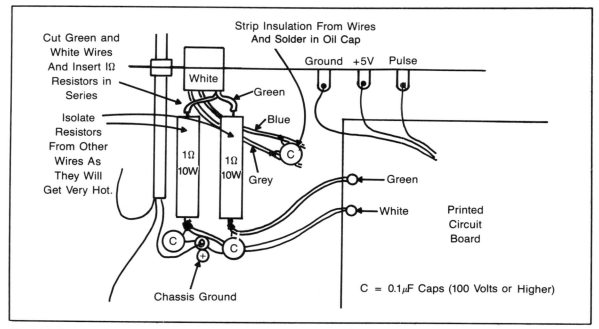

Fig. 14-8. Drawing showing the positioning for adding in current limiting resistors and bypass capacitors to a 9534 controller.

momentarily short the feedback wires together several times. Do this until the number changes. Once it changes, count the number of times the wires can be shorted until it changes again. It should be 12 times. If this is the case, then the reed switch, cabling, or mechanical stability of the dish is at fault. If it can be done at the receiver but not at the dish, then the cabling is too small and should be increased by one gauge (i.e., from 22 AWG to 20 AWG).

If there is no change in the reading, then check RA11 and RA12 in the controller. If RA11 is burned up there is your problem. Replace it with a 180 ohm, 1/2-watt resistor. If RA12 is a 33 ohm resistor, replace it with a 100 ohm resistor. This applies to the Luxor only.

When the 9534 is first hooked up, or after it loses its memory because someone forgot to change the batteries, the actuator will seem to move slower. This is normal and is due to the lack of memory data in the processor. Once it has been programmed, its motor speed increases. To increase the slow speed, adjust the pot located on the main circuit board (it's the only one).

Memory loss is indicated by a blinking 00 on the 9534 display. Usually, this is due to weak batteries or poor contact between the battery and the terminals. Check the voltage on the batteries, it must be 2.9 Vdc or more. If it is not, then the batteries are weak and must be replaced.

If alkaline batteries are used they should last about two years. Be sure to change batteries with the unit turned on or your memory will come up scrambled.

Memory loss can also be caused by the delay power-down circuit in the microprocessor being defective. Try changing out the micro. If that does not help, check that the EPROM is an ACT-03. It was designed to eliminate memory loss during receiver turn on and turn off. Contact Luxor about EPROM replacement.

If two of the three digits on the 9534 go out, then the microprocessor has stopped. To reset it, turn the receiver off and then on. An occasional occurrence is normal. It is often due to loading down of the +5 Vdc line by the Polarotor or some internal chip overheating. This can be caused by high temperatures and line spikes. Be sure that when the cable is routed to the actuator that it does not go near pumps, transformers, 220 Vac lines for compressors, etc. The video clean-up modification can cure this problem also (see Fig. 14-8). If it happens almost every day in the afternoon, check to see if the sun is shining directly onto the controller. This has been known to cause this problem. Either close the drapes or move the controller!

If the motor stops working, the 6 1/4 slo-blo fuse located inside the 9534 may have blown. If it has, then check the dish for binding, ice build-up, wind loading, etc.

TECHNICAL NOTE
 # 1

Subject: Lateral strain on actuators resulting in bending of the inner tube.

Solution:It is the responsibility of the installer of the power jack to the dish to **ensure** there is no *Lateral* strain on the actuator. The installer may find it necessary to install knuckle kits (which may be obtained through Distibutors or from factory at a nominal charge) or modify mount by changing triangulation or both so that the actuator will obtain *FULL* extension and *FULL* retraction without lateral stress.

Failure to comply to the above will result in voiding of the **warranty** of the Actuator.

Fig. 14-9. MTI technical bulletin on lateral stress on actuator tubes.

There is also a 2 amp slo-blo fuse on the back panel. Be sure that both fuses are of the proper rating. If this continues, then the arm may be too short for the dish, contact the manufacturer of the dish for the recommended actuator tube length.

Pressing the ODD or EVEN buttons while the dish is moving will not change the polarity, only the channel numbers. Thus to actually change polarity be sure to wait the dish has stopped. It is best to hold the odd and even buttons down a little longer than other buttons so that the polarity circuit will listen to the command. Often a quick push will change the channel tuning and readout but not the polarity.

MTI Products

Pentec/MTI manufactures many different actuator controllers. Some of the models are the Starchaser (2100 series), the 2150, the Stardrive (2800 series), and the 4100 Programmable APS.

All of the MTI controllers are designed to use either an Acme or a Ball screw actuator that used 36 Vdc. Positional feedback is done by the pulse counting method and uses either reed switches or Hall Effect sensors.

There are two actuators that are supplied as part of the original system. These are made by Saginaw and Warner. The Saginaw ball screw has a gold tube and a 1,500 lb. load limit; the Saginaw acme screw has a silver tube and a 500 lb. load limit; the Saginaw 52 inch has a gold tube and a 400 lb. load limit; and the Warner ball screw has a black tube with a 1,500 lb. load limit.

The Saginaw actuators have an integral mechanical overload clutch to prevent damage if accidentally driven to the ends of its travel. The clutch will not prevent dish damage if an obstruction is encountered. On a drive that is completely extended, lateral pressures can cause bending of the inner tube, if this happens the actuator must be replaced, as motor overload, screw failure, moisture intrusion, and possible controller failure will be the eventual results.

On the Warner actuator, there is an adjustable internal end-limit switch for the extended position. This can be adjusted by using a small jeweler's screwdriver. Set it so that the actuator extends just beyond the last satellite. Note that the Warner actuator does not have a clutch and thus can be damaged by overextension.

MTI recommends that the motor be mounted to the dish (if at all possible) and the inner tube end mounted to the polar mount. This will prevent water from flowing down the tube into the motor housing.

The first one is before three Tech Notes are reprinted from the 4100 owner's manual and apply to any controller that uses a Saginaw actuator. Figure 14-9 is on lateral strain damage to actuator tubes. Figure 14-10 is about weatherproofing the actuator and draining water out of the housing. Figure 14-11 is about curing feedback miscounting due to motor noise.

MTI 2800. If the 2800 does not respond to the infrared remote control, the sensitivity could need adjusting, or the battery may need replacing. Replace the battery first. If that does not help, then remove the cover and adjust the sensitivity trim pot per Fig. 14-12.

TECHNICAL NOTE
2

Subject: Moisture Intrusion in Saginaw Drive

Could result in: Damage to unit due to freezing. Decreased positioning accuracy. Eventual damage due to corrosion.

Suggested corrective action: To prevent moisture intrusion into the gear housing we have found that a small amount of "RTV" or clear silicon sealant placed around the saginaw tube (sealing the tube to the gear housing) will greatly reduce and in most cases completely eliminate any moisture from entering the gear housing.

The inner tube is sealed to the outer tube by means of an "O" ring approximately 2" down inside the outer tube. To *guarantee* no moisture entrapment, drill a 1/8" to 1/4" hole on the down side of gear housing just *slightly* below bottom of outer tube for drain purposes.

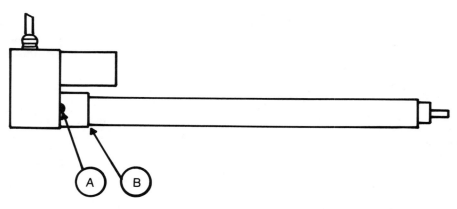

A) Drain hole must be drilled so that the hole faces down, this ensures proper drainage.

B) Encircle tube base with Silicon.

Fig. 14-10. MTI technical bulletin on moisture intrusion in Saginaw actuators.

If the control unit goes dead, check the internal 1 amp fuse. If the actuator does not move, check for 36 volts on pin #9 and #10 on the terminal strip. If there is no voltage, check internal fuse F2. Replace it with a 10 amp fuse.

To reset the display and counter to 000, push the small reset button on the read panel, lower left corner.

If the dish moves but the display does not change, check the wiring between the controller and the actuator. Check the magnet wheel for looseness. If it can be turned by hand, then it is too loose. Use lock-tite to fasten it to the housing.

If the unit uses a Hall Effect sensor, check the +5

Vdc on pin #5 of the terminal strip and on the small red wire that goes to the Hall-Effect sensor in the actuator. Check the blue wire for +5 volt pulses as the magnets move below the sensor. Also check terminal #6 for the same pulses. If there are no pulses, replace the Hall-Effect sensor.

If the unit uses a reed switch, then use an ohmmeter to check that the reed switch closes when the magnets pass below it and then opens back up as the magnet is moved away.

MTI 2100. The 2100 is basically a combination of the 2150 and the 2800. It features a memory to store several satellite positions.

The schematic and board layout for the power supply is shown in Fig. 14-13 and Fig. 14-14. There are three fuses (one external and two internal) labeled F1 (10 amp), F2 (1 amp), and F3 (2 amp).

Most of the troubleshooting tips for the 2800 can be used with the 2100, although pin numbers on the terminal strip will be different. There is one terminal strip and one jack to interface the motor drive and the controller with. The terminal strip has pins #1 and #2 connected to the motor. Pin #3 is the shield, pin #4 is the ground for the Hall-Effect sensor, pin #5 is the counting pulse, and pin #6 is +5 Vdc for the Hall-Effect sensor.

TECHNICAL NOTE

3

Subject: Mispositioning of MTI 4100 Remote Positioner due to miscounting. (Unit will not return to exact position every time.)

Could result in: Motor noise being fed inot Sensor Circuit.

Suggested corrective action: To prevent miscounting by the hall effect sensor or reed switch the shielding must not be removed from the smaller gauge wires and the shield drain wire must be attached to ground at the phillips head screw as shown in the drawing below. If the foil shielding is removed from the smaller wires, it must be replaced using aluminum foil. Failure to do so could result in miscounting. Be sure foil does *NOT* cause short between wires.

Reed Sensor (2 wires: black and blue, *NO* red.)
Or Hall Sensor (3 wires: black, blue and red.)

Slack in the red and black wire should be pulled through the grommet and kept on the outside of the gear box assembly.

Black   Blue   Red

Blue
Purple
Shield drain wire
Green
Shielding

Shielding must *not* be removed from smaller gauge wires. If removed, it must be replaced by aluminum foil. Failure to do so could result in miscounting of the Hall Sensor or Reed Sensor. Be sure foil does *not* cause short between wires.

Red
Orange

Red and orange wires must be separated from the Shielded Sensor wires.

Red   Black

Weather Proof Clamp

Power Cable

Fig. 14-11. MTI technical bulletin on preventing false pulses from a Hall Effect sensor or reed switch.

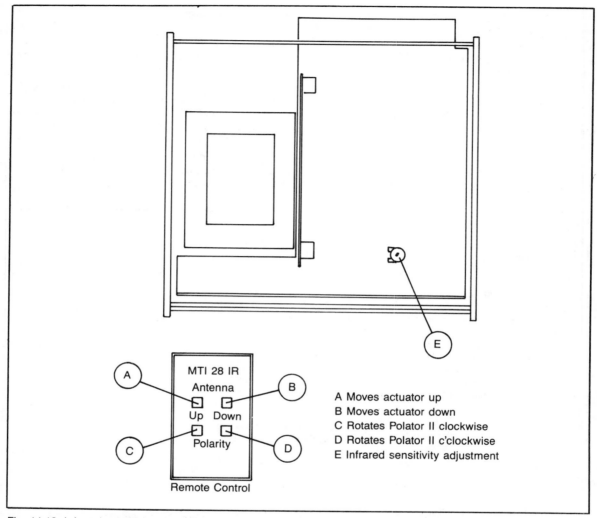

Fig. 14-12. Infrared sensitivity adjustment location for the MTI 2800.

On J1 the pins are : #1 is the motor drive turn on relay return (goes to ground to turn on relay), #2 is the shield ground, #3 is + 15 Vdc for the controller, #4 is the direction relay return (goes to ground to drive the dish to the east), and #5 is the counting pulses.

Draco Aimer II and III. Both models from Draco are designed for the same motor drive. The actuator is a linear type using an acme screw drive. It uses an opto-interruptor positional feedback mechanism, so that it is virtually immune from lightning, static, lawnmowers, and other ignition sources. The motor and drive are fully protected from moisture by a weatherproof cover and a

polyurethane seal on the end of the outer tube. The actuator features a unique gimbal bracket that allows up to 32 degrees of side load, so that it will allow the drive to fit on almost any mount, without binding or having a shim up the opposite end.

The Aimer III is a microprocessor controlled actuator controller that has a built-in polarization controller circuit for a PR I. It interfaces with any receiver that has a PR I controller circuit for automatic format reversal. In most cases, this is accomplished by using their Interface II module. It is hooked up as shown in Fig. 14-15. Installation is easy. Just plug the interface into the tuner

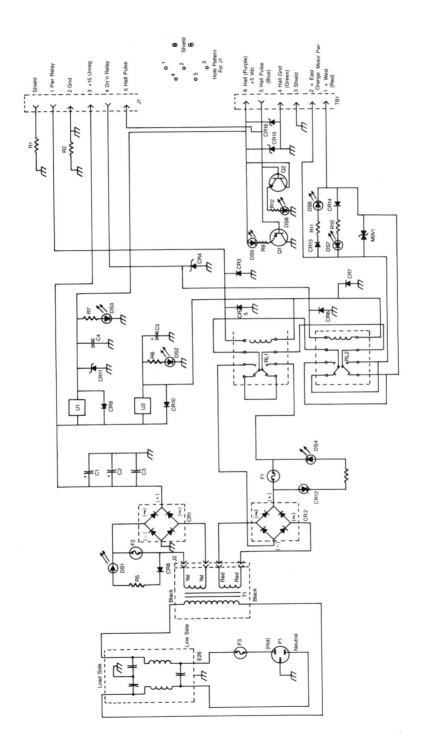

Fig. 14-13. Schematic of the MTI 2100 power supply. It features an integral ac line filter as well as separate secondaries to provide the power to the controller and to the motor itself. Courtesy of Pentec/MTI.

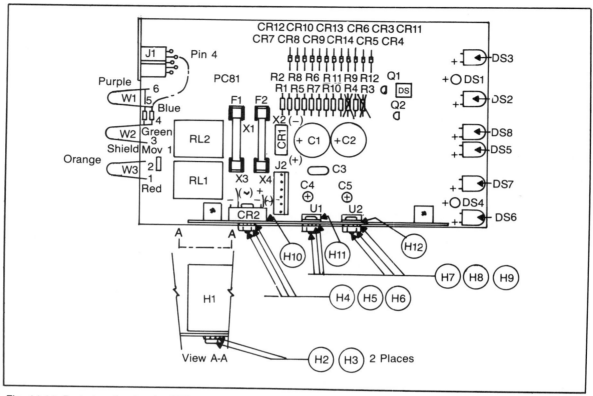

Fig. 14-14. Parts location for the MTI 2100 power supply.

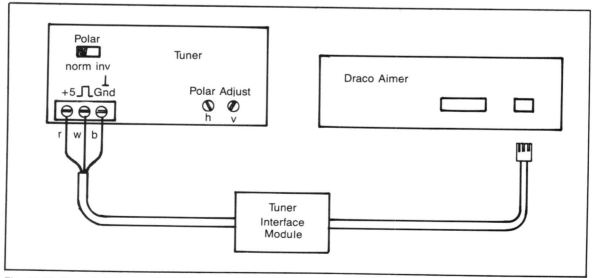

Fig. 14-15. Tuner interface module for interfacing the Draco Aimer III to most receivers with built-in polarity control.

interface slot in the back of the Aimer III and connect the red, white, and black wires to the receivers PR I terminals, or splice in the cable on models without a terminal strip.

If the Aimer readout says E on odd channels and O on even channels then the format switch must be changed on the receiver. When the receiver is turned on, the even/odd button on the Aimer is disabled. To check that the PR I is working, turn off the receiver. The button should function properly. For receivers with horizontal and vertical outputs instead of PR I outputs, the red and white wires should be tied together and hooked to the horizontal terminal. The black wire is hooked to the ground. If the polarity indication on the Aimer is wrong (i.e., O for even channels and E for odd channels), then move the red and white wires to the vertical terminal instead of the horizontal. For Drake receivers with even and odd terminals, only the black and the white wires are used. Hook the white wire to the even terminal and the black wire to the ground. See Fig. 14-16 for the two hook-ups.

The Aimer has several self-diagnostic features built-in. If the readout says 'FUSE', then it may mean that the motor fuse on the back panel is blown, that there is a short or open in the molex connector wiring, that the motor is plugged in wrong, or that the PR I wiring is shorted. If the word 'JAM' appears in the display, it is indicative of the actuator being stopped by the mount not being set properly and overtraveling, by the end limits being set incorrectly, or by the actuator shifting in the mount. If the word 'LIM' appears in the display, then the actuator has reached a preset end limit. If it is not correct, then pull the actuator all the way into the outer tube, or to its full in position. Turn off the Aimer. Press the SETUP button while it's off. This erases all of the Aimer's memory. If the SETUP button is pressed while it is on, it unlocks the parental lockout.

If all the satellites are off, then the actuator may need to be reset on the arc. This is done by selecting a preset satellite and pressing the GO TO button. Once the actuator stops, press the ALPHA button and the RESET B button. The AZ LED should be flashing. Rotate the

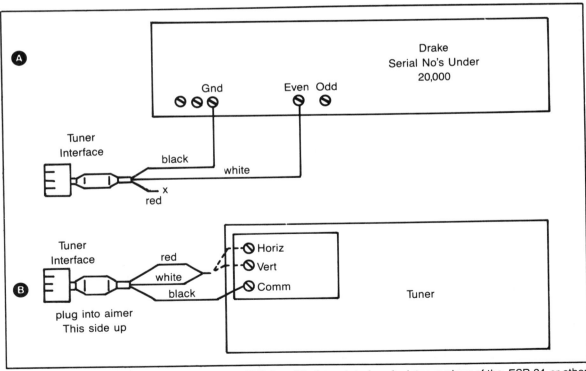

Fig. 14-16. (A). Draco interface using early Drake ESR-24s. (B). Draco interface for later versions of the ESR-24 or other receivers with separate horizontal and vertical polarization switching.

tuning knob to fine tune the satellite for best reception. Press the STORE button. The Aimer should now properly track the arc.

When installing a system that has the dish more than 250 feet from the actuator controller, use Romex (117 Vac house wire) for the motor wiring and a shielded 4 conductor (18 gauge) cable for the feedback and Polarotor wiring.

Houston Tracker Products. Houston Tracker has several motor drives: the Tracker II, the II +, the III, the IV, the IV + and the IV Super +. The IV, IV + and IV Super + all use the same basic board. The only difference between the IV and IV + is the remote control. The Super + adds one board to the controller to create a tuning voltage that can be directly connected to almost any receiver. The next several paragraphs concern troubleshooting the II and II +. The IV + and the Super + follow.

Always use caution when reversing the motor direction on the II and II +. If the UP and DOWN buttons (on the II) or the FAST and SLOW buttons (on the II +) for the opposite direction are pressed before the dish has fully stopped, then the fuse or circuit breaker may blow.

If the right hand limit LED is continuously lit and the readout reads over 1000, yet the dish will move in one direction, then the sensor is defective or the wiring is open between the controller and the actuator. If the left hand LED stays lit and the readout shows a minus number, then there is a short somewhere in the system. The sensor is a ten-turn pot which is connected to the two blue wires. It should read between 140 and 850 ohms. *Note*: Be sure to turn the controller off before reading the resistance.

If the motor stops moving after a few minutes, then it is possible that the power transformer's thermal circuit breaker opened. It is also possible the back panel circuit breaker or fuse is opened. Turn off the controller, reset the circuit breaker, and check the fuse. Also check dish and actuator for obstructions. After approximately 15 minutes the transformer should be cooled down and the thermal breaker should have reset.

The two LEDs above the H and V buttons indicate which polarization has been selected. The LED in-between them however should never light up. If it does, then it is an indication that the PD is stalling (drawing excessive current) and is most likely up against its physical stop. If this happens, then the PDs motor could be damaged by overheating. If it is lit only on one polarity, then adjust the corresponding trim pot located on the bottom of the unit. If the polarization setting is off when the LED goes out, then the PD must be physically rotated

on the dish to peak up that polarization. The opposite polarization will probably have to be readjusted as well. Be sure to set the skew control in its mid position before readjusting the PD.

The IV + and Super + are discussed in the next several paragraphs.

Intermittent operation of the actuator when using the front panel EAST and WEST buttons, and yet the remote control works OK, is usually caused by motor noise being picked up by the UHF remote receiver which cuts off any data from the front keypad. The best solution is to move the wiring away from the antenna and point the antenna away from the wiring. If this still does not solve the problem, then an attenuator pad must be used between the antenna and the F connector.

If the remote control does not work, works intermittently, changes the wrong thing, or works only at a decreased distance, the most likely cause is a run-down battery. Replace with a 9-volt alkaline type for best performance. Another possibility is that the antenna on the back of the control box is not fastened tightly and in a vertical position. The UHF control signals should normally be usable up to about 200 feet. Intervening walls can attenuate the signals, depending on the material they are constructed from. Sometimes pointing the control in a different direction while pressing the buttons will solve the problem.

If the parental lock-out has been set and the code is forgotten, the control box must be unplugged from the ac wall outlet. Press the ON/OFF button and reinsert the wall plug at the same time. This procedure erases the parental lock-out information.

If a motor error indication should be displayed in the read-out, then most likely one or more of the motor control wires have been pulled loose. This most often will occur at the dish, although moving the control box can cause the same problem. Check both places for loose or shorting wires. When installing any equipment, always leave enough slack in the cabling so that the control box can be moved for cleaning. Usually five feet of movement is sufficient.

On a Super +, if the tuning is off or wrong, it is usually caused by temperature drift in the DC or LNC. Move the dish to Satcom 3R. Select channel 1 by pressing 8-0-1. Enter 9-2. Use the receiver's channel fine-tune control or its channel selector switch (or pot) to bring in channel 1. *Note*: During the daytime on F3-R, channel 2 is Nickelodeon.

If the message "SAT. NOT STORED" is displayed after entering a satellite, press 0-8 to get a listing of the programmed satellites and their codes to verify that the satellite was programmed.

To resynchronize the actuator to the memory, a satellite must be recalled and fine tuned for best positioning. Enter 0-5 twice. This will resynchronize all satellites. *Note*: If the satellite that the dish is fine tuned on is not the one originally selected, then all satellites will be incorrectly synchronized. This is easy to do if the selected satellite is G-1 or F-3, as they are so near to one another that a slight error in the actuator could position the dish onto the wrong one. And if, in turn, the dish is fine tuned to the wrong satellite then all satellites will be off by about 3 degrees.

Pressing 0-1 twice will clear all the memory. Pressing 0-4 and then the name and number of a satellite will delete it from memory. 9-9 sets the east limit. 0-0 sets the west limit. Pressing 0-9 will get a scrolling of Tracker IV features.

To store the polarization skew and format, press 0-2. Follow this by inputting the satellite name and number. To set polarization press the EVEN or ODD buttons and push the CCW or CW buttons until the polarization is correct. If the polarization will only get one polarity, then the PD needs to be rotated slightly on the dish. It should be set up so that the LNA is at a 45 degree angle. To reverse the format without changing the skew, press 0-3.

15

Setting Up a Test Bench

A TVRO TEST BENCH IS VERY SIMILAR IN ARRANGE-
ment and spacing to one used to repair stereo equipment, TVs and VCRs. The biggest differences will be
that there will have to be some type of 4 GHz source and
a test set for LNA and down converter testing.

FUNDAMENTALS

As a bare minimum, the bench should have a work area
just big enough to fit the largest receiver you'll work on.
Realistically, the smallest serviceable area is about 17″
deep by 22″ wide. An ideal area is more like 22″ deep
by 30″ wide. This size puts most of the test gear within
arm's reach without having to strain and without having
to move in your chair. Remember, these measurements
are the clear work area, there still must be room for your
test gear.

As a general rule, the upper shelf on the test bench
should be 24″ deep. The side areas should be a bit more,
26″ to 30″ wide. If you are constructing a workbench from
scratch, the best thing to do is to measure all your test
gear and design around their dimensions. The necessary
test equipment is covered in a later section.

Figure 15-1 is a well-stocked commercial test bench.

Already installed along the back are power strips for the
test gear, a shelf for test gear, a rack of test gear to the
side (which can be turned to face the bench along the side
wall). Out of the picture and below the bench, on the right
side, are drawers for test leads, adapters, tools, lunches
and everything else that needs a home.

Commercial test benches like this run from about
$150 for the bench top and legs, by themselves, to about
$400 for a full blown set-up with a test gear riser, built-
in power strips, and a set of drawers below the bench.
They are available in several depths and heights. For electronics work a plastic laminate is the best type of bench
top to get since it won't stain too badly.

The Work Surface

Whether buying or making your own repair bench,
it is important to have a surface that won't scratch the
unit under test. The best surface to use is indoor-outdoor
carpeting. A 22″ × 30″ piece will protect the unit under
test from sliding off the bench as well as from getting
scratched if it is turned over on its top.

The carpeting should be the short nap fiber variety,
not the plastic grass variety. If it is held in place with

200

Fig. 15-1. The well stocked test bench that I use to troubleshoot a Drake ESR-24. The test gear includes a DVM, a dual trace scope, a TV, a video monitor, a waveform monitor, a stereo system, a power meter, a frequency counter, a capacitance meter, a sweeper, and a 4 GHz attenuator and feed system.

velcro, it can be easily removed for cleaning. An inexpensive hair brush can be used to sweep off any solder splashes and dust accumulation.

Lighting

Overall repair shop lighting is usually fluorescent, which tends to wash out the video on TVs and monitors, as well as make reading scopes and waveform monitors harder. A better choice is indirect track lighting which is bounced off the walls above or to the side of the work area. Direct lighting for the actual work area can be done through a clamp-on swivel lamp or through a clamp-on magnifying lamp like that shown in Fig. 15-1.

Regardless of the type of lighting chosen, all scopes and video screens should be easily viewable without shading from reflections being necessary. Any windows should be covered with an antiglare grey material or be shaded from direct sunlight.

Power

The primary ac source must be a properly grounded three prong outlet. Ideally there would be a separate breaker for the shop. In most cases, one 20 amp circuit will be sufficient for all your power needs on a typical test bench.

A power strip can be run along the back of the bench to supply power for all the test gear and the unit you are working on. In commercial installations, the Fire Marshall will come to visit sooner or later to inspect your electrical system. Commercial installations cannot use 3-way ac adapters, so needless to say, the 6-way dual-outlet adapters are definitely out of the question. Extension power strips (not extension cords) are OK if each is run to a separate wall outlet, and not run together in a string. The best bet is to get an electrical contractor and have him install some conduit and plenty of outlets.

An ac line filter is also a worthwhile investment, as DVMs and other IC-based test equipment can be damaged by a sudden line surge or spike. Filters with built-in surge and spike protectors and a line noise filter are readily available through computer supply houses. All line filters have a maximum current rating so be sure to get a model that amply exceeds your maximum current draw.

HOOKING UP

An effective bench is one on which the unit under test can be hooked-up and be ready to test in less than two minutes. If it takes five or more minutes to open up the UUT and track down the proper adapter and then hook the receiver into the test system, for every 15 receivers you've lost an hour's time.

An effective bench has a screw gun positioned close-by to remove the screws holding the top and/or bottom covers and a dedicated place to set the cover and screws. A set of labeled cables to hook up the i-f input, the video output, the audio output, the rf output, and any unique controller cables in one movement. And it has all the test equipment test leads available.

Unfortunately, there are no standards for i-f, rf, or video connectors. About the only standard is that RCA jacks are used for the audio connectors on every TVRO receiver I've come in contact with. Video outputs may be BNC, type-F or RCA connectors. Rf outputs are usually type-F, although I have seen BNC and RCA used there also. Most if inputs are type-F, although UHF is also used. Therefore, a good bench will have a good selection of adapters.

Test Signal Source

To do effective troubleshooting of TVRO receivers, DCs, and LNAs requires a multilevel signal source. This signal can be a satellite simulator like the GBS 2600, GBS 1600, or Wavetek 1470, or an off-the-air signal from a dish. If the dish is not more than about 50 to 100 feet away, Belden 9913 or 9914 cable can be used to bring in a 4 GHz feed.

Ideally the dish will have a motor drive so that it can be moved to various satellites. This is to both simplify motor drive controller troubleshooting as well as to be able to find a transponder with color bars and an audio tone for off-the-air testing.

If you have a ten foot dish with 39 dB gain, and a LNA with 50 dB gain then the signal exiting the LNA is approximately − 41 dBm in level. This assumes a signal level of − 130 dBm hitting the dish.

9913 has an attenuation of about 1.2 dB per 10 feet at 4.2 GHz. So, after running 50 feet of 9913, the signal is around − 47 dBm. If we split the signal by two, there will be a 4 dB loss, putting us at about − 51 dBm at the input to a down converter. A 20 to 30 dB attenuator

should be placed in the other line from the splitter to lower the signal level so that the LNAs, LNBs, and LNCs can be properly driven. A waveguide adaptor (coax to waveguide transition) is used to connect the RG-214 to the LNA, B, or C.

If the cable run is longer than about 60 feet, then a line amp is necessary to properly drive the DC of the unit under test. You can use an LNA in line to act as a line amp if you balance the cable losses to LNA gain. Again a coax to waveguide transition would be used. Figure 15-2 is a schematic of an idealized repair bench, set-up.

Most down converters will accept a signal level from − 55 dBm to about − 30 dBm without being under or over-driven. However, there is usually a sweet spot between being underdriven and being overdriven where the DC performs best. Ideally, you would have an adjustable 4 GHz level at the test bench to check DCs. H-P makes a variable 4 GHz attenuator for $600. The practical alternative is to use several different lengths of RG-213 cable to attenuate the signal, or to have on hand several values of in-line attenuators. Using pads of 1, 3, 6, 10, 20, and 40 dB, you can get virtually any attenuation value.

In addition to attenuators and various adapters, there are several other pieces of paraphernalia that are required for a TVRO test bench. Some of these are shown in Fig. 15-3. These are just some of the special pieces of gear which don't have much use anywhere outside of the TVRO world.

A. A coax cable to waveguide transition. It is used to connect 9913 type cable using a male N connector to the female N connector that is mounted on the bottom of the transition. The N connector's center pin sticks into the waveguide opening and acts just like an antenna, radiating the microwaves into the waveguide. It is used to put LNAs in-line or to adapt an LNC or LNB system to a multiple receiver system that also uses DCs.

B. Power inserter/dc block. It is used to insert dc voltage onto the center conductor of the cable to power the LNA. It can also be used to measure the current draw and voltage going to an LNA. If no dc is hooked up to its feedthrough it can be used as a dc block.

C and D. A two way and an eight way splitter from Merrimac. The eight way is useful for showroom applications, and the two way is perfect for a one person repair shop.

E. A 15 dB attenuator for 70 MHz lines. It is a homebrew attenuator that also passes dc for receivers that have the tuning voltage going up the rf cable. The schematic is written on the box. It consists of a 10 μH choke that passes dc but stops rf, four 100 ohm resistors,

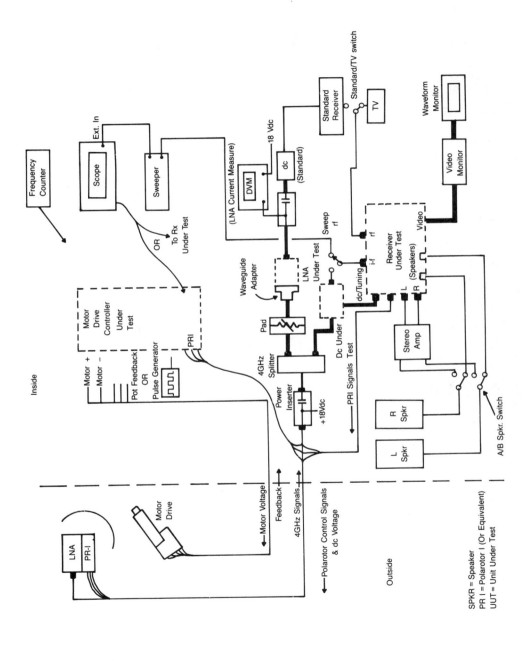

Fig. 15-2. Schematic of the well stocked test bench. It is set up to troubleshoot and repair motor drive controllers, LNAs (or LNBs and LNCs), down converters, and receivers from any manufacturer.

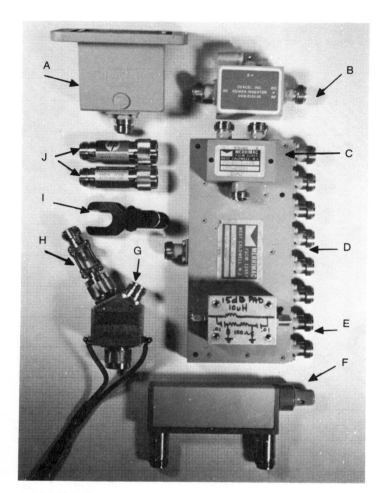

Fig. 15-3. Various interfacing equipment for hooking up all types of TVRO equipment. Included is a waveguide adapter, power inserter, a two and an eight way splitter, a 70 MHz pad, a 4 GHz step attenuator, a coax switch, a dc block, an LNA clamp and two inline attenuators.

and two .01 μF caps: The resistors attenuate the signal about 15 dB, and the caps block the dc from going through the resistors. Table 15-1 lists the values of the resistors for various attenuation from 1 to 50 dB.

F. Hewlett-Packard step attenuator. All $600 worth!

G. Hewlett-Packard coaxial relay. Used to choose between horizontal and vertical inputs. Notice the fitting on input #1, labeled **H**. It is a custom dc block for 4 GHz using a chip capacitor. It can be inserted in-line to stop any dc from damaging the step attenuator, spectrum analyzer, or other dc sensitive test gear.

I. Great temporary clip for holding LNAs to waveguide adapters.

J. Two attenuators from H-P. The upper one is 6 dB, the lower one 20 dB. They come in sets. Common values range from 3 to 40 dB. They can be inserted before a DC or an in-line LNA to optimize the signal level.

Table 15-1. Resistor Values for 75 Ohm Pi-Type Attenuators of 1 to 50 dB Loss.

Loss in dB	R2	R1 + R3
1	10Ω	1.2K
2	18Ω	680Ω
3	27Ω	470Ω
6	56Ω	220Ω
10	100Ω	150Ω
15	220Ω	100Ω
20	330Ω	91Ω
25	680Ω	82Ω
40	3.8K	75Ω
50	12K	75Ω

Symmetrical PI Attenuator Pad

R1   R2   R3

HOOKING IN LNAs AND DCs

Figures 15-4, 15-5 and 15-6 show different methods used to attach a dish feed signal to a unit under test. Figure 15-4 shows how to attach a DC to the incoming feed. The photograph shows a STS DC (D) under test. It has been hooked up to the coax switch (B) through the dc block (C). This prevents the + 18 Vdc LNA supply voltage from causing any problems. Notice that one of the signal inputs has a pad or attenuator (A) in the line. This is to more evenly balance the incoming signal levels between horizontal and vertical. The 15 dB, 70 MHz pad (E) is also being used to prevent overdriving the receiver. The pad is inserted after a 150 foot coil of cabling so that the level that enters the receiver is closer to the real level it will experience.

Figure 15-5 shows an Amplica LNA (F) undergoing testing. In this case the signals are coming into the coax switch (B) and going through an attenuator (C) before entering the coax transition (D). The coax switch setup is the same, with an attenuator in line (A) to match up the levels more evenly between vertical and horizontal. The quick clamp is being used (E) to secure the LNA to the waveguide. Another type of clamp (G) and its special

removal tool (H) is also shown. This type of waveguide clamp makes a very tight fit, but it also takes longer to hook-up the LNA using this type of clamp.

Figure 15-6 shows the same LNA on a Polarotor I test fixture. In this setup, two coax transitions are used to attach the horizontal and vertical signals to an orthocoupled feedhorn. This feedhorn is coupled to a standard PR I which is controlled by the receiver under test. The LNA is fastened to the output waveguide of the PR I, where it will receive either vertical or horizontal signals depending on the probe's position.

MAKING A DC TEST GENERATOR

To easily determine if an LNA, down converter, and receiver are working, a 4 GHz test generator like the Newton GBS-1600 could be used (shown in Fig. 15-7). But if you don't have the money to buy one or you don't happen to have the generator at hand, a quick go/no go test of an LNA can still be made. The cheapest and easiest way to get a 4 GHz signal is to use a single-conversion down converter as the signal source.

All single-conversion down converters without isolators, leak some LO signal out the input port. This

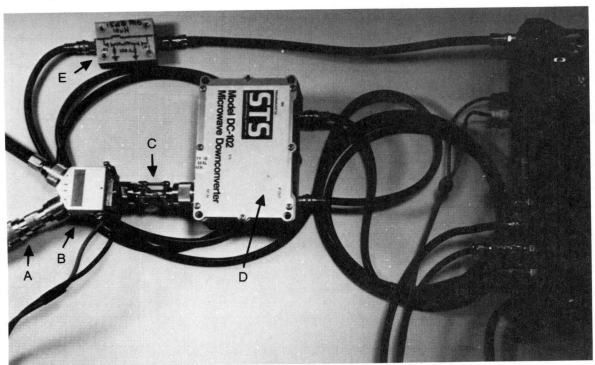

Fig. 15-4. Down converter and receiver testing showing the use of the coaxial switch, the dc block, and the 75 ohm pi-filter.

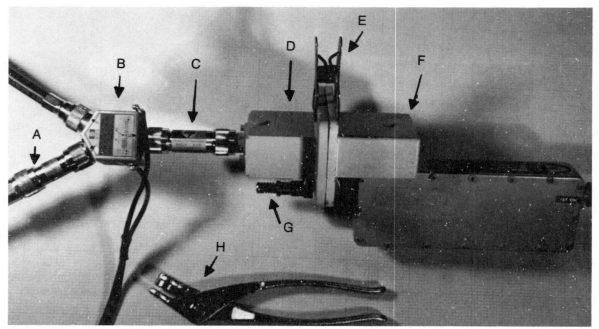

Fig. 15-5. Test setup for an LNA (or LNB or LNC). It consists of a vertical and horizontal input, a 4 GHz coax switch, an in line attenuator and a waveguide adapter. Two types of waveguide clamps are shown, the clothespin type (E) and the spring-loaded type (G). The spring-loaded type needs a special tool (H) to clamp and unclamp it.

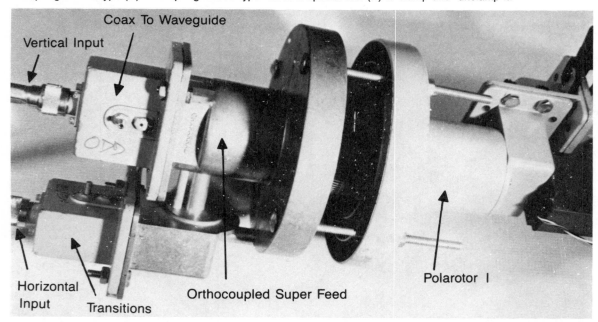

Fig. 15-6. A method of bringing in dual polarity signals that also checks the polarity control circuit. Two PDs are used, a PR I and an orthocoupled feed. They are clamped together. The LNA mounts to the PR I, and the horizontal and vertical signal lines are mounted to the orthofeed through waveguide adapters.

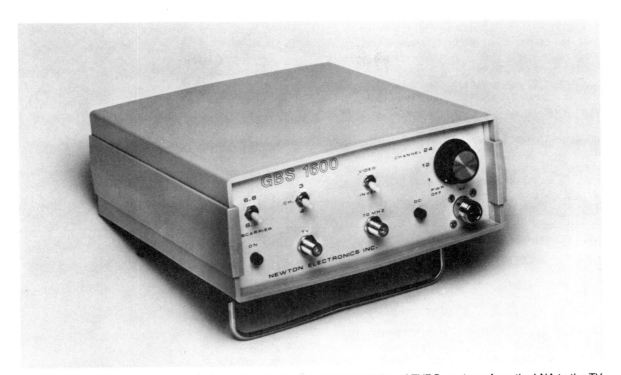

Fig. 15-7. The GBS-1600 portable test generator. It is for in-the-field servicing of TVRO systems from the LNA to the TV set. It has a 4 GHz color bar output on transponders 1, 12, and 24, a 70 MHz output, and an rf output on channel 3 or 4. It will also detect the presence of dc on the rf cable to verify power and/or tuning voltage. Courtesy Newton Electronics, Inc.

Fig. 15-8. When using a down converter in a multiple installation or as a test generator, be sure that the dc line is cut so that no dc is present at the N connector. Both an isolator and a waveguide adapter are dc short circuits.

LO signal is tuned 70 MHz above (or below) the channel that the receiver is set on. In most cases, this is a nuisance, but you can make use of this leakage to create a 4 GHz signal.

As an example, you can make a compact and portable test generator out of an older KLM down converter. Use one from a IV, V, VI or 7X receiver package. You could use the receiver to power and tune it, but all you really need is +18 Vdc and a +2 to +8 Vdc tuning voltage. The DC has a terminal strip with the ground on pin A, the power supply voltage on pin B, and the tuning voltage on pin C. If the trace that supplies the dc voltage to the LNA is cut, then you can put a coaxial waveguide

transition directly to the input via some RG-214 (See Fig. 15-8).

The tuning voltage can be obtained from the dc supply voltage via a voltage divider and 10K pot. You could also adapt the scan circuit from the Sky Eye 7X to generate a tuning voltage sweep to sweep the LO across the satellite band. See Fig. 15-9 for the circuit diagram. If you use two rechargeable 9 volt Ni-Cad batteries, then it will be fully portable and should be good for about an hour's use between charges.

To use the DC Test Generator, merely attach the waveguide transition and about 15 feet of 9913 cable to the DC, turn on the supply voltage, and place the

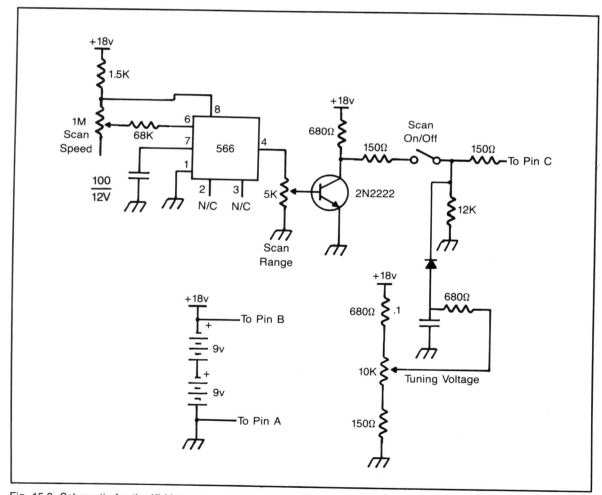

Fig. 15-9. Schematic for the KLM test generator controller. The circuit supplies dc to the down converter as well as a variable tuning voltage that can also be swept for a continuously variable signal. Use Ni-Cad or rechargeable batteries.

waveguide so that it points into the dish or feedhorn. Placement is not critical since the LNA is looking for signals that are at about -90 dBm, and the output of the generator is about -40 dBm.

It can be manually tuned or swept across the band. By observing the signal-strength meter on the receiver under test, you can readily see if it is working. No meter movement indicates that there is definitely a problem. A small indication means that there is something getting through. A pegging meter is an indication that the electronics are probably working, but the dish is off.

If the meter barely moves or doesn't move at all, then the generator could be directly attached to the input of the DC under test. Again look for meter indication. If it reads OK at this point, then something is wrong with the LNA or the connecting cable (RG-214).

Reconnect the LNA cable to the DC and connect the generator to the LNA end of the cable by using a female to female type-N adapter. If the meter reads almost the same as in the DC test, then the cable is OK. This would leave the LNA as the suspect component.

By using the Test Generator on known good units, you can familiarize yourself with the meter indications and picture indications to be expected by putting its signal into the dish, the LNA, the LNA cable, and the DC.

By using the DC Test Generator, you can easily determine if it's the LNA, the RG-214 cable, the DC, the receiver cabling, or receiver that has a problem. And if it's found that the DC has a problem, then the Test DC could be substituted for the system DC to substantiate that the down converter is indeed where the problem lies. In this case, the channels would be tuned by the test generator itself, but the pictures should be good regardless of what receiver is being used as long as it has a 70 MHz i-f.

Troubleshooting a System

USUALLY THE INITIAL TROUBLESHOOTING ON A TVRO system must be done in the field. Very few customers are going to take off their feed and disconnect their receiver, and then bring it all in for repair. Even if they do, the chances are good that the bad component won't be brought in! Thus, the almost extinct TV repairman's *house call* is making a comeback, both because of TVRO installations as well as after-the-sale repairs.

Because of the nature of TVRO equipment, almost every system will be installed by professionals (hopefully they will be anyway). Most customers aren't willing or able to install a system by themselves. Those that try, the so called cash and carry customers, will often end up calling for help sometime during their installation. On the average, most installations, and thus most house calls, are located between 10 and 50 miles away from the showroom or repair shop. But before you complain about that 75 mile round trip to the customer's house, think of Bob Behar and Hero Communications. They have to be the record holder for long distance installations. They've installed systems on every continent in the world. So next time you have to do a service call, just be glad you don't have to contend with any 9,000 miles away. It'll make the drive a little easier.

TEST BENCH ON WHEELS

Anyone trying to install or repair TVRO systems without a dedicated service van or truck is wasting both his time and his customer's time. Time is expensive and thus forgetting some piece of test gear, an adapter, or a small tool just because you thought it was in the toolbox, and it's not, is doubly expensive and should be uncalled for. A properly set up service vehicle will increase your available time by having all that you need efficiently placed so that only a quick check is needed to ensure that everything is where it should be. One trip to and from a customer's site is enough, especially if it's a warranty call.

There are many different ways to configure a house call vehicle. But whether it's a van or a truck or even a station wagon, it must be able to carry every tool that could possibly be used during an installation or a service call. And while carrying all the tools, test gear, and technicians, the vehicle must be able to make it to any customer in any weather. Because it never fails, the day starts out beautiful, you go 8 miles up a paved backroad and solve the customer's problem, but by the time you're ready to leave, it's pouring down outside and the road back looks like a river. At moments like that, you'll wish you'd spent the extra bucks and got a four-wheel drive instead of the economy import station wagon.

So learning from experience, go out and get best heavy duty van that you can afford. You've designed a great logo to be painted on the doors or maybe the side walls, but now what else needs to be done to turn the van into a rolling TVRO shop?

First, unless you're in an area where the crime rate is zero, get rid of the logo for the van. The logo is screaming "Steal me, I've got untold thousands of dollars of expensive electronics equipment inside." With that advice in the back of your mind, console yourself with a basic white cargo van with a top-dollar security system.

Test Gear

For simple sorting, tools and test gear necessary for a shop vehicle can be broken down into three piles: 1) tools used only during installation, 2) tools and test instruments used during installation and for troubleshooting and repairs, and 3) specialized test gear and tools for troubleshooting and repairing a system. The van can be organized to reflect its main purpose.

If it's mainly used as an after-the-sale repair vehicle, it could have all the test gear rack mounted and have a work bench and a spare parts cabinet. Such common items such as connectors, replacement cable, spare LNAs, swap-out down converters, and even swap-out receivers should also be readily available.

If it is mainly an installation vehicle, then ample storage space for reels of cable, spare connectors, and adapters is important. Enough horsepower to pull a trailer will be important also, especially if the vehicle has to carry several people, a couple reels of cable, and hundreds of pounds of tools and spare parts. Power tools and hand tools could be stored in a locked cabinet especially fitted to the vehicle to allow easy removal so that it could be carted to the dish location with all tools being carried in one trip. Regardless of its main purpose, if the vehicle is set up correctly and efficiently the first time, it will serve all your purposes efficiently.

Site Surveys

One important troubleshooting task is doing proper site surveys. Here is where a lot of future headaches can be avoided. If improperly done, a site survey, even using a dish on a trailer, will tell you that 1) the satellites are still up there (surprise, surprise) and 2) that in the driveway, street or front yard there's no, some or a lot of TI.

A site survey, to be effectively done, must be done from the exact site that the dish will be placed, preferably with the same type of dish, in both material and f/D ratio. Of course, in most cases this is out of the question, so why do a site survey at all?

A good site survey is a necessity, but using a dish isn't a necessity. In most cases, if one has done a few installs in one area, the actual obstructed viewing area can be determined by using a compass and an inclinometer, or even by estimating the Clarke Belt's location across the sky. This is usually sufficient unless there are trees or buildings close by. Even then a compass and an azimuth/elevation chart can be used to find the exact heading to any particular satellite. In some cases, moving only three or four feet will be all the difference between getting no signal at all and getting clear reception.

The second reason to do a site survey is to look for TI. Here a dish isn't necessary either. An LNC and a test receiver are all that's necessary. The best place to check for TI is to get up on the roof and slowly move the LNC around in a circle while observing the signal strength meter as the receiver is tuned just below and just above each channel. Holding the LNC both horizontally and vertically will give you a good idea of what TI is there and how strong it is. In most cases, if the signal meter doesn't peg, the TI won't be strong enough to cause problems. If it does peg, then record which channels it was found on, and recheck those channels back on the ground where the dish will be located. If the dish is going to be placed on a tall pole, those channels on either side of the TI will probably be affected. It would be advisable to factor in the cost of TI filters in the system cost if this is the case.

Probably the best way to do a sight survey is to let the customer do it for you, or at least start it anyway. I've found that half your time is spent explaining to the customer why the dish can't go here, or there, or even over there. Most customers can't fathom how big the dish really is, until it is actually placed in their backyard. Figure 16-1 is a handout that can be given to the customer to help him decide where the dish might be placed. It also asks several questions about interfacing the equipment with existing video equipment. Not only will this make a more satisfied customer, but it will also generate added revenue. Figure 16-2 is the drawing that accompanies the handout so that the customer can see for themselves that they have a clear line of sight from their proposed dish location.

As you can see, several potential problems can be

SITE SURVEY QUESTIONAIRE

The following questions will help you to get ready for your upcoming satellite system installation. To help us blend the system into your house and media center, we will be doing a site survey on:

M T W TH F SAT SUN,____________________at____________________ AM PM.

Please look over the questions and consider your answers before the survey date.

1. Consider your property in directional terms. Determine the approximate location of due south. Go to where you feel the dish should be located. Read over the information on the 3rd page (with the drawing). Refer to the drawing for assistance. Notice that the drawing has numbers assigned to the satellites. They refer to the amount of TV programming that each satellite has. Number 1's are the main satellites.

2. Try to visualize the physical size of the dish you selected. Now try to place that same object into your back yard. Where does it seem to fit? Remember that the dish will be angled back about 45 degrees and, as it rotates about its axis to change satellites, it must have room to move. There should be a protected area four to five feet to the east and to the west of the dish.

3. Where will the satellite receiver be placed? Will a TV be used or will a video monitor be used? Will a VCR be hooked up? Will more than one TV be hooked to the satellite receiver? If your receiver choice is remote controlled, do you want extra controls at the other TV's location? Let the survey person know your answers, so that we can bring all the necessary materials during installation.

4. Look over the sample programming schedule we gave you. Which satellite services do you think you're most interested in? Check off those services and be sure to show this to the person doing the survey.

5. Any system can be set up for more than one receiver. If you think two or more receivers are desired, let the survey person know so he can discuss such a system with you. It's more cost effective to do it now, than it is to do it later.

Fig. 16-1. Site survey questionnaire and typical installation charge sheet.

headed off even before you start. Knowing more about your customer's desires will enable you to satisfy them, which will foster good will and possibly a sale or two in the future.

Checking Components in the Field

During installation, it's a good idea to check the actual current draw of the LNA, LNB, or LNC. The reasoning behind this is to record the reading for future reference, just in case there's a system problem. Usually each unit is within 10% of the spec sheet, so this is not absolutely necessary, but many times problems can be diagnosed by the current draw of the device, which is its heartbeat, so to speak. Comparing the starting figure with

Installation Charges

The installation charge is based upon several factors which will be set during the site survey. Our prices are as follows:

1. Basic installation charge $400.00

This price includes; travel charges for systems within a 25 mile radius, setting up the dish on level ground using a standard 7' pole mount, running up to 100 feet of cable into the house, aligning the dish, receiver and motor drive (if used) for proper reception and operation. We will supply the A/B switch box and cable for hooking up the system to one TV set. We will also instruct you on using your new system. If need be, one follow up visit is also included in this price.

2. Hooking up a second TV, a VCR, a video monitor or a stereo system
. . . . $35.00/each

This price includes a pre-assembled 10 foot interconnecting cable. The attached component will be checked for proper operation with the satellite system. If the length of cabling is longer than 10 feet, then the following charges apply:

.20/ft for RG-59 coax cable
.32/ft for audio and video cable
.18/ft for speaker wire
.35/ea for type-F connectors
.45/ea for RCA phono plugs

Some TV's require a balun transformer (to go from 75 ohm coax to 300 ohm twin lead terminals), they are $3.50/ea.

3. Installations Surcharges . . . . per the following

If your location is more than 25 miles from our showroom, then there will be a onetime .41/mile (one-way) charge. If your dish is more than 100 feet from you receiver, then an additional .95/ft will be charged for cabling beyond the first 100 feet. If your location requires the dish to be mounted on a tall pole, on a roof or on a steep slope, then there will be a surcharge of up to $300.00 added to the basic charge depending upon your needs. All of these charges will be set forth during the site survey.

Remember the pre-paid cost of the survey is non-refundable, but it will be credited to the cost of the system at the time of purchase.

---

a later measurement will tell you a lot about the health of the microwave component.

If a component in the LNA, LNB, or LNC fails, the result may be reduced gain, excess sparklies, fluctuating signal levels, or even no signal at all. But these same problems could be caused by a faulty cable or receiver. A quick way to check if the LNC is faulty is to check the current draw and supply voltages.

If the current draw is zero, than the LNA, LNB, or LNC is effectively disconnected. This could be caused by a broken power supply wire, bad power inserter, bad voltage regulator in the receiver, or a bad microwave component.

If it's 20 to 25 mA less than it was at installation, it's

a good chance that one stage of the LNA has died. This holds true even for LNBs and LNCs, since they still have an LNA inside the same housing. If it is an LNB or LNC and it is off by more, then the down converter i-f amp may be bad, the VTO or LO may be bad, or more than one gain stage may be bad in the LNA.

If it draws the same current as it did during installation, there could still be a broken connection that is blocking the signal, but that is not affecting the current draw, so the current reading will only tell you that the bias circuitry is still OK.

The next thing to check is the power supply and tuning voltages. These voltages should be checked at the LNA end of the cable. That way the cable is checked as well. A current/voltage checker can be made up to ef-

fectively allow measurement of the current and voltage while the unit is operating. It is shown schematically in Fig. 16-3.

Substitution

The easiest method used by novices in the field is substitution. Namely, before making the service call, the technician rounds up all the same components used in the system and tests them at the shop. Then at the customer site, they are substituted one at a time. This will usually solve the problem if it's due to component failure and not dish mispointing.

Of course, this requires an extra set of all the brands and models that you service. And it does mean that these

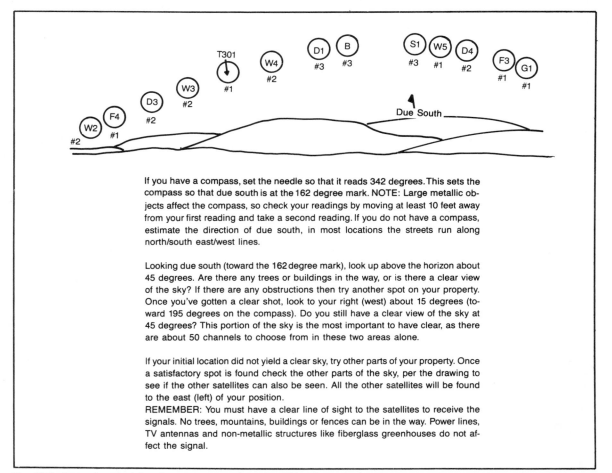

If you have a compass, set the needle so that it reads 342 degrees. This sets the compass so that due south is at the 162 degree mark. NOTE: Large metallic objects affect the compass, so check your readings by moving at least 10 feet away from your first reading and take a second reading. If you do not have a compass, estimate the direction of due south, in most locations the streets run along north/south east/west lines.

Looking due south (toward the 162 degree mark), look up above the horizon about 45 degrees. Are there any trees or buildings in the way, or is there a clear view of the sky? If there are any obstructions then try another spot on your property. Once you've gotten a clear shot, look to your right (west) about 15 degrees (toward 195 degrees on the compass). Do you still have a clear view of the sky at 45 degrees? This portion of the sky is the most important to have clear, as there are about 50 channels to choose from in these two areas alone.

If your initial location did not yield a clear sky, try other parts of your property. Once a satisfactory spot is found check the other parts of the sky, per the drawing to see if the other satellites can also be seen. All the other satellites will be found to the east (left) of your position.
REMEMBER: You must have a clear line of sight to the satellites to receive the signals. No trees, mountains, buildings or fences can be in the way. Power lines, TV antennas and non-metallic structures like fiberglass greenhouses do not affect the signal.

Fig. 16-2. Second part of the site survey questionnaire. This allows the homeowner to site out their property for the best place to place the dish.

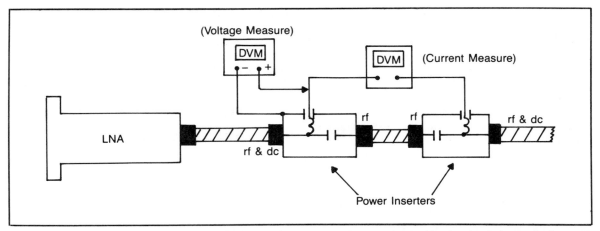

Fig. 16-3. Field checking the LNA current and voltage using two power inserters hooked back-to-back. It can also be used on LNBs and LNCs if the proper connectors (or adapters) are used.

units cannot be sold as new units, since they will undoubtably get scratched and damaged in the course of substituting out components. But it is a simple, expensive way to check out systems.

A simpler and cheaper way to go, is to have a test system. This is a complete electronics system that can be used to check any brand or model receiver, DC, or LNA. With the proper adapters, almost any brand or model receiver, DC, and LNA can be used.

A test system would consist of a 120 degree LNA, a down converter, and a receiver. Your basic setup, right? But why a 120 degree LNA you ask? The purpose of a service call should not only be to troubleshoot and hopefully repair a system, but also to tweak-up a system for max performance. By using a 120 degree LNA, any dish mispointing or feed degradations will be much more readily apparent than if you have the hottest 75 or 85 degree LNA in the shop. What's more, a 120 degree LNA is downright cheap these days.

As long as all the right adapters were available, just about any single-conversion system could be checked by using the Test DC (as outlined in Chapter 15), an LNA, and a simple continuously (pot) tuned receiver.

TEST PROCEDURE FOR DEXCEL LNCs

Dexcel manufactured two LNCs; the DXS 1000 and the DXS 1200. The DXS 1000 was manufactured for over 4 years. It is used with the Dexcel DXR 1000 and DXR 1100 receivers, and with early Boman and MTI receivers. The DXS 1200 was manufactured for about two years and is used on the DXR 900, DXR 1200 and DXR 1300 receivers. The DXS 1000 is painted light blue, and the

DXS 1200 is painted tan.

Both LNCs use a type-N for RG-59 cable and a Cannon connector. The Cannon is a 4-pin military style connector, originally made by Bendix and is usually referred to as the Bendix connector. The pins are labeled A through D. On the opposite end of the cable is a 4-pin Mike (or Mic) Plug. Its pins are labeled 1 through 4. The cable goes from pin 1 to pin A, pin 2 to pin B, pin 3 to pin C and pin 4 to pin D. Even though the connectors are the same, the cabling is different between the DXS 1000 and the DXS 1200, and like the LNCs, is not interchangeable.

The DXS 1000's power supply voltage is +28 Vdc. It will be found on pin A. The tuning voltage will be found on pin D. It should range between +10 and +20 Vdc (depending on which channel is selected). Pin C is the ground or shield.

The DXS 1200's power supply is +12 Vdc and is found on pin B. Ground is supplied by the shield on the coax cable's connector. Pin A is not used, while pins C and D are shorted together in the LNC. When the Bendix connector is plugged into the LNC, the cable shield (from pin 3) is connected to the black wire (on pin D) that supplies the ground for the polarity control device. The red (pin 1) and white (pin 4) wires are connected directly to the polarity control device. Tuning voltage is sent up the center conductor of the coax cable and enters the LNC through the type-N connector.

Checking the current draw requires a current meter inserted in series with the wire from pin 1 of the Mic Jack to pin A of the Bendix connector on the DXS 1000. On a DXS 1200, the wire from pin 2 to pin B would be used.

This is the Black wire in the standard Dexcel 1000 or 1100 cable, or the orange wire in the standard 900, 1200, or 1300 system. There are two ways to check the current: put an in-line fuse holder in the line or make up a special current checking cable.

Putting an in-line fuse holder in the line adds some over-current protection to the LNC, since it would be left on the customer's unit. Use a 1/4 amp fast-blow fuse. To measure the current, remove the fuse and hook the current meter to the two contacts.

Inserting the fuse in-line requires that the 4-pin Mic Jack be disassembled. Unscrew the two strain relief screws and remove the strain relief. Unscrew the small screw that holds the metal cover to the connector. Slide the cover up the cable. Unsolder the wire from pin 1 (normally colored black). Run both wires from the in-line fuse holder through the metal cover and solder one wire to pin 1, the other to the black wire on the DXS 1000. On a DXS 1200 LNC, the fuse holder's wires would go to pin 2 and the orange wire. Be sure to use shrink tubing or tape on the wire splice to prevent shorting.

A current/voltage checking adapter could also be made up. It requires that a male and a female mic plug and jack be purchased. These are available from Radio Shack (part #274-001 and #274-002). To make up the Checking Cable, solder approximately 6 inches of #20 wire between pins 3 and 4 of the two connectors. Attach 3 inch wires to pins 1 and 2 on both connectors. Put four female bannana jacks, to use as test points, on the ends of the 3 inch wires. It is easiest to use two red and two black banana jacks. Put the red jacks on the wires coming from pin 1 of each connector. Put the black jacks on the wires coming from pin 2 of each connector. Use shrink tubing on the male connector and the banana jacks to prevent shorts. The whole cable may be wrapped for appearance, leaving just the four jacks showing. It is illustrated in Fig. 16-4.

To use the Current Checking Cable, insert it between the standard LNC cable and the receiver. Connect a current meter, set to read up to 200 mA, to the two red or black jacks. The red jacks would be used for checking DXS 1000 LNCs, while the black jacks would be used to check DXS 1200 LNCs. Turn on the receiver and read the LNC current draw. The meter reading should be around 140 mA to 160 mA for most LNCs.

TROUBLESHOOTING GUIDE

The rest of this chapter is a troubleshooting guide that covers the various parts of the typical TVRO system. But before we can narrow our search down to one component in the system we must first determine if the dish is actually pointed at a satellite. If the dish is off as little as 3 to 4 degrees, then the satellite that it is aimed at will not be seen.

An azimuth and elevation directory is helpful in determining if the dish is pointing correctly. It typically lists the azimuth and elevation for each satellite for most latitudes and longitudes in the United States and Canada. Of course, once you've done several installations in one location, you will almost instinctively be able to eyeball the dish and determine if its position is within the ballpark. Once you know the dish is close, then the following troubleshooting guide should be helpful.

Noise on All Channels, Both Video and Audio

Possible Causes. Dish pointing error, electronics failure, or in the case of fiberglass dishes, someone forgot to add the reflective metal to the dish when it was manufactured.

To determine if dish pointing is the cause, one must first determine if the electronics are functioning. This is most easily done by using the DC Test Generator (from Chapter 15) or a commercial microwave generator. Point

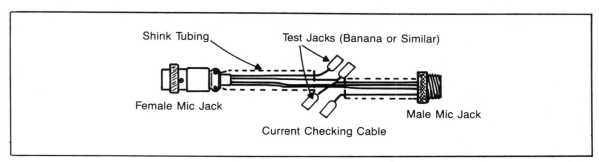

Fig. 16-4. Test cable specifically made up for Dexcel LNCs. The two banana jacks are used to measure the current draw on the two types of Dexcel LNCs and DCs.

it into the dish and observe the receiver for an indication that it is receiving a signal. If it is not, then isolate the LNA, DC, and receiver to find the defective component. If the electronics check out, then the only remaining possibilities are dish positioning, feed positioning, or the dish itself.

As far the dish itself being the problem, there was one manufacturer that I heard of (who's been gone a long time now), that made fiberglass boats, and who thought that making TVRO dishes would be a piece of cake. So they started to manufacture dishes, only they didn't always remember to put a reflective surface in them! In some cases one or more panels would be without metal while the others would be OK. You got pictures but boy were they noisy, and the gain was not anywhere close to what it should have been, depending on how many petals had metal.

The reason that I mention this is that these dishes are still out there somewhere. If somebody drives up with a big trailer full of fiberglass dishes of unknown origin and offers them to you at below wholesale cost, buyer beware. Get your ohmmeter out and start scraping some test points to check for continuity before deciding if you've got a bargain.

Checking Dish Positioning. The first checks that I make once the electronics have been OKed, are that the pole is set perfectly, that the polar axis is set correctly, that the offset angle is correct, and that the dish axis is parallel to the Earth's axis. All of these together only take about 5 minutes to check with the proper tools and charts.

The required tools are a compass, inclinometer, and carpenter's level. Enough string to stretch across the dish is handy.

Since most installations use a pole mount, that is what I will be discussing. The pole must be perfectly vertical for the dish to properly track the Clarke Belt. Use a carpenter's level to check the pipe. Check it twice, 90 degrees from each other to ensure the pole is vertical in all axis. If it is not vertical, then the dish will not be able to properly track all the satellites. If it is not too bad, then the end satellites can be optimized and the center satellites (like Anik) will only be slightly off.

Also check the dish mount pipe. It slides down over the pole and is usually clamped to it by several bolts. If the bolts are not tightened in the correct sequence then it could be biased off while the pole could be correct, which would result in the same problems. Sometimes a pole that is offset slightly can be compensated for by biasing the mount pipe in the opposite direction.

Next check the polar axis. It should be set to the latitude of the dish location. Since inclinometers, like those that Sears sell, are only accurate to ± 1 degree, rounding off the latitude to the nearest degree is all that's required. Make sure that the surface you are setting the inclinometer on is even and that it is parallel to the axis points that the dish is actually rotating on. Once the setting is found, tighten all bolts relating to the polar axis setting to the dish manufacturer's specifications. Recheck the angle to make sure that tightening the bolts didn't affect anything.

Next, check the dish declination offset angle. It can be measured by setting the dish to true south and then running a string from the top of the dish to the bottom. Lightly place the inclinometer on it to read the offset angle. On some dishes, the offset angle is read on an angle indicator designed specifically for that dish. On some dishes there is a backplate that can be used to measure the angle on. Regardless of the method, the measurement should equal the sum of the latitude angle and the offset angle. Appendix B includes a declination chart that covers latitudes up to 70 degrees.

The final check is to set the axis for true north. The best procedure is to use an accurate compass and to set two stakes about 10 feet apart on a true north-south line that runs through the pole. Be sure to add in the magnetic deviation to the compass reading or else you could be up to 10 degrees off. Also be sure that you are at least 10 feet from the dish or other metallic structures as they will affect the compass reading. Appendix B includes a Magnetic Deviation map of the U.S. It will give an approximation of the magnetic deviation in your area. Call your nearest airport or harbormaster and ask for the magnetic deviation for the immediate area.

At this point, moving the dish about the Clarke Belt should produce pictures if the electronics are working correctly. I use transponder 11 if the receiver does not have a scan control, as it is active on almost every satellite. Remember to use reverse polarity for Galaxy, Westar, Spacenet, and Anik satellites and normal polarity for Satcom, Comstar, and Telstar.

Poor Picture Quality, Low Signal Strength, and Excessive Sparklies

Possible causes. Dish pointing errors, feed positioning errors, LNA degradation, moisture in LNA cable, poor connector contact, or LNA power supply problems.

Go through the dish positioning check-out steps per the previous problem. If the pictures are still found to be below par, then the next check is the feedhorn focusing and centering.

Rotate the dish for easy feedhorn access, and use a

Focal Finder to determine if centering and focusing are correct. A Focal Finder is a device with a rubber stopper that fits into the circular waveguide opening of most polarization control devices. It has an extendable rod that points directly below the feedhorn's opening. If the point where the extended rod hits the dish is not exactly in the center of the dish, then the feedhorn is not centered. Adjust the feedhorn until the rod is centered on the dish and is perpendicular to the dish's axis.

Another method is to measure from the edge of the dish to the feedhorn's circular waveguide at three equidistant points around the dish. If all measurements are the same, then the feed is centered. If they are different then the feed is offset and should be adjusted until all three measurements are the same.

To determine if the feedhorn is parallel to the dish axis, set the dish so that it is pointing due south. Place an inclinometer on the back of the scalar rings or on the circular waveguide opening. In either case, the angle should equal the angle of the dish. If the dish was measured by stretching a string across it, then the angles should be equal. You could also subtract the dish's offset angle from the reading. The result should be equal to your latitude (see Appendix B for an Offset Angle Chart).

If the reading equals your latitude, then the feedhorn is set correctly (if the rod was centered). If it's not, then the feedhorn is not parallel to the dish axis. If the feed uses three or four legs, then there is usually some length adjustment to each of them that is used to center and focus the feedhorn. If it is a J-hook feed support, then it may need to be bent slightly or it may just need to be rotated around to find the correct setting.

If the rod is carefully marked with the proper focal distances of the dishes you use, then you can tell at once if the feed is focused properly. Otherwise, measure from the center of the dish to the end of the feedhorn's circular waveguide. This distance should be 1/4 inch less than the focal distance for the dish. In other words, the actual focal point is 1/4 inch inside the circular waveguide, not at the end of it.

Determining the Focal Distance. If the focal distance is unknown, it can be determined in two ways: by measuring the diameter and the depth of the dish and calculating it, or by using the f/D ratio and the diameter to calculate it.

When measuring the diameter, be sure to measure from the reflective surface. Don't measure any outside support ring, as this could add two or more inches to your measurements. Also be sure that the tape measure is not allowed to droop as this will also add extra nonexistent inches to your dish.

To determine the depth, fasten a string across the antenna so that it crosses the center. Measure the distance from the string to the center of the dish, this measurement is the depth of the dish. To determine the focal distance for the dish, square the diameter and divide the total by 16 times the depth.

To use the f/D ratio method, it is also necessary to find the actual diameter of the dish. If you multiply the f/D ratio by the measured diameter you end up with the focal distance.

Once the feed is centered and focused and the dish is able to track the Clarke Belt, then the pictures as well as the signal strength should be excellent. If there is still excess sparklies, then the surface accuracy of the dish should be checked. This can be done by stretching two strings across the dish. They should cross exactly in the center of the dish and should be at a right angle from one another. At their crossing point, they should just touch. If they don't, then it is an indication that the dish is warped. Move the strings clockwise 45 degrees and do the same check. If they still don't touch then the dish must be tuned up. If they touch at both locations then the dish's outer perimeter is even. On a screen dish, visually sight across the dish looking for bumps or ridges in the metal, anything more than a 1/4 inch should be smoothed out.

Fiberglass and plastic strutted dishes have a tendency to sag at the bottom. This will become more pronounced with age and ice and snow build-up. Screws and bolts can loosen as the dish is moved back and forth by the motor drive and by the wind blowing into and across the dish. Movements up to 2 inches can be seen in many unsupported J-hook feeds as the wind blows into the dish. This oscillation will cause fading and unstable pictures. The dish itself can also vibrate back and forth after a very short time if the mount and motor drive are not correct for the dish. Dishes that are driven back and forth across the arc can also develop loose bearings, especially in sandy and dry climates like the Southwest desert.

Unfortunately, many of the mounts and antennas manufactured for TVRO use are not adequately weatherized. Most do not have anodized hardware nor are they painted with the proper polymer paints to withstand ultraviolet light. There is a real problem with fiberglass dishes cracking and allowing water into the lower laminations. This causes further cracking and glazing and can corrode the reflective metal that was sprayed or laminated to the fiberglass. Already, some of the screen dishes installed only two or three years ago are starting to rust through where the J-hooks, screws, and pop rivets are located.

Aluminum corrosion due to salt spray and air pollution is also becoming more noticeable. So far testing has not shown any drastic reduction in efficiency for dishes or feeds that are corroded, but as the corrosion increases there will surely be problems due to lowered gain.

Cabling and Connectors. Read through Chapter 4. If the connectors were not weatherproofed during the installation, then that is most likely your source of problems. Water in coax cables is a real problem. The foam dielectric will act like a wick, drawing moisture up into the cabling. This will short out 4 GHz signals, pull down the supply voltage, or change tuning voltage going to the DC (if the voltage is run up the center conductor of the coax). In most cases, heating up the connector and the three or four inches of cable with a heat gun or hair dryer will usually be enough to steam out the moisture. If there is moisture in the coax, then it will show up as a resistance between the center conductor and the shield.

Check the depth of the N-connector center pin. It should be slightly below the level of the inner ring. If it is farther down, then the connector may not be making good contact with the female connector. Check the female connector for bent or missing sections on the center conductor contact.

If the connectors were packed with any water resistant substance, clean them out. Silicon grease or lube is not designed to work in connectors that have dc present. Impurities in the grease can cause the 4 GHz signals to short to ground or the voltage to be pulled down. If the voltage drops below the advertised minimum voltage for most LNAs, then the gain drops off rapidly. For most 15 volt LNAs, somewhere around 11 or 12 volts is where the LNA dies. Between 12 and 14.5 volts it will function but with very reduced gain.

Secondary Image in Picture Background

Possible Causes. Cross polarization from an adjacent satellite, incorrect setting of polarization control device, image-reject mixer problem in the DC, video being broadcast on terrestrial carrier, rf modulator or TV set picking up local TV station.

If the problem only occurs on some satellites it may be due to cross-polarization from an adjacent satellite. This is especially common to smaller dishes, and it will be most obvious on F3, due to G1's stronger power and F3's 20 degree skew.

If the modulator or TV is being affected by a local TV channel, disconnecting all inputs to the TV set will probably result in a picture being watchable on the screen from the local channel. If the picture is still there when the modulator is changed to its other setting, then an auxiliary modulator for a higher or lower channel may be necessary.

If the image is not seen on channels 18 and above, but it is noticeable on channels 17 and lower, then it is probably in the DC, as the image falls 7 channels above the channel tuned, thus channel 17 may have an image from 24, 16 would have the image from 23, etc. This holds true for a high side LO. If the DC has a low side LO then channels 1 through 6 would be clean, but 7 through 24 may have the images floating in the background. If either of these cases appear, then the mixer circuit in the DC has a problem. Often a diode will be open or shorted in the mixer, or it will have become unbalanced in some way. In most cases, it is best to take it back to the shop. Setting the image rejection is not something one does with a Heathkit scope and the VITS.

Scrambled Picture

Possible Causes. Incorrect setting of the video polarity switch, defective video polarity switch, encrypted broadcast, PLL chip lock-in range off, video amplifier section failure, rf modulator failure, TV fine tuning offset from receiver's output frequency.

If all channels are scrambled then try switching the video polarity switch. Often just pressing on it slightly will stop the problem. Since the video goes through the switch in most receivers, then the contacts may be bad. Use a scope to check for video on the output. If the switch is changed the sync pulses should go from positive to negative and vice versa. If this is the case, then the video polarity switch is OK and so is the previous circuitry.

PLL chips have a trimmer cap that sets up the chip to lock in on the correct frequencies. Often an unstable picture that drops into a scrambled picture will be caused by a flakey trimmer cap. Try pressing lightly on the cap while watching the picture. If it clears up then the trimmer caps is the source of the instability.

Wavy Vertical Black Bar in Picture

Possible Causes. Incorrect setting of the video polarity switch, encrypted broadcast.

Cannot Receive All Satellites

Possible Causes. Misalignment of dish, incorrect

setting of format switch.

In some receivers the afc range is fairly narrow. This prevents the receiver from being pulled off by a TI carrier. It also prevents the receiver from locking on the next channel, which is offset 20 MHz. In such receivers, if the satellite format switch is incorrectly set then the receiver will be tuned for an even channel and the PD will be set for an odd channel or vice versa.

Black or Grey Screen on Some Channels

Possible Causes. TI or LO leakage from another down converter.

If there is another satellite system within a few hundred yards, then LO leakage is possible from one dish to the other. To check for LO leakage, tune the neighboring dish through all the channels while keeping the test system tuned to a midband channel like 11. Tune the neighboring receiver to channels 6, 7, 15, and 16 while observing the receiver under test. If its meter jumps or the picture goes grey or black then stops as the other receiver is tuned to another channel, then the LO from that system is leaking into the system under test. An isolator in between the LNA and DC of the interfering system should clear it up.

Signal Meter Jumps and Picture Changes Quality

Possible Causes. TI, LO leakage from another down converter, moisture in the waveguide, LNA or cabling, intermittent connection in the LNA, DC, or i-f strip.

Motor Drive Only Goes One Direction

Possible Causes. Defective limit switch, defective direction switch, power supply failure, improperly programmed end limit, feedback circuit failure.

Refer to Chapter 14's Troubleshooting Guide.

Polarization will not Change, is Sluggish, or Intermittent

Possible Causes. Wire gauge too small for the cable run, defective PD, polarity fine-tune adjustments set up incorrectly, moisture in the PD, polarity format incorrectly set-up during programming.

Refer to Chapter 6's Troubleshooting Guide.

Audio Buzz

Possible Causes. Ground loop, misaligned audio

detector, misaligned rf modulator, video level to modulator set too high, video high-frequency adjust peaked too high.

See Chapter 11's Troubleshooting Guide.

Audio Hum

Possible Causes. Ground loop, power supply failure (diode or filter capacitor problem), shorted amplifier.

See Chapter 11's Troubleshooting Guide.

Offset Center-Tune Meter

Possible Causes. The afc misadjusted, TI, misaligned i-f strip, channel tuning misadjusted.

If the offset is the same for all channels, then it would seem to be a misalignment in the receiver or a missetting of the afc centering control. Usually the afc control is adjusted with the DC connected but without any signals coming in. The meter should be centered. If it is not, then the afc should be adjusted so that the meter is centered. Be sure to check that the meter itself is not biased off by its physical centering adjustment. This is done with the receiver turned off and with the meter terminals shorted together.

Video Jitter

Possible Causes. Video level incorrect, afc misaligned, video detector misaligned, clamping circuit defective, open filter or bypass capacitor.

Video problems usually stem from a video level that is too high or too low. Video jitter is usually caused by a video level to the modulator that does not deviate it enough. A similar symptom is caused by the PLL circuit not being aligned properly.

If the video seems to pulsate or flicker, especially during bright scenes, then the clamp circuit may be defective or may be turned off. Check the clamp on/off switch. A bad bypass capacitor could be allowing feedback to get into other parts of the circuit. This can also cause motorboating audio.

Jumping or Drifting Channels

Possible Causes. Moisture in cabling at DC or LNC, intermittent tuning voltage cable, defective afc circuit, defective VTO or VCO in DC or LNC.

No Audio or Video and no Signal Indication

Possible Causes. Defective coax cabling, mispointing of antenna, defective LNA or DC, defective i-f strip.

Use the Test DC (Chapter 15) to determine if the DC, LNA, and cabling is OK. The signal-strength meter comes off before the limiter circuit, so the limiter and the rest of the receiver is probably OK. Check the coax cable as it enters the rear of the receiver, and as it connects to the board. If it is connected to a plug, try unplugging and reseating. Check coils for broken leads and defective ferrite cores.

Black Screen but Audio is OK

Possible Causes. Video amp defective, defective video polarity switch, shorted video cable.

If the sound gets through, then the video detector circuit is OK and the problem lies in the video amp, filter, or clamping circuit. Again the video goes through the polarity switch, so that is the first place to check. If transistors are used, check the bias voltages. If ICs are used, then checking the input and output signals may be all that you can check.

Channels Off by One Number

Possible Causes. Polarity format in wrong setting, PD misadjusted, channel tuning voltage off, channel display decoder or driver is defective.

If the polarity format switch is set wrong or the APS is programmed wrong, then the receiver could be set for an odd or vertical channel when the PD is actually set on horizontal. If the receiver is fine you will see the adjacent channel above or below the desired channel.

If the receiver has not been matched to the cabling and DC then the tuning voltage to the DC may be wrong. This can lead to all channels being off or only one or two, depending on the tuning scheme used. Each receiver has its own matching procedure. Some models are given in Chapter 18, but for ones not listed, contacting the manufacturer for instructions is the best procedure to follow.

If the receiver can be stepped through all the channels and yet the channel readout is wrong, then possibly the display driver circuitry has a problem. If one of the input lines are shorted or pulled high, then either bit 1, 2, 4, or 8 would always be on. This should be easy to tell if the receiver is set on channel 1. With the receiver on channel 1, then a readout of 3 would indicate the 2 bit is high, a readout of 5 would indicate the 4 bit is high and a readout of 9 would indicate the 8 bit was high. If the display blanks out on all channels but 1 and 11, then that is another indication that bit 8 is stuck high. If bit 4 was stuck high, then channel 1 would read out as 5,

channel 2 as 6, channel 3 as 7, channel 4 as 4, while channels 5, 6, and 7 would read normally and 8 and 9 would probably be blank.

Horizontal Bars Float through Picture

Possible Causes. Ground loop, using too small of a gauge wire for PD, power supply filter cap defective.

If the bars disappear when the PD is disconnected then the wire size is probably too small. This can sometimes be compensated for by adding a 100 to 1,000 μF electrolytic capacitor at the PD. It should be connected between the B+ wire and ground.

If the bars remain with the PD disconnected, then there may be a ground loop between the receiver and another component. Use a ground relief adapter to lift the chassis from the wall ground. If the bars go away, then there is a difference between the receiver's ground and the dish ground. The dish ground is the most important ground for lightning and shock protection. It should be left intact, and the receiver, stereo, VCR, etc. should be lifted above ground with the ground adapter.

If doing both procedures does not change the bars then the power supply output should be checked for ac ripple on the dc voltage. A capacitor of equivalent or higher voltage than that used in the receiver can be tack soldered across the main filter cap to see if the filter cap is the culprit. Filter caps will degrade, reducing their capacitance. If this happens the problem will gradually get worse.

Fuse on Receiver Blows

Possible Causes. Shorted power supply, shorted LNA, LNB, LNC, or DC, shorted speaker cables (if receiver has integral amp).

Disconnect the LNA power, DC tuning voltage line, and polarity controller from the receiver. If the fuse still blows then there is definitely a short in the receiver. The most likely causes are a shorted transformer, shorted rectifier diode, shorted or leaky filtering capacitor, shorted regulator circuit or IC regulator. If the fuse has been replaced with one of a higher rating, then any manufacturer's warranty is void and there will probably be much more damage than a simple component short.

Fuse on Motor Drive Blows

Possible Causes. Motor size (torque) too small for the dish, motor mounted incorrectly, rust or other build-up on tube.

Typically there is an internal protection fuse on most

motor controllers. Do not use a fuse of a higher rating if the internal fuse keeps blowing. The most common reason for continually blowing fuses is because the motor is binding or is not strong enough for the type and size of dish. Refer to Chapter 14 for differences between motor drive types and the current/load ratings.

If it is winter, then the most likely cause is a freezing up of the shaft. Try wrapping heat tape around the gear housing and lower tube. Often just putting expandable covers over the extension tube and the motor will stop motor freeze ups because water will not be able to penetrate the housing. Try repacking the gears and extension tube with a molycoat or other low temperature grease.

Loud Audio Hum and
Dark Horizontal Bars through Screen

Possible Causes. Filter capacitor or diode in power supply defective, bad ground loop.

The most likely cause is a power supply filter that has opened up. This allows an ac voltage to be passed along with the dc, resulting in a 60 Hz hum in a half-wave rectifier or a 120 Hz hum in a full-wave bridge rectifier. If one of the diodes opens up or shorts, then the filter caps cannot maintain the dc level and will sag as the ac input voltage changes, resulting in the same type of sound.

A ground loop with a high-voltage differential can also cause the same symptom. If the motor voltage is shorted to the case at the dish, it can cause this to occur, although in most cases the motor fuse will blow. If it is the culprit, then unplugging the motor drive controller should make it disappear.

Channels Will Not Change

Possible Causes. Tuning circuit in receiver defective, VTO or VCO defective in DC or LNC, tuning voltage cable disconnected or defective.

Check the tuning voltage at the DC or LNC. If it is there, then chances are the receiver and cabling are OK. If there is always the same channel, even if the receiver is turned off and then back on, then the LNA and DC are getting a supply voltage but the VCO or VTO in the DC is not reacting to the tuning voltage. If the DC is field serviceable, then the VTO should be replaced. If it is necessary to return the unit to the manufacturer, it is usually best to return both the DC and the receiver so that they can be repaired and rematched as a unit.

If the tuning voltage is not present at the DC, then check the connector on the rear of the receiver that outputs the tuning voltage. If it is not present there, then the tuning voltage amplifier is probably dead. It is typically a garden variety op-amp that should be easily replaced in the field.

Motor Drive Dead

Possible Causes. Open wire between controller and drive, open motor winding, worn brushes.

Read over Chapter 14. It covers Antenna Positioning Systems in full detail. As a general rule, mechanical binding, loose bolts and screws, and disconnected wiring are all common problems. If the motor drive is over a year old and is used a lot, then suspect worn brushes if it is intermittent or sluggish.

Specialized Components

T O EFFECTIVELY TROUBLESHOOT AND REPAIR TVRO systems and receivers, it is necessary to understand the operation of the various major components that make up the typical receiver circuit. These components range from specialized ICs that have been adapted from TV and FM radio circuitry to unique products especially designed for TVRO receiver use. Also included in this chapter is information on the more common parts found in any electronics equipment, such as diodes, FETs, transistors and ICs.

DIODES

Diodes are two lead, polarized devices that change an ac voltage into a pulsing dc voltage. This process is called rectification. The schematic symbols for various diodes are shown in Fig. 17-1. Diodes are used in power supplies, AGC circuits, demodulation circuits, meter circuits, tuning circuits and clamp circuits.

Diodes will only conduct on either the upper half or the lower half of an ac waveform, depending upon their polarity. Figure 17-2 shows the signals that are found in a typical rectifier circuit. By adding a filter capacitor, the pulsing dc voltage is smoothed into a steady dc voltage.

Diode use in power supplies is covered in more detail in Chapter 8.

A similar type of circuit is used to demodulate the video in some receivers. In a delay line demodulator, there are two or sometimes four diodes in a bridge arrangement. These are usually Schottky diodes but may be common signal diodes. The diodes must be matched as close as possible to one another to ensure getting an undistorted video waveform out. Figure 17-3 is a typical delay line demodulator. Video demodulation circuits are covered in more detail in Chapter 10.

Faster Schottky diodes are used in the clamping circuitry. A typical part is the H-P 5280-2800. This is a low barrier Schottky diode, which has a very fast turn on and turn off time. Thus it can stop the 30 Hz dithering waveform by holding the average dc level constant, while letting the higher frequency video signals pass on through. An example of a clamping diode circuit is shown in Fig. 17-4.

A very unique diode that is actually part capacitor and part diode is the varactor diode. This component is used to tune various circuits by changing its capacitance as the applied bias is changed. An example of a varactor diode being used to tune a PLL circuit is shown in Fig. 17-5.

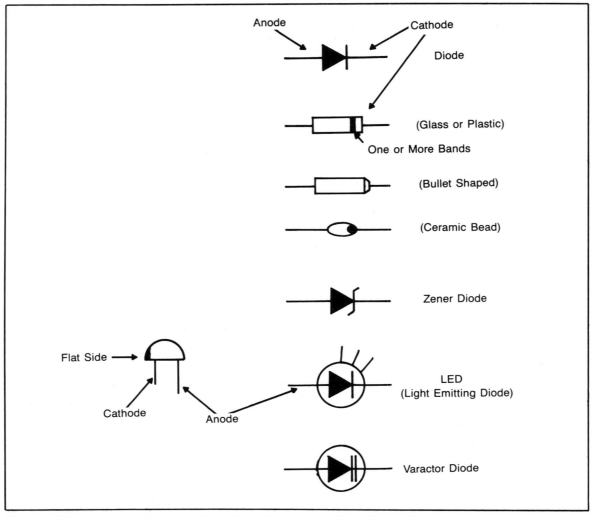

Fig. 17-1. Various types of diodes and their schematic symbols and polarity markings.

TRANSISTORS

Transistors are three lead devices that can be used to amplify, limit, or rectify the signals that pass through them. The three transistor leads are called the base, emitter and collector. These three leads cannot be switched around in a circuit or it will not work properly. A transistor is basically two diodes that have a common connection (the base). The schematic symbols and physical examples for transistors are found in Fig. 17-6.

All transistors can be divided into two families, npn and pnp. The n and p stand for negative and positive. It

is an indication of the normal voltage polarity found on the collector, base and emitter leads respectively. The easiest thing to remember about transistor circuits is that an npn transistor is turned on, or becomes conductive between emitter and collector, by a more positive voltage on the base (in relation to the emitter), while a pnp transistor is turned on by a more negative voltage (or ground) being applied to the base.

Figures 17-7 and 17-8 show the two types of transistors and the typical biasing voltages that may be expected. The npn transistors are more universally used than are pnp transistors.

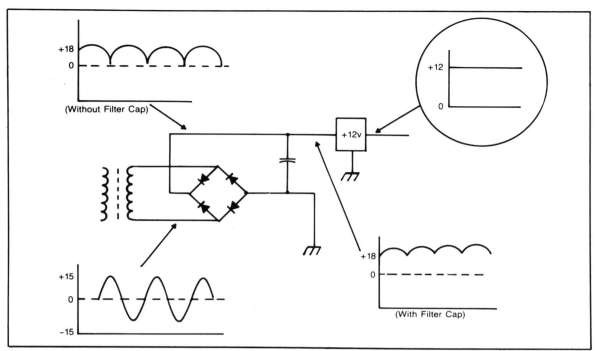

Fig. 17-2. The signals that are found in a typical power supply using a bridge or full-wave rectifier circuit. The input from the transformer is ac (in this case 30 Vac). The output of the bridge is pulsating dc. By adding a large value filter cap, the dc is smoothed out. Putting it through an IC regulator assures an accurate, almost pure, dc voltage as long as the input voltage is more than 3 volts higher than the output voltage.

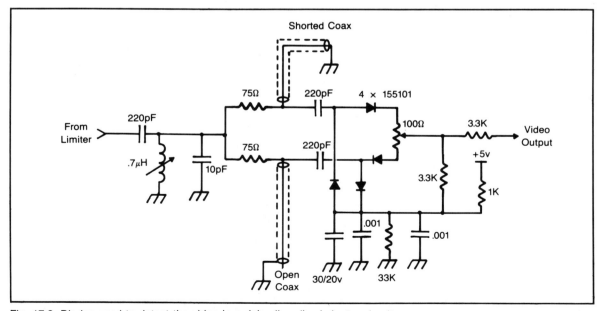

Fig. 17-3. Diodes used to detect the video in a delay line discriminator circuit.

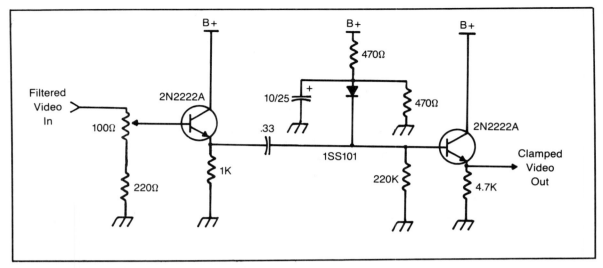

Fig. 17-4. A typical clamping circuit that uses the same diodes as in the discriminator circuit.

Transistor Cross-Reference Books

Probably the most valuable reference source on transistors is the cross-reference books put out by ECG, RCA, IR, and Radio Shack. In these books, thousands of transistors, FETs, and ICs and their cross to common replacement part numbers are listed. Also listed is the specifications for the replacement parts. Thus, if you have a transistor, but don't know the type (whether it's an npn or a pnp), but it's listed in the book, then its replacement part can be looked up. From it, you can determine whether the transistor is an npn or a pnp.

As you look up transistor part numbers, you'll find that in most cases only a few values of npn and pnp transistors are used to replace hundreds of common and in-house numbered transistors. For TVRO use, there are a handful of transistors that should be kept in stock. Two small-signal transistors that will substitute for most applications are a 2N3904 (npn) and a 2N3906 (pnp). For video circuits a 2N2222A in the metal can package will replace almost any npn transistor. For i-f amplifiers 2SC2498, 2SC2876, and 2N5179 transistors are interchangeable in most circuits.

Watch lead placement as the 2SA, 2SB, and 2SC series are often different than the 2N series cross reference part. Figure 17-9 shows the difference between two equivalent parts which have similar bodies but different lead designations.

For voltage regulation circuits, higher power npn transistors are commonly used. The Texas Instruments

TIP series of transistors in the TO-220 package are often used. A few types, such as the TIP-31A or the TIP-41A should be stocked. A 2N3055 is sometimes used in the TO-3 package as a pass transistor, so a handful of them should also be stocked.

FIELD EFFECT TRANSISTORS

Another three lead device is the FET. It is similar to the transistor in that it is a solid state device, and that it can be used as an amplifier or as a switch. But it acts differently than a transistor. For one thing it is a voltage controlled device, whereas the transistor is a current controlled device. It also has a very high input impedance and very low internally generated noise.

The schematic symbols for the different types of FETs is shown in Fig. 17-10. The three leads are named the drain, gate and source and are listed as D, G, and S. There are four types of FETs: the p-channel FET, the n-channel FET, the p-channel MOSFET (metal oxide semiconductor FET) and the n-channel MOSFET. MOSFETs usually have four leads and a 3N prefix. Note that MOSFETs are also static sensitive devices and thus must be handled properly. See the section on static later on in this chapter.

FETs and MOSFETs are used as mixers in audio demod circuits, as voltage amplifiers in both audio and video circuits and as fast switches in clamping circuits. A typical mixer circuit is shown in Fig. 17-11. Testing FETs and transistors is very similar. Basic ohmmeter

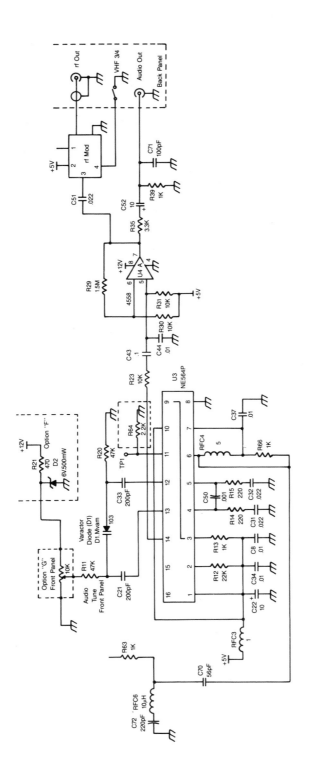

Fig. 17-5. Varactor diode being used to tune the audio channel on a PLL detector. As the voltage across the audio tune pot changes, so does the capacitance of the diode. This changes the lock-in frequency of the PLL.

227

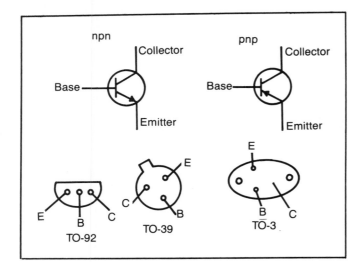

Fig. 17-6. Pin outs and schematic symbols for npn and pnp transistor.

checks can be run to determine if the junctions are good, open or shorted. If in circuit, checking the bias voltage is the best test.

INTEGRATED CIRCUITS

There are many families of ICs used in TVRO receivers. By family, it is meant that a group of ICs all use the same type of signals and voltage levels and thus can be classified together as one family. Some of the families used in TVRO receivers and motor drives are TTL, CMOS, ECL, and Linear.

All ICs can initially be divided into two large super-families, called digital and analog. Digital means that the chips respond to only two voltage levels, a high state of typically +5 Vdc called a logic 1, and a low state, usually ground called logic 0.

Analog ICs respond to an analog signal, one in which continuously varying signal levels are present. Analog ICs

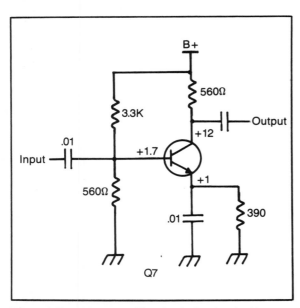

Fig. 17-7. Typical npn amplifier circuit showing the bias normally associated with an npn transistor.

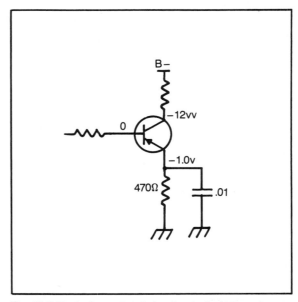

Fig. 17-8. Typical pnp circuit showing typical bias voltages. PNP transistors are not used very often in TVRO circuits.

228

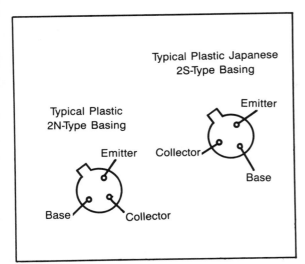

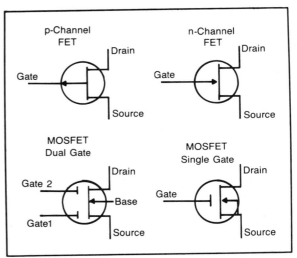

Fig. 17-9. Even though replacement transistors may cross between 2N numbers and 2S numbers, often their pin outs will be different. If the transistor on the left (the 2N substitute) was replaced for the 2S-series transistor on the right without interchanging the base and collector leads, then the transistor may be destroyed.

Fig. 17-10. Schematic symbols for FETs, including p-channel and n-channel FETs, and single and dual-gate MOSFETs.

are usually referred to as Linear ICs. There are occasional exceptions, where the families are used in reverse, i.e., a digital chip being used to pass an analog voltage or a linear chip being used as a comparator to sense a high or low voltage.

At this time there are very few chips made specifically for TVRO usage. So TVRO circuit designers sometimes push the limits and use chips designed for some other elec-

tronics discipline in what amounts to completely new applications (some of which push the ICs beyond their specification limits, unfortunately resulting in early field failures).

Chips are borrowed from just about every electronic field. From the computer world comes the ECL (emitter coupled logic), TTL (transistor transistor logic) and CMOS (complementary metal oxide semiconductor) logic chips. From FM radios comes some of the special purpose linear LSI (large scale integration) chips that contain an entire demod and multiplex decoder on one chip.

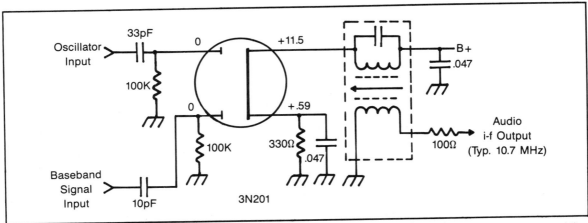

Fig. 17-11. Typical dual-gate MOSFET used as an audio mixer. The baseband video signal is input on one gate and the LO is input on the other, the output is the desired audio subcarrier typically centered at 10.7 MHz.

From the TV and VCR world come linear and digital tuning chips and video amplifier chips. From video games comes the rf modulator chips and channel number insertion chips. From audio cassette technology comes the noise reduction chips, both Dolby and DNR (dynamic noise reduction).

Each family has certain attributes which must be understood in order to properly troubleshoot it. To fully cover every IC in every family that is used in TVRO is impossible for an overview type of book like this, but a general discussion of each type follows with specific examples of its usage.

Transistor Transistor Logic

TTL is a digital family, which uses a +5 Vdc power supply. The signals associated with the TTL family are square waves with a change in state occurring at about +2.3 Vdc. A voltage above that point is high and a voltage below that point is low. Figure 17-12 shows the voltage relationships between high and low. TTL chips draw a fair amount of current, so a subfamily was created called the LS or Low-power Schottky family, which draws much less current. Another subfamily is the S or Schottky family, which is faster than the standard TTL family of chips.

TTL parts are identified by the 7400 series numbering, while the low-power Schottky chips are known as the 74LS00 series and the Schottky IC's are known as the 74S00 series. Do not substitute between families, if a 74S00 is used it must be replaced with a 74S00. The same thing goes if the chip is a 7400 or a 74LS00.

One usage for a TTL chip is in the divide by two

NE564 PLL demodulator. The divide-by-two is sometimes accomplished by a 74S74 dual D flip-flop. The 74S74, shown in Fig. 17-13, is used to divide the 70 MHz down to 35 MHz before being fed into the 564. Because the output is a squarewave, the divider also functions as a limiter.

The incoming i-f signal (at 70 MHz) enters the IC on pin 3, which is the Clock Pulse input of flip-flop #1. The signal is conditioned for the proper dc level by trim pot RV10, and by clamping diodes D16 and D17. The flip-flop changes state once for every two state changes on the input, thus the output signal is divided in half and is centered around 35 MHz. Since the maximum clock frequency is 125 MHz for the 74S74, it is running at a little over half speed and thus works very nicely.

Complementary Metal Oxide Semiconductors

CMOS (pronounced see moss) has supplanted TTL in the newer computers because of its low current requirements. It can also be powered on almost any voltage from +3 to +15 Vdc. Speed suffers as the power supply voltage is lowered, so most designs run CMOS at +12 to +15 Vdc. Because of its low current draw CMOS is rapidly becoming the choice of almost every circuit designer for anything requiring digital logic. CMOS chips are identified by their 74C00, 4000, 4500 and 14500 series part numbers.

CMOS chips range from simple digital or analog switches like the 4066, to LSI microprocessor chips like the Z80 (which is actually technically NMOS). There is a CMOS chip that duplicates the logic functions of almost every TTL 7400 series chip.

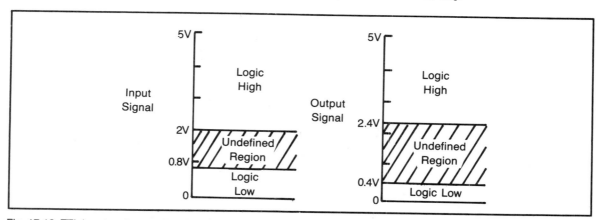

Fig. 17-12. TTL input and output logic levels. A signal in the undefined region can cause either a high or a low depending on clocking signals and the type of chip involved. CMOS circuits have similar undefined regions between high and low states, but they are dependent upon the chip's operating voltage.

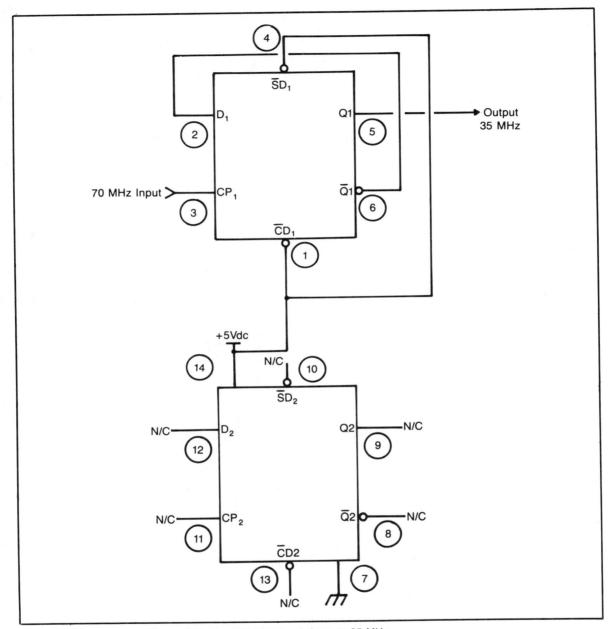

Fig. 17-13. TTL flip-flop used to divide the 70 MHz signal down to 35 MHz.

Emitter Coupled Logic

ECL is an older logic family that is well-known for its speed. It has found extensive use as limiters, dividers and i-f amplifiers in many TVRO receivers because of its speed and pseudo-analog behavior. It uses a +5 Vdc power supply. An ECL chip is identified by the 10,000 or MC1600 series numbering.

The MC10114, 10115, and 10116 are line receiver ICs designed for computer bus systems. They are used as 70 MHz amplifiers and as limiters in many satellite

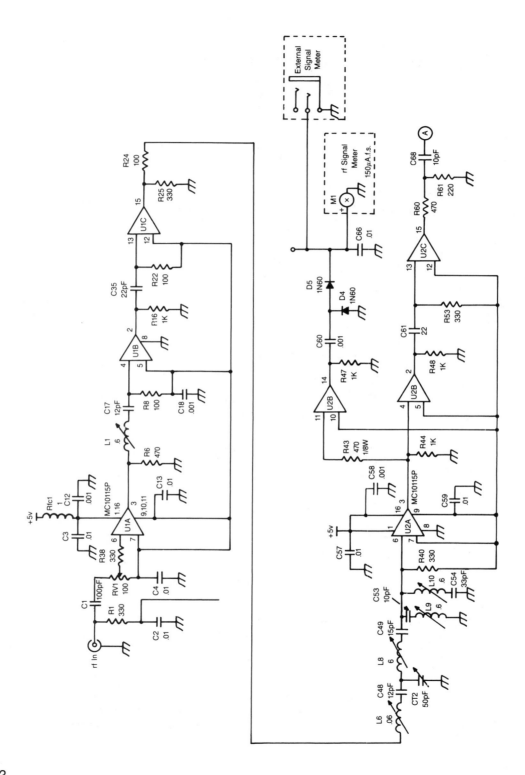

Fig. 17-14. Examples of an ECL chip (10115) being used to provide i-f amplification (U1a, U1b and U1c) as well as limiting (U2b and U2c). Courtesy of Gould/Dexcel.

receivers. One example is shown in Fig. 17-14, which is the i-f section of a Gould/Dexcel DXR-900. Here a MC10115 is used to drive the i-f filter and the signal meter. A second 10115 is used as the limiter.

Another common ECL chip is the MC1648, which is a VCO (voltage controlled oscillator). It is used in the R20 receiver from Amplica, as the tunable LO for the balanced audio demodulator. Figure 17-15 shows the audio circuit from an R20. Baseband video is input on pin E3, which is fed into pin 1 of the MC1496 (U7). A tuning voltage is input to pin 3 of the 4558 (U12). It is mixed with an afc voltage from pin 7 of the 3089 chip (U9). The resulting voltage is amplified and sent to the MV2209 varactor diode, which is used to tune the VCO. The VCO's output is 10.7 MHz higher than the desired channel's frequency. It is fed to pin 8 of U7. The resulting i-f is fed to U9 via the 10.7 MHz ceramic filter. U9 is a complete FM radio demodulator. Audio is output from pin 6 of U9. It is deemphasized and amplified to drive the rf modulator and audio output.

Linear chips

Linear chips are used throughout almost every TVRO receiver. They are used to amplify the i-f, detect the video, amplify the video and audio. They function as matrix decoders, as tuning voltage generators, as display comparators, and as voltage regulators.

There is no one numbering system applied to linear chips. Some of the identifying prefixes are the XR200 and XR400 series from Exar, the CA3000 series from RCA, the LMxxx series from National, the uAxxx series from Fairchild, the RCxxx series from Raytheon, the MCxxx and MC1xxx series from Motorola, the NExxx series from Signetics and the TLxxx series from Texas Instruments. In most cases chips with similar numbers but different prefixes will work as replacements, thus a uA747 would be equivalent to a LM747, RC747, or a MC1747.

One thing to look out for in ordering linear ICs is that many are available in many package types. The package is usually identified by a suffix. A K suffix means that the case is a TO-3, a T suffix means a TO-220 case, an AC suffix equals a TO-92 case, an H is a TO-5, and an N is the standard DIP (Dual Inline Package).

Some ICs will have two suffixes, as in LM733CN. The C specifies that the operating temperature is from 0 to +70 degrees centigrade instead of the normal −55 to +125 degrees centigrade specification that would be implied without the secondary suffix. The N still means that the IC is in a plastic DIP package.

Some of the more commonly used linear ICs that should be kept in stock are the LM733 video amp (or NE592, which is a pin for pin replacement), LM4558 (or equivalent 1458) and LM747 dual op amps, LM723 adjustable regulator, LM7805, 7812, 7815 and 7818 positive voltage regulators, LM7912 and 7915 negative voltage regulators, NE555 timer, LM1496 balanced demodulator, MWA120 i-f amplifier, LM1889 rf modulator, 741 op-amp, and the NE564 PLL demodulator.

Appendix A contains pin out drawings for commonly used ICs found in the most popular TVRO receivers. In some cases, a block diagram of the chip is also given.

Static Protection

Some MOSFETs and some CMOS chips are not gate-protected, meaning that they may be destroyed by static. They must be handled with care. Keep all MOSFETs and CMOS parts in static resistant bags (identified by their pink color), or keep the leads shorted together by inserting them into anitstatic conductive material, shorting the leads together with aluminum foil or by twisting them together until they are ready to go into a circuit. Before handling static sensitive parts, discharge any static charge that may have built up on your body by using a grounding strap. Use only soldering irons that have a grounded tip. Table 17-1 summarizes how to handle static sensitive parts.

HYBRID COMPONENTS

Hybrid components are made up of a combination of discrete parts like transistors, ICs, resistors and capacitors, which have been encapsulated into one package. These are used for i-f amplifiers like the MC5801 and i-f filters. They are basically tested like ICs in that they must be treated like black boxes. All one can do is check for the proper supply voltage and bias voltage, and that the input signal is correct. If everything checks out OK but the output signal is incorrect, then the component should be replaced.

SURFACE ACOUSTIC WAVE FILTERS

There is one main source for the SAW components used in TVRO receivers, CTI (Crystal Technology). They manufacture many different SAW filters and resonators that are used as i-f filters, delay lines, and oscillators in rf modulators and down converters.

SAW filters are available in many bandwidths. Table 17-2 is a cross listing of CTI part numbers to i-f bandwidth. These part numbers are usually stamped on the top of the metal case. To decrease or increase the i-f band-

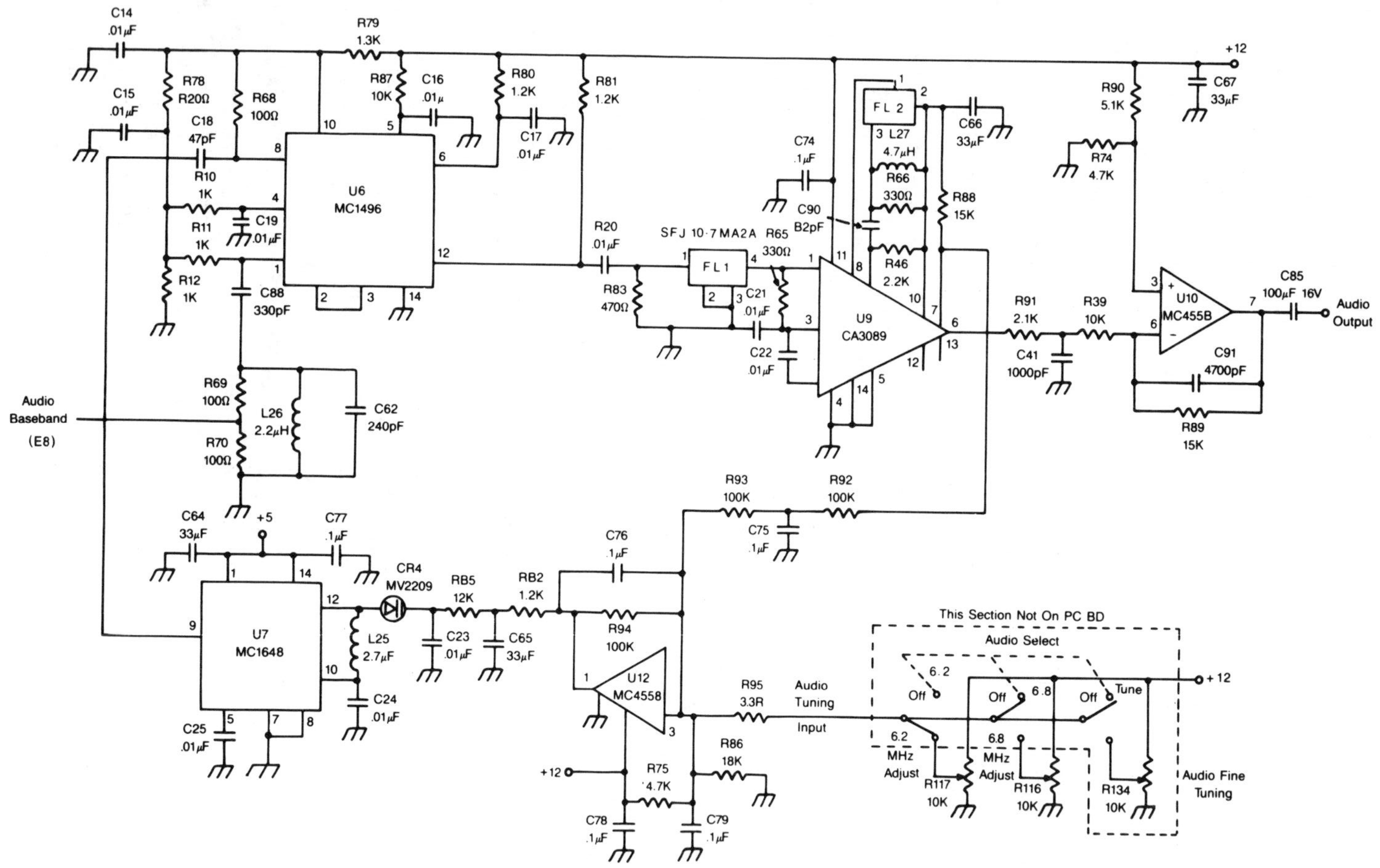

Fig. 17-15. Example of an ECL voltage controlled oscillator, the MC1648, being used in an audio tuning circuit. Courtesy of Amplica.

Table 17-1. Procedures for Handling CMOS and MOSFET Parts.

Handling Procedure	Use Conductive Material	Ground to a Common Point
Handling equipment	X	
Metal fixtures and tools		X
Handling trays	X	X
Soldering irons		X
Table tops	X	X
Service personnel		X *
General device handling		X *

*Use grounding straps in series with a 470K ohm resistor to ground.

Note: In dry environments (less than 30% humidity), static accumulation is much greater and the above precautions take on greater importance, but they should be strictly adhered to regardless of the humidity. Even though most of today's MOSFET and CMOS devices are gate-protected, they can still be damaged if improperly handled. By following the above precautions, you are ensuring that the devices will not be damaged by your handling.

width, the SAW filter need only be changed, but in most receivers the i-f strip is optimized for a specific pass band and for a specific SAW filter and thus may not react properly to the different component.

SAW filters consist of a thin quartz, lithium niobate, or berlinite wafer that is cut to enhance the piezoelectrical qualities of the crystal. A thin metallic layer is deposited onto the surface of the crystal. Transducers, in an interdigital pattern or in a multiphase unidirectional pattern, are then engraved into the metal by microphotolithography and chemical processing. The transducers are connected to the input and output pins of the device either directly or through a matching and phasing network. Figure 17-16 contains drawings of the two types of SAW filters.

A SAW filter works by converting the electrical input signal into an acoustic wave that is propagated across the surface of the crystal. The pattern that is laid down on the crystal, as well as the crystal itself, effects the frequency response of the acoustic wave, resulting in a bandpass filter. The signal is converted back into electrical energy by the second transducer.

The first generation of TVRO SAW filters had insertion loses of up to 30 dB and thus needed an extra gain stage in the i-f strip. Present designs keep the insertion loss to only 3 dB to 10 dB. There are also many new filters now available for higher i-f frequencies (in the 500 MHz to 600 MHz range).

Table 17-2. CTI SAW Filter Numbers and Parameters.

CT 1 Part #	Center Frequency	Saw Filter Width (−3 dB Points)	Loss in dB
139	70	16	25
138	70	21	25.5
134	70	25	26.5
132A	610	28	28
133	70	31	28.5
132B	610	35	30
131	70	36	30

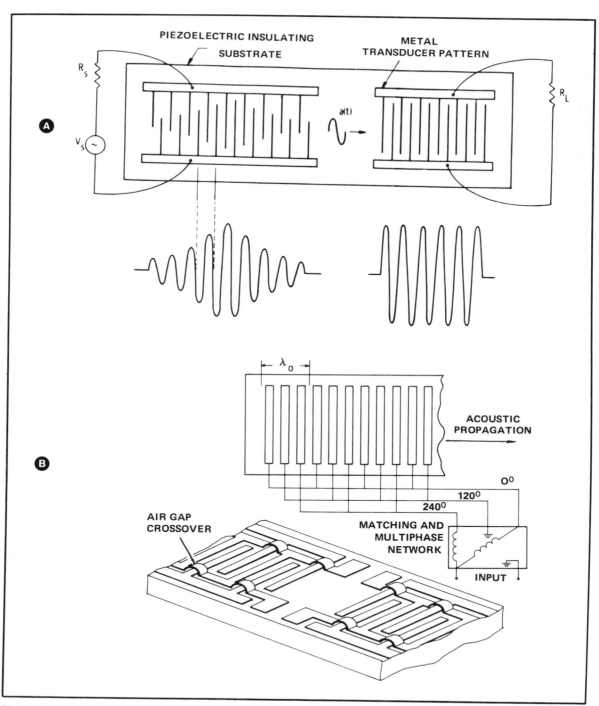

Fig. 17-16. A SAW or Surface Acoustic Wave of the first generation that used bidirectional couplers is shown in A. A second generation SAW device with unidirectional transducers is shown in B.

Product Information

T HIS CHAPTER CONTAINS INFORMATION ON PROD-
uces from Amplica, AVCOM, Channel Master,
Conifer, Dexcel, Drake, Luxor, STS, Sat-Tec, and
Winegard.

AMPLICA

Amplica's first receivers were the models R-10 and R-20.
They were sold in various formats including as a total
system with a dish, feedhorn, LNA, and DC.

R-10 Receiver

The R-10 receiver has one main board (830086) for
the i-f strip, video detection and processing, and audio
detection and processing. There is also a board that the
front panel switches are mounted on (830106). The wiring diagram for the R-10 is given in Fig. 18-1. The R-10
block diagram is shown in Fig. 18-2.

The Amplica receiver has a very wide input range
because of the automatic gain control (agc) circuit which
feeds some of the filtered signal into an amplifier that
drives the signal strength meter as well as the pin diode
on the input (CR9). This diode lowers the signal level coming into the receiver to compensate for short cables or
high-gain LNAs. On longer cable runs or when there is

less gain in the system, the diode is turned off and the
signal is not attenuated. From the i-f strip the video is
detected by a divide-by-four PLL circuit using a NE564
chip.

The i-f schematic is shown in Fig. 18-3. The 2SC2876
transistors are used for i-f gain. There is some fine tuning of the i-f filter by C107 and C108, but it is pretty much
a fixed i-f strip. The agc level (R111) and the meter level
(R115) are independently adjustable.

The limiter, demodulator, video processing, and tuning circuits are shown in Fig. 18-4. Limiting is accomplished by running the i-f signal through the U3, a 10116
ECL chip. Since the 564 chip (U2) is running at 17.5 MHz,
the 70 MHz signal must be divided by four. This is done
by the two 10131 ECL dual flip-flops. The video circuit
is pretty standard. To unclamp the video, CR2 is removed
and a 47K resistor put in its place. R67 is the video level
control. The video frequencies cannot be adjusted without changing circuit values.

The circuit's most unique feature is the 566 oscillator
circuit. It puts out a variable voltage to scan the channels. Main channel tuning is accomplished through a pot
(R128). There are two internal tuning voltage presets for
setting the top and bottom channel's tuning voltage.
These are R112 and R113.

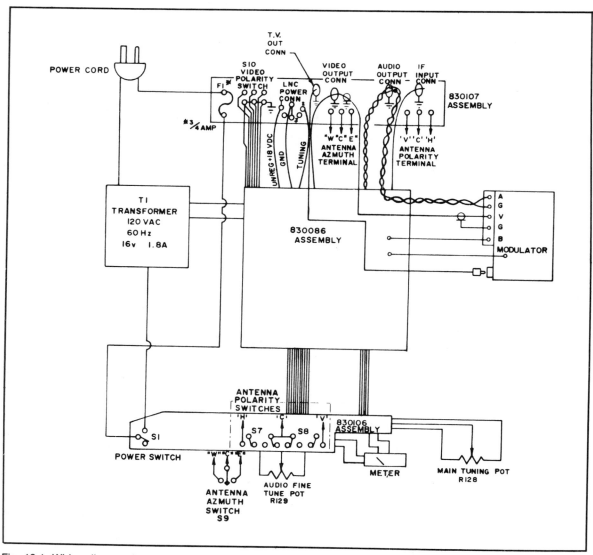

Fig. 18-1. Wiring diagram for the Amplica R-10 receiver. Courtesy of Amplica.

Figure 18-5 is the audio demodulation and processing circuit. It uses the CA3089 i-f amplifier and FM detector chip as its main component. The input to the CA3089 is one audio channel centered at 10.7 MHz. The channel is tuned by varying the frequency on pin 8 of U6 (the 1496 balanced demod chip). This frequency is mixed with the baseband signal coming into pin 1, resulting in the 10.7 MHz output.

The 1648 is an ECL VCO chip. Its frequency output is controlled by the capacitance of varactor diode CR4,

which in turn is controlled by the voltage from U10. U10's output is a combination of the audio afc voltage from pin 7 of U9 and the voltage from the audio tuning pots. R116 and R117 are presets for 6.2 and 6.8 MHz, although in most installations, the 6.2 MHz preset is set to 6.62 MHz (for MTV). R134 is the variable audio tune control.

The receiver uses a simple full-wave rectifier, IC regulator power supply that outputs three voltages: unregulated +15 to +18, a regulated +12, and a regulated +5.

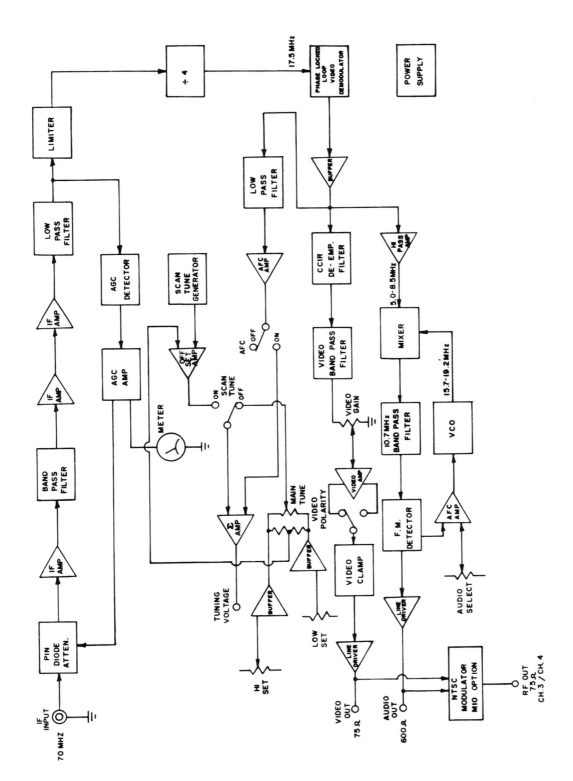

Fig. 18-2. Block diagram of the Amplica R-10 receiver. Courtesy of Amplica.

239

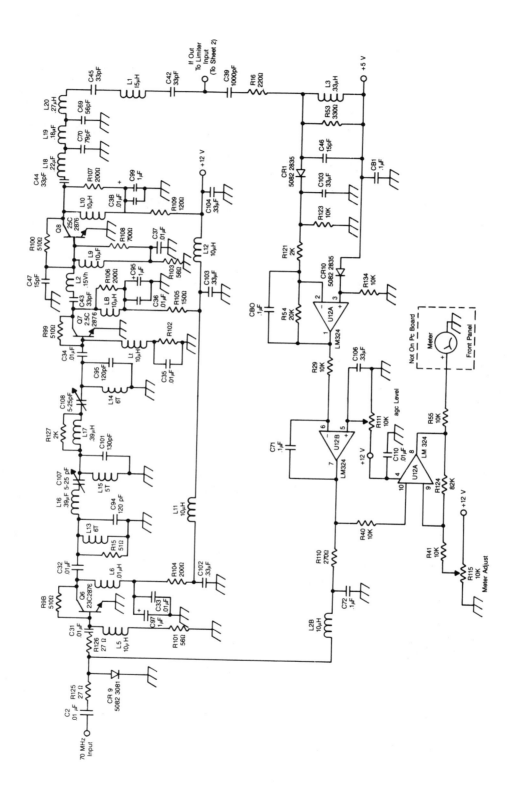

Fig. 18-3. R-10 i-f amplifier and filter schematic (sheet 1). Courtesy of Amplica.

240

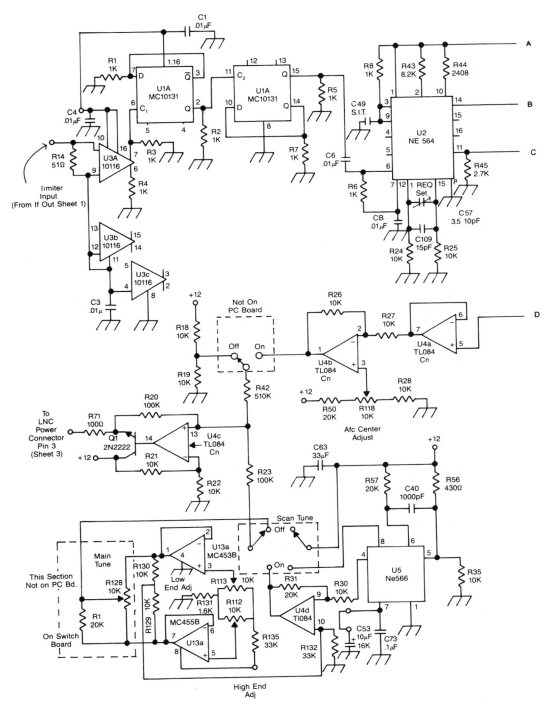

Fig. 18-4. R-10 limiter, demod circuit, video processing, and tuning schematic (sheet 2). Courtesy of Amplica.

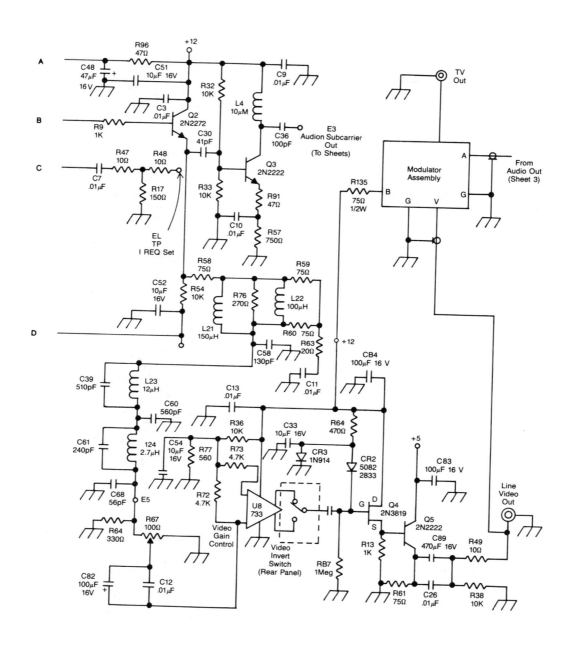

Fig. 18-4. (Continued from page 241.)

242

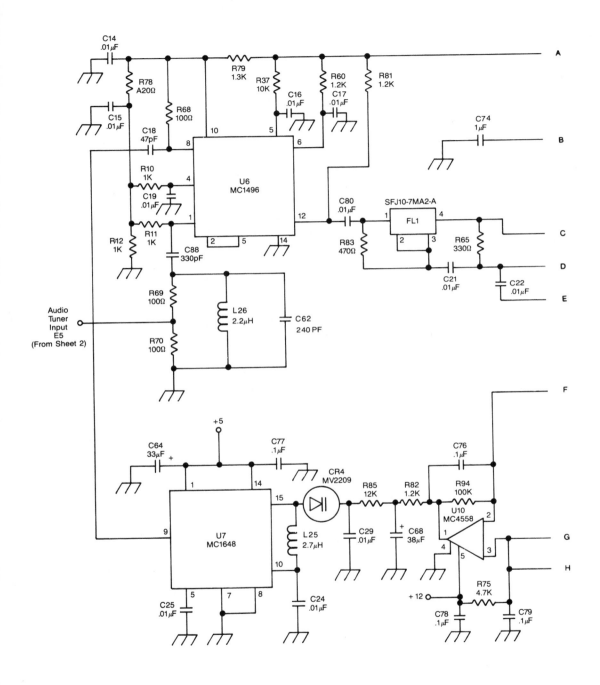

Fig. 18-5. R-10 audio processing and power supply (sheet 3). Courtesy of Amplica.

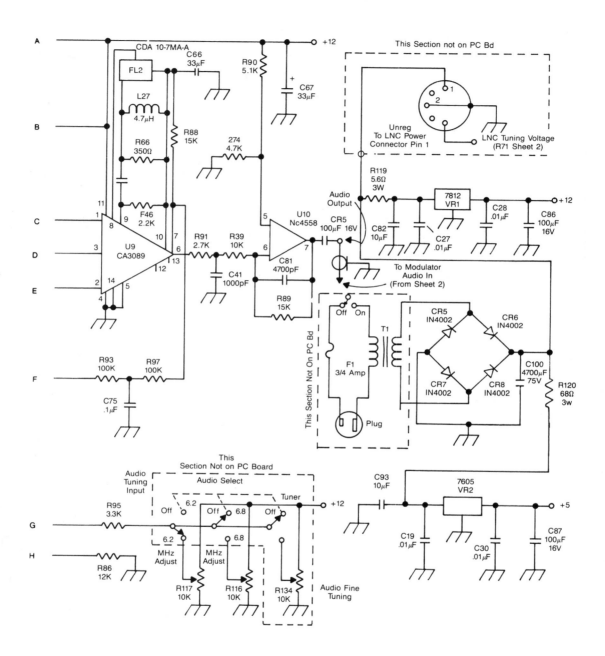

Fig. 18-5. Continued from page 243.

R-20 Receiver

The R-20 receiver featured an infrared remote control and a built-in modulator. Figure 18-16 is the wiring diagram for the R-20. It consists of five PC boards. The 830191 drawing contains the power supply, i-f, video, and audio sections. The block diagram in Fig. 18-7 is very similar to the R-10's in the i-f, video, and audio sections. The R-20 uses a Pulse Averaging discriminator at 17.5 MHz instead of the PLL circuit that is used in the R-10. The other difference is the additional circuitry that is used to detect the infrared remote, derive the tuning voltage, and display the channel number.

The schematic is shown in Fig. 18-8 through Fig. 18-11. The infrared detector circuit (830194), the LED readout board (830189), and the channel selector switch assembly (830190) are shown in Fig. 18-12. The audio tuning and meter board assembly (830193) is shown in bits and pieces throughout Fig. 18-8 to 18-11.

The i-f section, shown in Fig. 18-8 is almost identical to the R-10's except for the i-f filter, which uses a 25 MHz wide SAW filter. The only adjustments in the i-f are the agc level (R161) and the meter sensitivity (R160).

The video detection and processing circuitry is shown in Fig. 18-9. The limiter and divider circuit is the same as that used in the R-10, but the method of detecting the video is slightly different. It is a pulse averaging discriminator. It used to be known as a gated-beam detector (for those that are familiar with tube circuitry).

Its output is a pulse-width modulated signal. The width of the pulses are related to the frequency of the incoming signal. As the signal goes above 17.5 MHz, the pulses get wider, and as they go below the nominal center frequency they get narrower. These pulses are run through an integrating circuit, the output of which is a dc voltage whose average amplitude varies with the pulse widths. The narrower the width the lower the voltage, and the wider the pulse the higher the voltage. After the signal is low-pass filtered by the inductors and capacitors between Q7 and Q8, it is essentially the same as the original video signal. The rest of the video processing circuit is virtually identical to the R-10.

The audio detection circuit and power supply are shown in Fig. 18-10. The main difference is in the power supply. There is an extra + 12 volt regulated source (VR3) to supply the remote-control detector circuitry. VR3 stays active at all times, even when the receiver is turned off. The other parts of the power supply, VR1, VR2, and the LNC power cable are turned off by Q16. If the base of Q16 is pulled low, then it turns on. It is pulled low by

Q15, which is part of the remote-control circuit shown in Fig. 18-11.

To cut down on turn-on temperature drift in the LNC or DC, it is a good idea to leave them powered on at all times. This can be easily done if the LNC power wire is moved from its connection to R29 to the output of the rectifier bridge. This way when Q16 is turned off and the receiver is off, the LNC will still be powered, since that point is always active as long as the receiver is plugged into the ac line.

Figure 18-11 is the tuning, remote sensor, and display part of the R-20. It is radically different than the R-10 and in fact is probably more complicated than necessary. Basically, its purpose is to provide a tuning voltage to drive the two LED readouts to give a channel number display and to provide an interface for video fine tuning and audio tuning.

The heart of the logic circuit is the 555 oscillator (U6). It clocks all the logic together, and also pulses the LED readouts at a very fast rate so that they consume less current. The left side of the schematic is a scan tune circuit made up of U18 and U10d. U18 is a 566, which is a VCO that outputs a triangle wave at pin 4. The triangle wave is buffered by U10 and rectified by the four diodes, CR5-CR8, so that a positive going ramp voltage is produced. This drives pin 6 of the analog to digital converter IC (U19). As the voltage rises, the digital word that is output on pins 11, 12, 13, and 14 increases. This in turn drives U20, which is the remote-control interface IC.

U15, U16, and U17 make up nine SPDT switches (3 per package) that are controlled by inputs A, B, and C. The A input controls the X switch, the B input controls the Y switch, and the C input controls the Z switch. If they are high, then the X, Y, and Z outputs are connected to the X1, Y1, and Z1 inputs. Conversely, if A, B, and C are low, then the three outputs are tied to X0, Y0, and Z0. A is pin 11, B is pin 10, and C is pin 9. The outputs are on pin 14 for X, pin 15 for Y, and pin 4 for Z. The six inputs are: X0 = pin 12, X1 = pin 13, Y0 = pin 2, Y1 = pin 1, Z0 = pin 5 and Z1 = pin 3.

To give an example of how they work, let's turn the receiver on and off and follow what happens. If we use the front panel switch, then the Z0 input (pin 5) to U16 will be grounded (or low). If C is low, then Z is connected to Z0. This will only occur if the local/remote switch is in the local position. If it is in the remote position, then C will be high, which will mean Z will be connected to the Z1 input (from U20). Thus we could get no response from turning on the receiver's front panel on/off switch if the receiver is in the remote mode.

When you flip the remote switch to local, the C in-

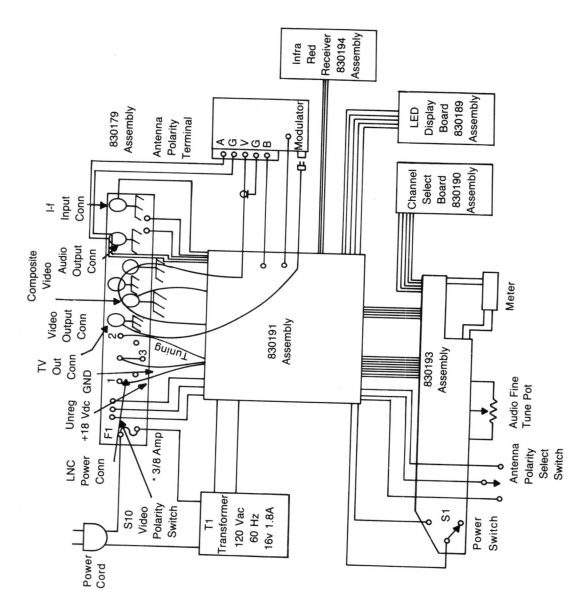

Fig. 18-6. R-20 wiring diagram. Courtesy of Amplica.

246

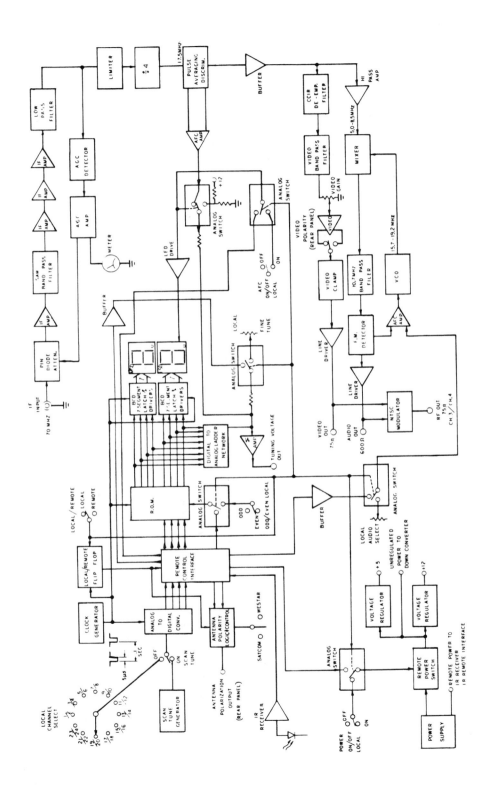

Fig. 18-7. R-20 block diagram. Courtesy of Amplica.

247

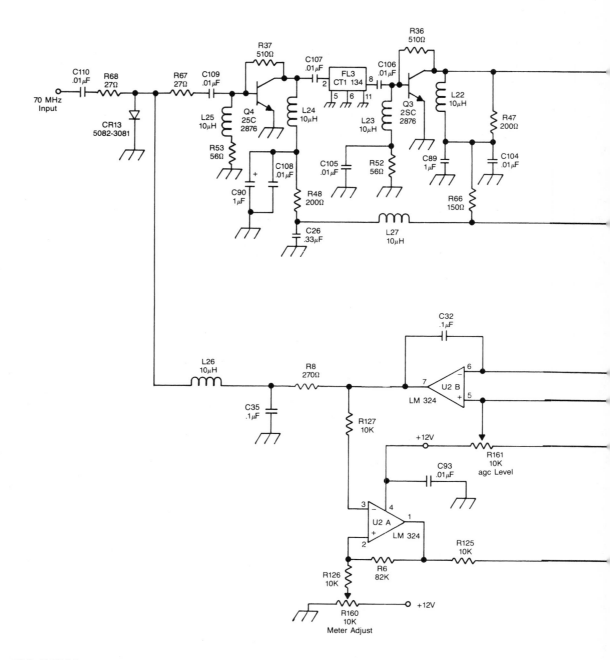

Fig. 18-8. R-20 i-f amplifier and agc schematic (sheet 1). Courtesy of Amplica.

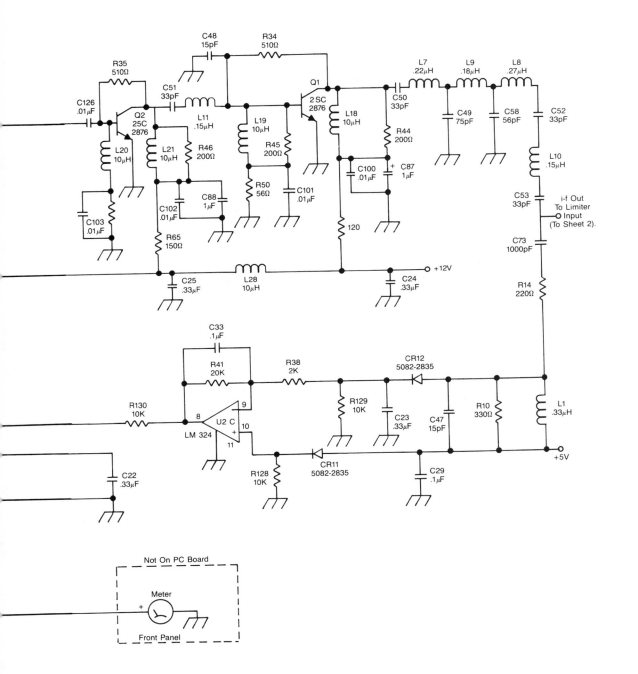

249

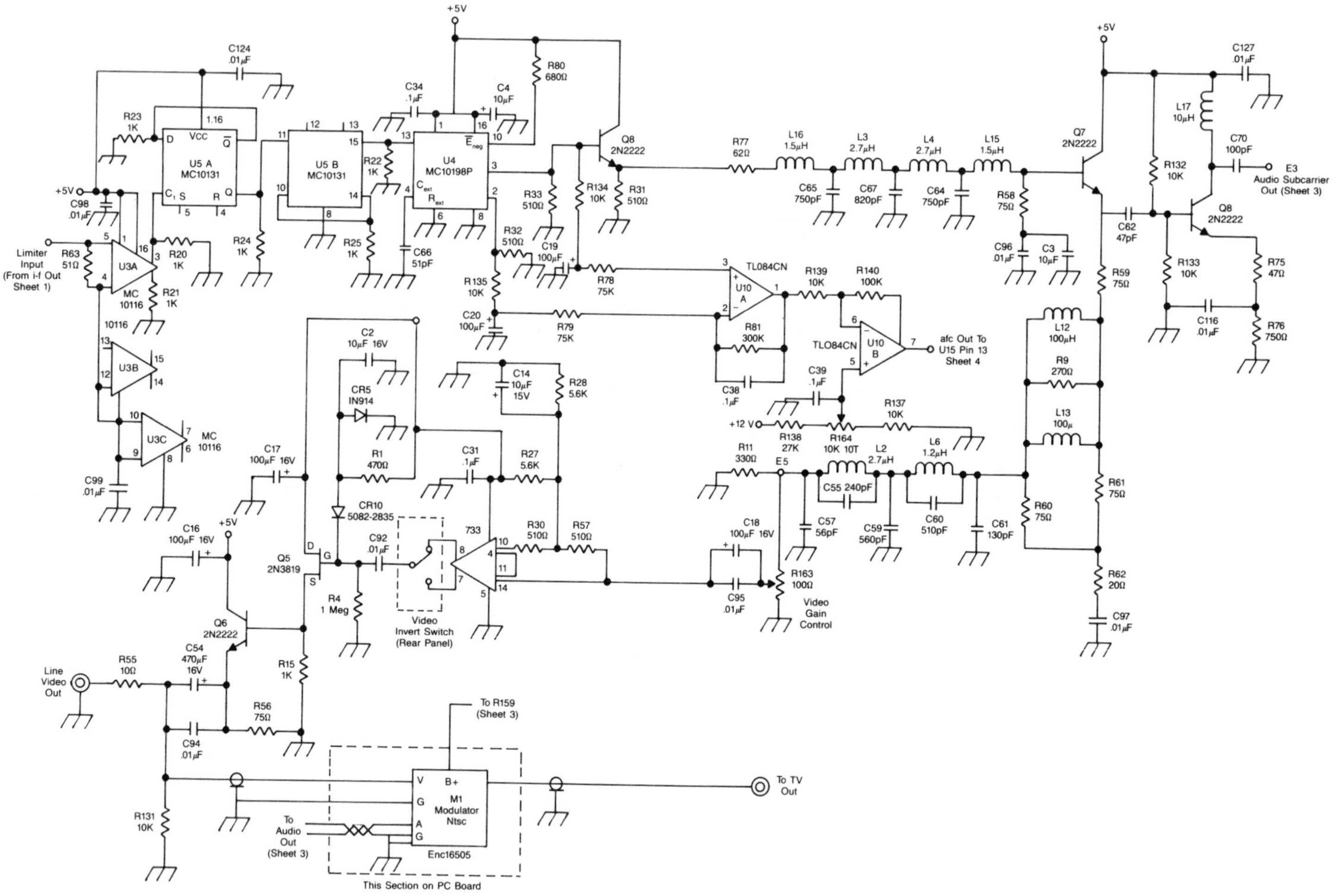

Fig. 18-9. R-20 video processing schematic (sheet 2). Courtesy of Amplica.

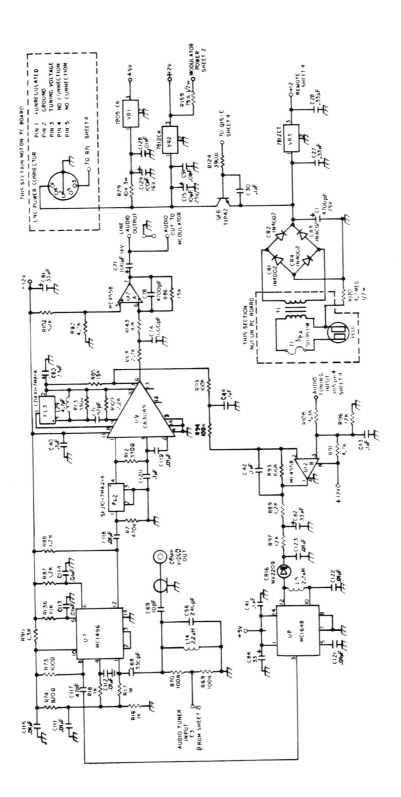

Fig. 18-10. R-20 audio tuner and power supply (sheet 3). Courtesy of Amplica.

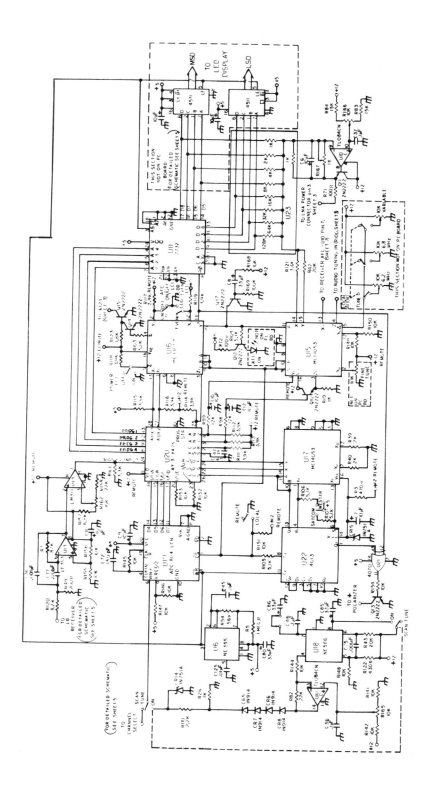

Fig. 18-11. R-20 logic, display, tuning, and remote control schematic (sheet 4). Courtesy of Amplica.

put (pin 9) goes low, which shorts pins 4 and 5 together. Since 5 was low, now 4 is also low, being that they are shorted together. This will pull the base of Q14 low, turning it off. This in turn causes the base of Q15 to go high turning it on. The collector of Q15 is now at ground potential which turns on Q16 (from Fig. 18-10).

To turn off the receiver, the power switch is opened so that it applies a high signal to pin 4. This turns on Q14, which in turn shorts out the bias on Q15, turning it off. Q16 is then turned off, stopping the voltage from going to the LNC, to VR1, or to VR2.

The tuning voltage is derived from the digital lines coming from U11, a 2732 ROM (Read Only Memory) chip. Lines D1 to D8 are the data lines. They output up to 256 different values, depending on which lines are high and which are low. These are fed to the display circuit through the 4511s and also to the tuning voltage amp through the 8 resistors that make up U23. They scale the voltage to the digital word. They are a digital-to-analog converter (ladder network in Fig. 18-7). The output voltage is summed with the afc voltage and fine-tune voltage and is amplified by U10c and sent to the LNC. The highest and lowest tuning voltages are determined by R166 and R167.

Figure 18-12 shows the channel selector switch, the digital readout driver circuit, and the infrared detector and amplifier circuit. Figure 18-13 explains how to set the high and low fine tuning pots on the R-20 and R-20A receivers.

AVCOM

AVCOM was started by Andy Hatfield, an engineer who designed a homebrew receiver for himself back in the late 70's. AVCOM is not a big name, but it is still considered the cream of the crop in home TVRO receivers. They build basic industrial strength equipment without frills but with a custom-made quality that is hard to beat anywhere else.

COM-2A or 2B Installation Tips

1. When installing the RDC-11 down converter use RG-214 between the LNA and DC if the cable run is less than 45 feet. Use RG-217 for runs of up to 100 feet. Hardline is required for longer runs. A signal level of −55 to −25 dB is required to properly drive the DC.

2. The DC that is used in similar to that made by Drake in that it uses a −7 to −12 Vdc for tuning. In a pinch, a Drake DC can be substituted for test purposes. As in the Drake DC, the box is not waterproof and so must be placed in a weatherproof enclosure to protect it from moisture. Be sure to use drip loops on all cables entering the box or its purpose will be defeated.

3. The DC can pass the dc voltage up the LNA cable by connecting a jumper between the +23 and LNA terminals on the terminal block. If the DC is used in a multiple receiver installation or if an isolator is used between the LNA and DC, then dc must NOT be allowed to pass on through to the TO LNA connector. In this case, there should not be a jumper between the +23 and LNA terminals.

4. There should be at least 3 dB and not more than 10 dB of loss between the DC and the receiver. The receiver's threshold extension circuit performs best under these signal levels. This can be achieved by using between 120' and 275' of RG-59 between the DC and the receiver. Longer cable runs will require RG-6 or even RG-11 to be used. Shorter cable runs will require an attenuator that can pass dc voltage to make up the difference. An inexpensive attenuator is a 2 way splitter that is modified to pass dc. One output port is terminated into 75 ohms. A 2 way splitter loses about 3 dB.

5. Use a shielded #18 gauge wire for the +23 Vdc line. It is best to include two extra wires in this cable so that an external meter can be used at the dish for dish and feed peaking. A Radio Shack meter movement (270-1751 or other 100 microamp meter movement) in series with a 50K ohm pot, can be connected to the SIGNAL and GND terminals on the back of the COM-2A (or B) with up to 300 feet of #20 gauge wire. Adjust the pot to keep the meter reading between half and maximum.

6. For multiple installations, isolators and extra shielding between down converters will be required to prevent LO crosstalk.

Field Adjustments For The COM-2A And 2B

i-f Gain. Adjusts the signal level coming into the receiver. The control is located on the circuit board in the right rear corner of the receiver (near the coax jumper). Ccw increases the i-f gain. Normal position is mid range. Adjusting this control will also influence the signal meter reading.

Signal Strength Meter Adjustment. The Meter Sensitivity control on the front panel of the 2A (rear panel of a 2B) allows the meter to be used as a full scale peaking instrument. If properly set, the meter will allow signal variations of 1/10 dB to be observed. If the receiver is being properly driven, then the meter should smoothly follow every change in dish positioning. If the meter does not, then it is probably being overdriven. Adjust the i-f

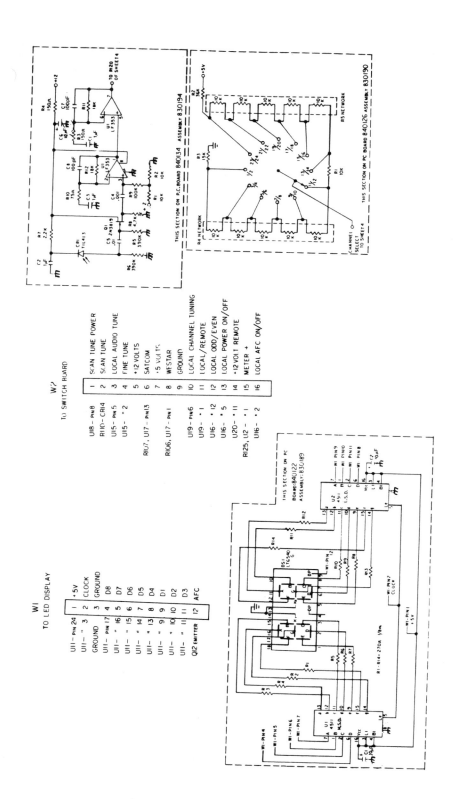

Fig. 18-12. R-20 LED readout, channel selection, and remote receiver circuit (sheet 5). Courtesy of Amplica.

Amplica, Inc. 950 LAWRENCE DRIVE, NEWBURY PARK, CA 91320 (805) 499-2621 • TWX 910-336-1291

R-20, R-20A SATELLITE RECEIVER HANDBOOK ADDENDUM

Section: 5 - 1 Receiver Adjustment Procedures.

End Point Adjustment Procedure.

The purpose of this adjustment is to center the remote fine tuning and the front panel fine tuning while calibrating the end stops all in one action.

1. Plug the receiver into 120 VAC wall receptacle. This sets the internal fine tuning to the center point.
2. Push the Remote Button.
3. Push EVEN; Push 24. DO NOT USE VIDEO FINE TUNE. (We assume at this point, the system is operating and on a 24 channel satellite).
4. Push AFC - OFF (DO NOT USE VIDEO FINE TUNE).
5. Adjust Hi-end control for best reception on transponder 24. See Fig. 3 in Section 5-1 of the Handbook.
6. Push 2. (AFC still OFF). DO NOT FINE TUNE.
7. Adjust Lo-end control for best video on transponder 2.
8. At this point both the internal and external tuning center points are calibrated to the Hi-Lo end points.

BP:DAK22

Fig. 18-13. R-20 alignment procedures. Courtesy of Amplica.

gain control to lower the meter reading (by turning it cw) or insert more attenuation between the down converter and the receiver.

Video Level. Using a waveform monitor the video should be set to 1 volt P-P. If it is high or low, then the video level can be adjusted by the trim pot that is located on the video board in the center of the receiver about five inches behind the front panel.

afc Balance. Trim pot located on the tuning board (the 3″ × 3 1/4″ circuit board located at the center rear of the receiver). It must be adjusted following this procedure: 1. Set the three position switch (scan tune, operate, afc off) to the afc Off position. 2. Tune in a weak transponder and adjust for equal salt and pepper sparklies. A below threshold signal level is necessary for proper adjustment. (If your signals are too clean, the dish can be moved off the satellite slightly to increase sparklies) 3. Switch to the Operate position without changing the tuning. If the picture changes at all, adjust the afc Balance control until the same picture (equal salt and pepper sparklies) is tuned in.

Tuning Dial Calibration. (For receivers SN #8849 and up) Set the receiver on channel 14. If the knob does not point midway between 12 and 16, loosen the knob's set screw and set the pointer to that position. Tighten set screw. Tune to channels 2 and 24. Note if the knob's pointer is positioned correctly. If it is not, then there is a calibration pot to widen or narrow up the overall span. This pot is located behind a hole on the bottom of the COM-2A remote. On the 2B, the cover must be removed. Adjust so that when both channel 2 and 24 are tuned in, the 2 and 24 marks line up with the pointer.

i-f Strip and other Internal Adjustments. Contact the factory if any other adjustments are suspected to be out of alignment. They will inform you of the proper test equipment necessary as well as the proper procedures required to maintain high quality video and audio performance.

CHANNEL MASTER

The following technical bulletins (Fig. 18-14 to Fig. 18-21) are from Channel Master. Some of them are generally applicable to all receivers, while some of them can be used for any of the Drake manufactured receivers (which for CM is the model 6128). The bulletins dealing specifically with the Drake-built receivers are Fig. 18-17, Fig. 18-19, and 18-21. The rest of the bulletins can be used for any system (if model numbers are thought of in generic terms).

The 6128 is virtually identical to other Drake-made receivers and thus most troubleshooting, circuit tips, and schematics from Conifer, Drake, and Winegard will apply to it. *Note:* The parts will be the same value but they will probably have different ID numbers.

CONIFER

The RC-2001 receiver is virtually the same as the Drake 24 (below serial #20,000) and the Winegard 7032, and so the receiver circuitry won't be covered here. Refer to the sections on those models for more information.

The biggest difference lies in the integral motor drive controller. The schematic of this circuit is shown in Fig. 18-22. It contains the polarity control circuit as well as the motor drive controller and positional feedback circuitry. It has a typical polarity control circuit consisting of a 555 IC in an astable oscillator configuration to drive a Polarotor I or equivalent PD.

The motor drive uses a potentiometer for feedback. This creates a voltage that is dependent upon the actuator's location. This is sent to the − input of the 741 IC. This is compared with the satellite selection voltage that is applied to the + input. The voltage difference between the + and − inputs is output on pin 6 of the IC. This voltage can be a negative voltage, zero, or a positive voltage depending on whether the desired position, as determined by the satellite select voltage, is equal to or is to the west or the east of the present location.

If the voltage is more positive than 0.6 volts, then Q1 (a 2N3904 npn transistor) is turned on. If the voltage is more than − 0.6 volts, then Q2 (a 2N4402 pnp transistor) is turned on instead. If it is almost zero (less than 0.6 Vdc yet higher than − 0.6 Vdc), then both transistors are turned off and the motor stops.

When the transistors turn on they supply a ground path for relays RY1 and RY2, which in turn switch power to the motor. If RY1 is turned on, then a positive voltage is applied to the upper motor terminal and a negative voltage is applied to the lower terminal, which retracts the actuator. Conversely, if RY2 is activated, then the positive voltage is applied to the lower motor terminal and the negative voltage is applied to the upper terminal, which extends the actuator. When the relays drop out, the motor terminals are shorted together which causes the motor to stop faster.

One difference in the motor drive circuit from most other simple controllers are the two preset locations that can be called up. These are labeled SAT A and SAT B. They put a preset voltage onto pin 3 of U1 whenever they are called up.

Figure 18-23 shows the motor-control board and part of the main video board in the Conifer receiver. The four diodes for the motor drive are located just below the two large filter caps. They are, left to right, D4, D7, D6, and D5. RY2 is the upper relay and RY1 is the lower relay. The four controls at the upper left, from top to bottom, are the horizontal skew (R2), vertical skew (R1), SAT A preset (R8), and SAT B preset (R11). Also notice the coiled delay line and i-f filter board located on the main board.

Troubleshooting

Most of the troubleshooting tips listed here also apply to most Drake receivers and to the Winegard and Channel Master receivers that have been built by Drake. All of the down converters, except for the very early ones from Drake (pre 1/83) use the same circuitry. The early models from Drake used two MWA110s for 70 MHz amplifica-

INFORMATION BULLETIN
CHANNEL MASTER SATELLITE RECEPTION EQUIPMENT

BULLETIN No. 1   DECEMBER 11, 1981   SUBJECT: **TROUBLE SHOOTING**

SYMPTOM	POSSIBLE CAUSE	SUGGESTED CORRECTIVE MEASURES
NO SOUND **NO PICTURE** 1. Gray screen changing channels	a. +15 VDC to receiver out	Measure +15 VDC in receiver @ +15V regulator out (pin #3 to ground = +32 VDC on control box)
	b. Receiver bad (Video demodulator defective)	Unit to be repaired or replaced
2. Light or Grey background with white sparkles	a. Antenna off satellite	Peak system on satellite—optional tuning meter will speed up adjustments
	b. No modulator output	LED on front panel should be on; substitute another modulator; run video and audio from receiver to a VCR or TV monitor.
	c. No LNA power +24 V regulator LNA cable open or shorted	Measure for +24 VDC @ pins 7 or 8 to ground (pin 9) Check for +24 VDC at RF out of receiver (where LNA cable connects) observe voltage at LNA cable right at LNA input (+24 VDC). If current can be measured, the LNA should draw approximately 120 to 140 Ma when operating normally.
NO PICTURE **SOUND OK** Picture noisy or completely gone	a. Video tuning voltage problem	Receiver/control repair or replace
	b. Bad video output cable	Cables should be tested for continuity- repaired or replaced.
	c. Bad modulator	Substitute modulator (see no modulator output above)
NO SOUND **PICTURE OK**	a. Audio cable bad	Test cable and repair/replace as necessary
	b. Modulator problem—4.5 MHz audio frequency out of alignment	Substitute modulator or adjust 4.5 MHz trimmer until audio is clear
	c. Audio not tuned to an active subcarrier	Insure that the audio tune on control box is adjusted in an effort to pick up sound
	d. Audio tuning voltage problem	Receiver/control repair or replace
	e. Bad audio circuit in receiver	Receiver/control repair or replace
SOUND DISTORTED **PICTURE OK**	a. Audio incorrectly tuned	Adjust audio tuning
	b. Modulator problem—4.5 MHz audio frequency out of alignment	Substitute modulator or adjust 4.5 MHz trimmer until audio is clear
	c. Defective audio demodulator in receiver	Receiver/control repair or replace
WEAK SIGNAL **SPARKLES**	a. Antenna not peaked	Adjust antenna for best signal or strongest signal on tuning meter
	b. Polarization incorrect	Adjust for strongest signal/best picture
	c. Poor view of satellite	Any obstacles blocking the line-of-sight path from antenna to satellite (trees, bldgs., fences, etc.) will degrade system performance.

Fig. 18-14. Technical Bulletin #1. Courtesy of Channel Master.

SYMPTOM	POSSIBLE CAUSE	SUGGESTED CORRECTIVE MEASURES
WEAK SIGNAL SPARKLES (Cont.)	d. Focal distance incorrectly adjusted	This is factory set but if in doubt, contact CMSS for proper length.
	e. LNA too noisy	LNA has high noise figure. The system may require an amplifier with a lower noise figure—such as going from 120°K to 100°K LNA.
CHANNEL TUNING OFF (Unable to pick up low or High End channels)	Tuning calibration off	Recalibrate using recommended procedure contained in manual.
CANNOT CHANGE CHANNELS	a. Channel tuning voltage problem	Defective cable—check all cables in receiver for good connection.
		Receiver has AFC Loop problem—requires repair/replacement
	b. Receiver has no tuning voltage to mixer.	Receiver need repair
UNIT JUMPING CHANNELS ON ITS OWN	a. Excessive or intermittent power line fluctuations	This problem can be alleviated by isolating the control box. Plug the control box into an isolating transformer and the transformer into the house voltage. This is a 120 VAC isolating transformer.
	b. Receiver VCO unstable	Receiver needs repair
ROTATOR DOES NOT TURN AT ALL OR ONLY ONE WAY	a. Lost voltage to rotor	Voltage at the rotor can be easily measured: From #1 or #2 to #3 of rotor should read approximately 24 VAC under the rotor load. This is with the H-V switch activated in H or V position. If no voltage trace back to pins 1, 2, 3 of receiver, then to control box (measure from pin #4—common to 12 or 14)
	b. Defective H-V Switch	Switch in control box needs replacing
	c. Microswitch type rotor has switch problem	Check microswitches in rotor for proper operation —have unit replaced if necessary.
	d. Defective rotor or rotor capacitor	Unit needs replacing
ROTATOR FAILS TO STOP	a. Microswitch type rotor: switches not engaging properly loose or bad connection	Check switches in rotor for loose connections, defective switch, etc. Replace if necessary.
	b. Control box has wiring short or shorted H-V switch	Test switch in control box for proper operation. Check for possible short in wires on switch
	c. Not functioning properly	Check position of "wiper blade" trigger for micro switches in bottom of rotator.
		Reposition and or tighten blade fastening screw.
RECEIVER CONTROL OVERHEATING	LNA shorting drawing excessive current. Should be drawing 120 ma to 140 ma.	Check current draw at LNA. Replace if necessary.

Fig. 18-14. Continued from page 257.

tion instead of the two 2SC2876 transistors that are currently used. The hybrid mixer circuit and LO circuit is the same on all designs.

If there are no signals or only weak signals coming from the DC, then the first thing to check is the input and output connectors. If moisture has entered any of the connectors and/or cabling, then the signal can be shorted out or the tuning voltage shorted out. If both connectors appear OK, and if the +15 Vdc line is OK, then the DC should be substituted. If the substitution of a known good unit solves the problem, then obviously the DC has problems.

INFORMATION BULLETIN

CHANNEL MASTER SATELLITE RECEPTION EQUIPMENT

BULLETIN No. 7	OCTOBER, 1982	SUBJECT: GENERAL TROUBLE SHOOTING GUIDE
SYMPTOM	**POSSIBLE CAUSE**	**SUGGESTED CORRECTIVE MEASURES**

SYMPTOM	POSSIBLE CAUSE	SUGGESTED CORRECTIVE MEASURES
Audio Buzz	1. No 75 ohm terminator in video base band output 2. Video drive level too high 3. 15V regulator inoperative	1. Place (replace) terminator in video out jack (see technical bulletin no. 17) 3160079 Terminator 2. Adjust R-94 to eliminate buzz without losing color (see technical bulletin no. 17) 3. Replace receiver
Distorted Audio (Hi Audio)	1. Audio output too high, or mistuned	1. Replace modulator. **NOTE: check output with VCR or spare modulator first, to be certain (see technical bulletin no. 17)** 5015005 Modulator
Low Audio	1. Audio output too low or mistuned	1. Replace modulator. **NOTE: check output with VCR or spare modulator first, to be certain (see technical bulletin no. 17)**
6.8 or 6.2 Audio Inoperative, Manual Tune OK	1. 6.8 and/or 6.2 audio presets detuned	1. Adjust 6.8 and 6.2 preset pots for correct setting (see technical bulletin no. 17)
Snowy Video	1. IF gain too low 2. IF coax wrong size 3. Low downconverter output 4. Low LNA gain	1. Adjust IF gain pot (rear of receiver) 2. Replace with correct size cable (see cable chart, information bulletin no. 5) 3. Replace downconverter (check voltage at downconverter for full 15 volts before replacing), retune channel pots (see technical bulletin no. 8) 5000161 Downconverter 4. Replace LNA
Center Tune Meter not Centered	1. Polarity reverse switch in wrong position 2. Cable compensation pot improperly adjusted 3. Channel pots off tune 4. Wrong size control cable	1. Set polarity switch correctly 2. Adjust cable compensation pot with fine tune control at 11 o'clock for center tune indication on meter (see technical bulletin no. 11) 3. Retune channel pots on affected channel (s) (see technical bulletin no. 8) 4. Replace with correct gauge (see chart, information bulletin no. 5)
Polarizer Intermittent	1. Bad connections	1. Check spade lugs at receiver, and connections at dish
Polarizer Sticks	1. Dirt, dust jamming mechanism	1. Spray lube (wd-40) in **probe side only** in small amounts around where probe protrudes through assembly.

USE TECH. BULLETIN No. 14 TO DETERMINE IF DEFECT IS IN CONTROL BOX OR JACK, BEFORE REPLACING ANY COMPONENTS.

SYMPTOM	POSSIBLE CAUSE	SUGGESTED CORRECTIVE MEASURES
Satscan Readout abnormal (ie: 000, 900, 999)	1. Sensor wires mis-connected 2. C1, C2, C11, C12 shorted, or leaky in control box	1. Check color code at both jack and control box and connect properly. 2. Replace C1, C2, C11, C12 (see technical bulletin no. 16) 5000174 Capacitor Kit
No Satscan Readout	1. Ribbon wire plug disconnected from LED readout 2. U1 defective (IC Chip)	1. Replug, add plug retainer assembly (see technical bulletin no.16) 2. Replace U1, (see technical bulletin no. 16) 5000175 IC Chip
Erratic or False Readout	1. Worm gear in Satscan jack loose 2. Sensor pot in Satscan jack loose	1. Replace and re-Loctite gear (see technical bulletin no. 15) 5000176 Worm Gear & Loctite Kit 2. Retighten pot screws (see technical bulletin no. 15)

NOTE: Channel Master does not suggest or authorize any other repairs than those contained in this bulletin. Any additional changes may void the warranty.

Printed in U.S.A. 10/82 (14067)

Fig. 18-15. Technical Bulletin #7. Courtesy of Channel Master.

INFORMATION BULLETIN
CHANNEL MASTER SATELLITE RECEPTION EQUIPMENT

BULLETIN No. 12   DATE: September, 1983   SUBJECT: Troubleshooting Procedures
                                                            Model 6127 Receiver

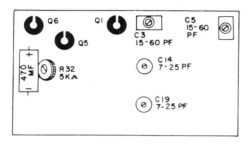

FIG. 1

RECEIVER
CHASSIS

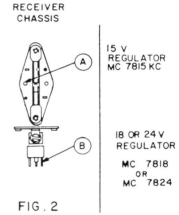

15 V
REGULATOR
MC 7815 KC

18 OR 24 V
REGULATOR

MC 7818
OR
MC 7824

FIG. 2

SYMPTOM

1. Vertical Jitter

2. Channel Drift

3. No Video
   No Audio

CORRECTIVE MEASURE

1. Increase video drive level (R32) in video demodulator (see Figure 1).

2. Adjust C14 and C19 in video demodulator while monitoring video (see Figure 1).

1. Replace channel select pot (R6) in control console. (Pot is 1K, 2 watt linear taper)

1. Check for +15 V output on Q1 (see figure 2, Point A). Replace if 15V is not present.

2. Check for +18V or +24V output of Q2 (see Figure 2, Point B). Replace if 18 or 24V is not present

3. Check for continuity of tuning voltage between receiver and control console. Voltage at center tap of channel select pot should be on pin 12 of 15 pin Joines plug at receiver. If continuity is not present 15 conductor cable is probably defective.

Fig. 18-16. Technical Bulletin #12. Courtesy of Channel Master.

4. Black Screen Audio OK	1. Check video amp (Q5) and video output amp (Q6) in video demodulator. Q5-C=15V B=13V E=12.3V Q6-C=15V B=12.3V E=11.7V (See Figure 1). Replace with 2N2222 if necessary. 2. Replace RF modulator
5. Intermittent Loss or Degradation of Video and/or Audio Quality	1. Check all cables: RG59, RG214 and 15 conductor control cable for possible moisture or bad connections. Replace if necessary.
6. Suspected Defective LNA	1. Put current meter in line at point B. Current should be approximately 120 mA. (See Figure 2). If current is high or lower than 120 mA, LNA is probably defective.

First remove the top cover and check for moisture damage. If there is any sign of moisture in the housing, then it is a good chance that the DC cannot be field repaired, as all the component's leads will corrode if left in water for any length of time. If moisture is not present or if it has not yet corroded the components (especially the resistors and capacitors), then hook it back into the system and check the internal power supply and tuning voltages.

The input supply must be at least + 15 Vdc at the input pin of the 7812. If it is lower, then the wiring is too small and it should be replaced with a larger gauge wire. If the wiring cannot be replaced, then unregulated voltage could be used to supply the DC. Move the wire that goes to the down converter dc voltage terminal to the input of the 7815 regulator instead of to the output. The voltage will be about + 22 Vdc, which will drop to about + 19 or + 20 Vdc by the time it reaches the DC. If there is no voltage at the output of U3 (the 7812 IC), then it is possible that the regulator was damaged by a lightning strike or by line transients. Luckily, it is typically the only failure.

Once the main supply voltage is verified, then check that there is no voltage on the hybrid circuit. If there is, then check that the 56 ohm resistor is OK and that it is properly soldered to ground. Check the bias on U3 and U4. It should be within a couple tenths of a volt as that listed on the schematic.

If the channels seem to jump around at random, then there is probably a cold solder joint somewhere in the LO circuit. Try resoldering U1 and U2 and the corresponding resistors. None of the holes on the board are plated through, so be sure to solder the leads on both sides of the board.

There should be a + 0.2 to + 0.3 Vdc bias on the junction of the 1000 pF caps and D1 and D4. If it is negative, then the 22K resistor is bad or there is a cold solder joint. D1 through D4 are 1SS99 diodes.

Typical tuning voltages are from − 4 to − 4.5 Vdc for channel 1 to − 9.5 to − 10.0 Vdc for channel 24. The tuning is tweaked by bending the horseshoe shaped coupling either toward ground (to lower the frequency) or away from ground (to raise the frequency). This is normally used only as a last resort.

On some receivers, there's no protection diode installed at the output of the 741 tuning voltage IC (U2). This is a 1N4005 diode that goes from the output (pin 6) and ground. The cathode goes to ground, thus shunting any positive voltage to ground. This should be added to all Drake-made receivers if it is not already present.

If the top channels tune in fine, but the lower ones do not, then the 741 needs replacing because the current output is not enough. It can be replaced by merely tack soldering another 741 chip right on top of the weak one. If that does not work, then the chip can be substituted by using a TL081, which is a higher current op-amp that is an exact replacement part.

If the antenna will only move in one direction and, on top of that, there's no picture or sound, then check the LNA fuse (1/2 amp slo-blo), as it's probably blown. If the antenna moves by itself, or starts and stops all by itself, then there is either a bad feedback connection or a bad feedback pot. Replace the feedback pot with a Bournes 3540S-1-102 or a Spectrol 534-7276. Before

TECHNICAL BULLETIN
CHANNEL MASTER SATELLITE RECEPTION EQUIPMENT

BULLETIN No. 11   AUGUST, 1982   SUBJECT: SIGNAL LEVEL METER
AND CENTER TUNE METER
ADJUSTMENTS FOR MODEL 6128
SATELLITE RECEIVER

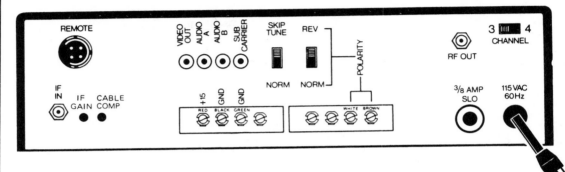

SIGNAL LEVEL METER ADJUSTMENT

Due to possible variations between units, it may be necessary to adjust the Signal Level Meter. The following method should compensate for these variations.

1. Locate Satcom 3R in the east, Satcom 4 in the west, and select the strongest channel—transponder that produces the maximum needle movement on the meter (deflections greater than full scale).

2. Using the fine tuning knob on the front panel, adjust for best picture.

3. Insert an insulated single-blade screw driver or alignment tool into the rear access port marked "IF GAIN."

4. By turning the screw driver, set the Signal Strength Meter on a reading of "8."

5. Now select the transponder that produces the weakest signal. Using the fine tuning knob, adjust for best picture.

6. Recheck all transponders for excessive meter movement. Transponder readings should vary between "2" and "8."

CENTER TUNE METER ADJUSTMENT

If the fine tuning control must be turned excessively to either end of its rotation to obtain a central indication on the Center Tune Meter, the following method should improve operation.

1. Position the fine tuning knob on the front panel at "11 o'clock."

2. Insert an insulated single-blade screw driver or alignment tool into the rear access port marked "CABLE COMP."

3. By turning the screw driver, set the Center Tune Meter on "0."

NOTE: In some extreme cases, adjustment of this control may not center the tuning meter. If this is the case, make sure that the ground connection between the down converter and the receiver is solidly made. If this checks out, it may be necessary to reset the individual channel calibrations. However, only an experienced technician should attempt this adjustment. See CMSRE Tech. Bulletin No. 8, Internal Adjustment Model 6128.

Printed in U.S.A. 7/82 (14019)

Fig. 18-17. Technical Bulletin #11. Courtesy of Channel Master.

Channel Calibration

If, after performing the compensation adjustment described in Section II-4 of the 6128 Installation Manual, there is a large variation in Fine Tuning control settings for various channels, or if the cable compensation potentiometer is at the end of its range, the channel calibration settings may have to be reset.

There are 13 potentiometers located on the main circuit board. Each one is marked with the channels it controls. One pot sets channel 1, another sets channel 24, and all the rest affect two channels each.

In order to set the adjustment, an accurate signal source is necessary for alignment. The best source is a satellite like SATCOM 3R. The person making the adjustments should be familiar with the programming on the satellite to make sure each pot is set-up on the right transponder.

If you use SATCOM 3R as a signal source proceed as follows:

1. Set the front panel Fine Tuning Control at about 11 o'clock, (pointing to the word "Fine"). This is the preferred position because it's closer to the electrical tuning center than straight up and down would be.

2. Using the UP or DOWN button, find Channel 24.

3. Adjust the channel 24 potentiometer for center reading on the Center Tune meter. NOTE: If centering the meter does not correspond to the best picture, refer to Center Tune Meter Calibration below, before proceeding with this section.

4. Select channel 22 and adjust the pot marked 22/23 for a center reading.

5. Select channel 20 and adjust the pot marked 20/21 for a center reading.

6. Proceed as above through all even numbered channels until all even channels have been set. Be sure you are setting each pot for the right transponder.

7. If any even numbered transponders were not transmitting, select the odd numbered channel which shares the same pot and set the pot for the odd numbered channel.

8. Select channel 1. Adjust the channel 1 pot for a center reading.

9. Go through all the remaining odd numbered channels and note that Fine Tuning should not have to be moved much to tune in any channel. You may wish to adjust a setting or two by splitting the difference between the odd and even channel for Fine Tuning centering. If the odd numbered channels are going to be used more often, reverse the proceedure above and set up the odd channels first.

Center Tune Meter Set-Up

1. Tune in a color bar test pattern or a picture with a deeply saturated color background.

2. Turn the antenna slightly off the satellite if necessary, to get some "sparklies" in the picture.

3. Carefully adjust the front panel Fine Tuning control for an equal number of black and white "sparklies" in the picture.

4. If the Center Tune meter does not read in the center of the meter area, adjust the potentiometer F50, located behind the meters on the P.C. board, until the meter is centered.

5. Recheck on several strong signals to confirm proper setting.

Audio Subcarrier — Preset Potentiometers

Pots R52 (6.8 set), and R54 (6.2 set) are factory set for the 6.8 and 6.2 MHz subcarrier frequencies. If the user desires, one or both of these preset frequencies may be varied. However, the two pots do not have the full 5.5 to 8.0 MHz tuning range like the front panel variable Audio Tuning does.

All Other Internal Alignment

The remaining internal adjustments require sophisticated laboratory test equipment for proper alignment. Consult the factory service department if you suspect alignment is required. It is extremely important that these adjustments be made only with the proper test equipment to assure that high quality video and audio demodualtion is maintained.

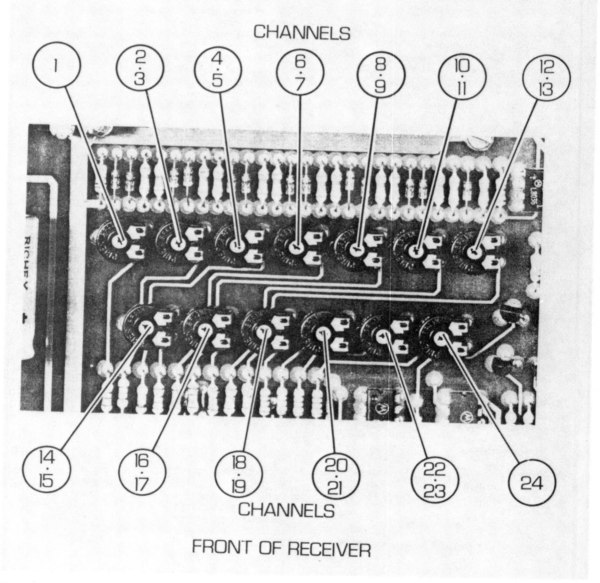

CHANNELS

FRONT OF RECEIVER

Fig. 18-17. Continued from page 263.

TECHNICAL BULLETIN
CHANNEL MASTER SATELLITE RECEPTION EQUIPMENT

BULLETIN No. 12   AUGUST, 1982   SUBJECT: DUAL LNA AND RECEIVER INSTALLATION
SCHEMATIC DIAGRAM

NOTES:
- Remove +15V Jumper on Down Converters
- Coax Switch: (H to N/C), (V to N/O), Contacts on CS 1 & 2
- Receiver: 1 — +15Vdc
              2, 3 — GND
              7 — +15 or 0Vdc (Polarity Control)

LNA	—	Low Noise Amplifier, CM 6111, 6112
DCB	—	D.C. Block, CM 6118
2-Way	—	2-Way Power Divider, CM 6152
C/S	—	Coaxial Switch, CM 6202
D1	—	Down Converter No. 1
D2	—	Down Converter No. 2
H	—	Horizontal
V	—	Vertical
I	—	Isolator, CM 6201
PS	—	Power Supply, 1 Amp., 15 Volt DC

DUAL RECEIVER INSTALLATION

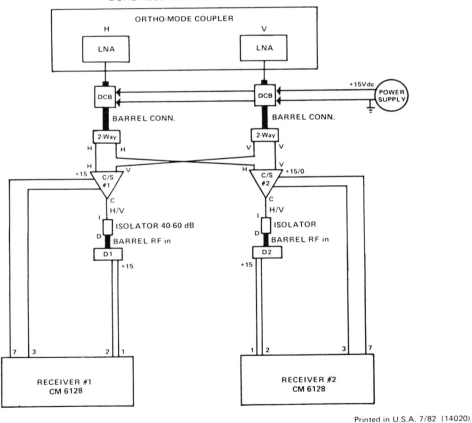

Printed in U.S.A. 7/82 (14020)

Fig. 18-18. Technical Bulletin #12. Courtesy of Channel Master.

265

TECHNICAL BULLETIN
CHANNEL MASTER SATELLITE RECEPTION EQUIPMENT

BULLETIN No. 17   OCTOBER, 1982   SUBJECT: RECEIVER (6128) INTERNAL ADJUSTMENTS, TERMINATOR AND MODULATOR REPLACEMENT

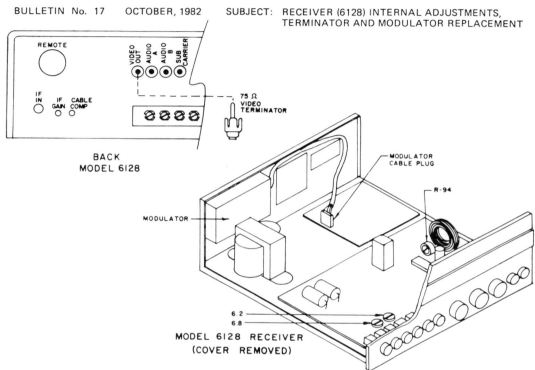

BACK
MODEL 6128

MODEL 6128 RECEIVER
(COVER REMOVED)

1. **75 ohm terminator:** Observe the rear panel on the receiver. If the video output phono jack is not in use for a VCR, etc. there should be a 75 ohm resistive terminator plugged into the jack. Without the terminator the video level may be slightly too high. This can cause audio buzz and possibly high picture contrast. *Note: Perform this step before adjusting R-94.

2. **Video drive level adjustment:** With unit operating, remove top cover from receiver and locate video drive level control R-94. While observing the picture and listening to the audio, carefully adjust R-94 for minimum audio buzz and best picture clarity.

3. **Replacement of Modulator:** A defective modulator can cause audio and video problems. To replace, unplug receiver and remove top cover. Remove nut holding modulator to back panel. Disconnect modulator cable from circuit board noting that the black wire goes toward the front of the receiver. Remove existing modulator and replace with new unit. Be sure to plug in cable with black wire toward front of receiver.

4. **6.8/6.2 Audio Adjustment:** If the audio sounds distorted or no audio is present when 6.2 or 6.8 is pushed in, but can be tuned with manual audio control, the audio preset pots may need realignment. With the unit operating, remove top cover. Set receiver to preset audio 6.8 and a transponder with 6.8 audio. Adjust 6.8 audio trim pot for clear audio output. Select a transponder which has 6.2 audio and push 6.2 audio in (ex.: Westar IV, TR.16 is broadcast on 6.2). Adjust 6.2 audio trim pot for clear audio output. These pots are located just behind the receiver front panel (see diagram).

Printed in U.S.A. (14073)

Fig. 18-19. Technical Bulletin #17. Courtesy of Channel Master.

TECHNICAL BULLETIN

CHANNEL MASTER SATELLITE RECEPTION EQUIPMENT

BULLETIN No. 19   DECEMBER 20, 1982   SUBJECT: Additional weatherproofing for SATSCAN
used in cold climates.

In order to reduce the likelihood of water entering the jack mechanism, run a bead of RTV sealant around where the jack tube meets the casting. Also run a RTV bead around where the rear cover joins the casting.

Model 6197
12 Volt SATSCAN™

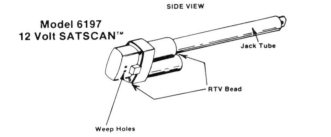

SIDE VIEW

Jack Tube

RTV Bead

Weep Holes

REAR VIEW

Weep Holes

On the 6197, make sure the plastic rear cover is mounted horizontal, as shown in the rear view above. Also make sure the weep holes are kept open and free of RTV.

In order to reduce the likelihood of water entering the jack mechanism, run a bead of RTV sealant around where the jack tube meets the casting.

Model 6250
36 Volt SATSCAN™

SIDE VIEW

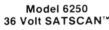

Jack Tube

RTV Bead

Weep Hole

REAR VIEW

On the 6250, the jack should be mounted so that the rear cover (casting) is vertical, as shown in the rear view above. There are two weep holes in the casting which should be kept open and pointed downward.

Printed in U.S.A. 12/82 (14143-#2)

Fig. 18-20. Technical Bulletin #19. Courtesy of Channel Master.

TECHNICAL BULLETIN
CHANNEL MASTER SATELLITE RECEPTION EQUIPMENT

BULLETIN No. 22   Date: September, 1983   SUBJECT: Channel Drift or Channel Jump in
6128 and 6128A Receivers

The following procedure will cure most channel drift or channel jump problems in Model 6128 and 6128A Receivers:

The majority of channel drift and channel jump problems originate in the RG59 coaxial cable connecting the down converter's IF OUT to the receiver's IF INPUT port. With the tuning voltage from the receiver being applied through the center conductor of the RG59 coax, any corrosion or other type of resistance will cause the tuning voltage to vary. In turn, channel drift will occur.

To correct, replace "F" type connectors on both ends of RG 59 coax or thoroughly clean any corrosion which may be present.

Bad connections on either end of the coax, whether it be center conductor or shield, will cause the above symptom.

NOTE: The LED channel display on the receiver does not change during channel drift.

Printed in U.S.A. 9/83 (14474)

Fig. 18-21. Technical Bulletin #22. Courtesy of Channel Master.

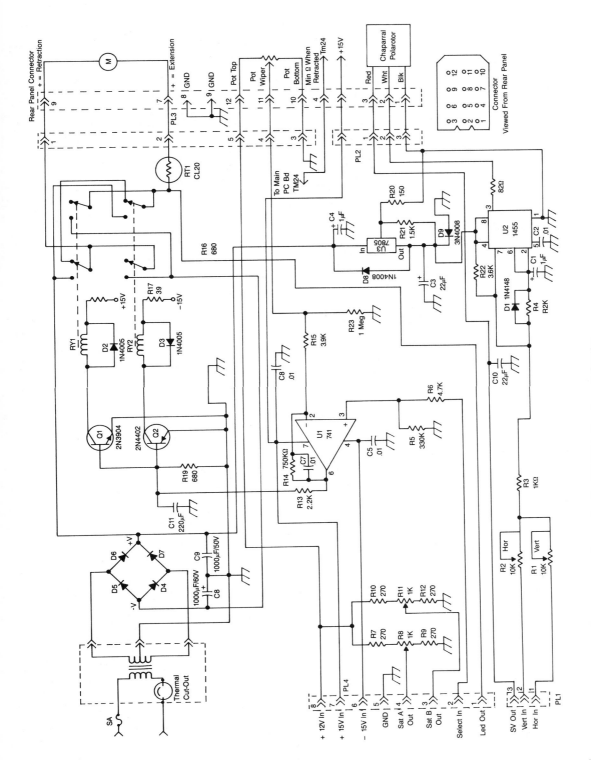

Fig. 18-22. Motor controller schematic from a Conifer RC-2001. Courtesy of Conifer.

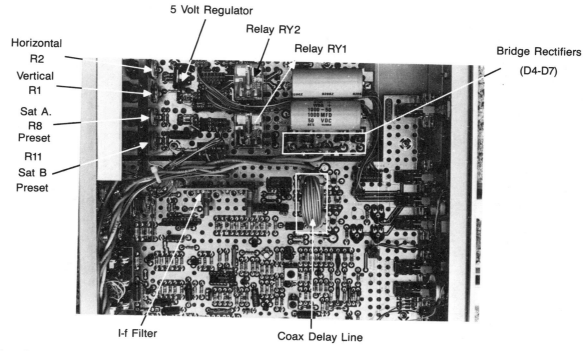

Fig. 18-23. The motor controller board and the i-f strip from an RC-2001. Courtesy of Conifer.

replacing the pot, be sure to check that the +5 volts is actually getting to the pot and that the ground and return wires are connected properly.

If the motor drives in one direction as soon as it is turned on, then Q1 or Q2 is shorted, or one of the relays is stuck in the on position. If it does not drive at all in one direction then Q1 or Q2 may be open, or one of the relays may be bad. If it won't go in either direction then check for ±18 volts. The rectifier bridge could have an open component, the transformer could have opened up, or the thermal cutout circuit breaker could have opened. Measure the resistance between the two input wires to the transformer. If there is no resistance at all, then either the fuse is blown or the thermal cutout has opened up. If the voltages are OK, check RT1, which may have opened up. If it is OK, then U1 may not be working properly. Check to see that the ±15 volts is present on pins 7 and 4.

If the rf modulator is causing problems, refer to the Drake modulator description and schematic shown in Chapter 12. If the i-f section is where the problem lies, then refer to Chapter 9 under the section on Drake i-fs. The video detection circuit is identical to both the Drake circuit described in Chapter 10, and to the Winegard schematic shown later on in this chapter.

Checking the Motor Controller

To bench check the motor controller build up the circuit shown in Fig. 18-24. It plugs into the Molex connector on the back of the receiver and allows troubleshooting of the controller without having the motor connected. It also allows you to preset the Polarotor for the proper 90 degree movement. The three LEDs indicate whether the controller is sending power to the Polarotor and whether the load resistor (in place of the motor) has any voltage on it and what polarity it is. The voltage from pin 1 to pin 3 should be +6 Vdc, while the voltage from pin 7 to pin 9 should be 18 Vdc when either LED is turned on. A three state LED could be used in lieu of the separate LEDs. Its connection is indicated by the dotted lines.

Before using the tester, it should be calibrated to a known good receiver. To do this, hook it up to the receiver and turn the receiver on. Press the SAT SEL button. Put switch S1 to position A and turn the front panel SATELLITE position control to 10 (8:00 position). Adjust the A trimpot until both LEDs are off. If the SATELLITE control is rocked back and forth, first one and then the other LED should light as it is moved away from position 10.

If both LED's work properly then set switch S1 to B, and adjust the SATELLITE control to 50 (12:00 position). Adjust trimpot B until both LEDs are off. Put the SATELLITE control to 90, and S1 to C. Adjust trimpot C until both LEDs go out. Use glyptol on the trim pots so that they will not change.

Press the H and V buttons while observing the Polarotor probe. It should move at least 90 degrees. Ad-

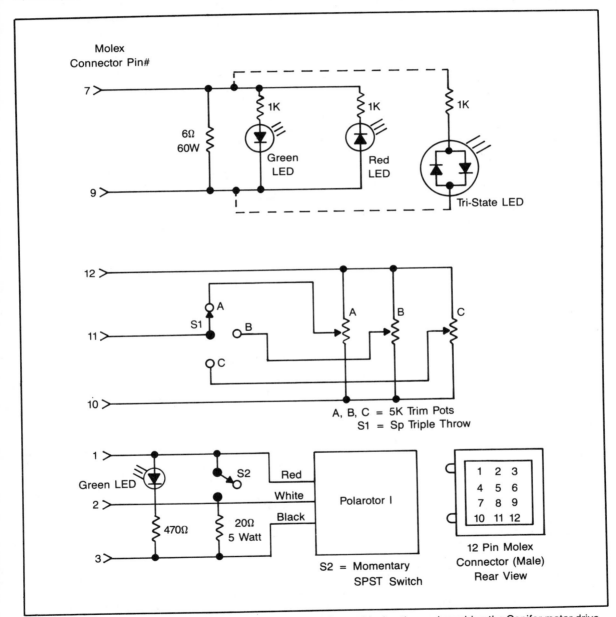

Fig. 18-24. Motor Drive and polarization device simulator. Used for troubleshooting and repairing the Conifer motor drive.

just the HORIZ and VERT controls on the back to get the largest probe movement possible. It should be at least 140 degrees. Mark the two positions as FCW (full clockwise) and CCW (counter-clockwise) on the side of the waveguide for reference. Center up a 90 degree movement evenly within the FCW and CCW marks by adjusting the HORIZ and VERT controls. Mark these two positions as H and V.

To use the simulator, connect it to the unit under test, put S1 into the B position, and push in the SAT SEL button. Turn on the unit and adjust the SATELLITE control until all LEDs are off. This should be around 50. Adjust the SATELLITE control toward 10. One of the motor LEDs should turn on. The front panel ANT LED should also be lit. Adjust the SATELLITE control back to 50 and both should go off again. Move the SATELLITE control toward 90. The other LED and the ANT LED should both come on. To check the feedback circuitry, set S1 to position A and adjust the SATELLITE control to 10. Do the same thing using position C and number 90. All LEDs should go out when the control is close to either number. If the LEDs do not respond properly then the previously listed troubleshooting tips should be followed to isolate the problem.

The polarization LED should be on, indicating that there is +6 Vdc present. To check the polarization, press in the H button and adjust the rear panel HORIZ control. The probe should move as the control is adjusted. Set the probe to line up with the H mark on the waveguide. Press in the V button and adjust the rear panel VERT control. The probe should likewise move. Set it to line up with the V mark. Pressing H and V should now rotate the probe 90 degrees.

The thermal shutdown capability of the 6 volt regulator can be checked by pressing S2 for about 20 to 30 seconds. After about 15 seconds the +6 volts should drop to zero, turning off the LED. Release S2, after about 15 to 20 seconds the voltage should come back. If not, try turning the receiver off for a few seconds and then turn it back on while observing the LED. If there is still no voltage, then the +6 volt regulator should be replaced.

DRAKE

The basic Drake-made down converter is shown in Fig. 18-25. All single-conversion down converters use similar circuitry. The input matching and hybrid circuit consists of a 1,000 pF cap, 56 ohm resistor, and the sideways H trace pattern. The outputs of the hybrid are the two traces that go straight up and down. The lower signal lags the upper one by 90 degrees at this point. These two signals

are sent to two mixers. The second input to the mixer is from the LO.

The LO is generated by U1 and U2. The frequency is controlled by the negative voltage that is applied to their emitters. The tuning voltage goes through a 2.7 μH choke which blocks the 70 MHz signals. The LO's output is coupled through the circular shaped *horseshoe* to the *antenna trace* on the board. It is applied to the mixer circuit along with the incoming signals. The output of each mixer consists of two signals that are at 70 MHz: the desired signal and the image signal.

The two mixer signals are sent to another transformer (or hybrid), that again delays the lower signals by 90 degrees. The result is that part of the bottom signal is 180 degrees out of phase with part of the top signal. And anytime two signals that are 180 degrees out of phase are combined, virtually zero signal if left. It just so happens that this out of phase signal is the image signal, which means it cancels itself out. Because of the frequency relationships between the desired channel and the LO, the desired signal in the bottom trace is actually delayed 270 degrees by the time it hits the second hybrid. This means that with another 90 degree shift it is back in phase with the upper trace.

The output of T1 is a 70 MHz signal which is applied to U3 through a .15 μH inductor and 33 pF cap. They form a wide bandpass filter that's resonance is centered at 70 MHz. U3 boosts the signal to drive U4, which in turn boosts the signal level to drive the connecting cable to the receiver.

Troubleshooting

There are more Drake receivers in the field than any other brand of receiver. They are not only sold under the Drake name but are also sold under an OEM agreement with Conifer (the RC-2001), Winegard (the SC-7032 and SC-7035), and Channel Master (the 6128). Typically the circuitry used in OEM receivers is virtually identical to the basic circuits from any Drake receiver (ESR-24, ESR-224, ESR-240, or ESR-324).

Figure 18-26 is the schematic for the Drake ESR-324. Along the top of the schematic is the i-f filter (L2 to L6), the i-f amplifier (Q3, Q4 and Q6), the signal-strength meter amplifier and detector (Q5, CR3, and CR4), and the limiter (U1).

The output of the second stage of the limiter is split into two signals. One is delayed by 270 degrees because it's sent through a delay line. The delay line is a length of coax cable that is coiled up on the board. The other signal is sent directly to a transformer with a grounded,

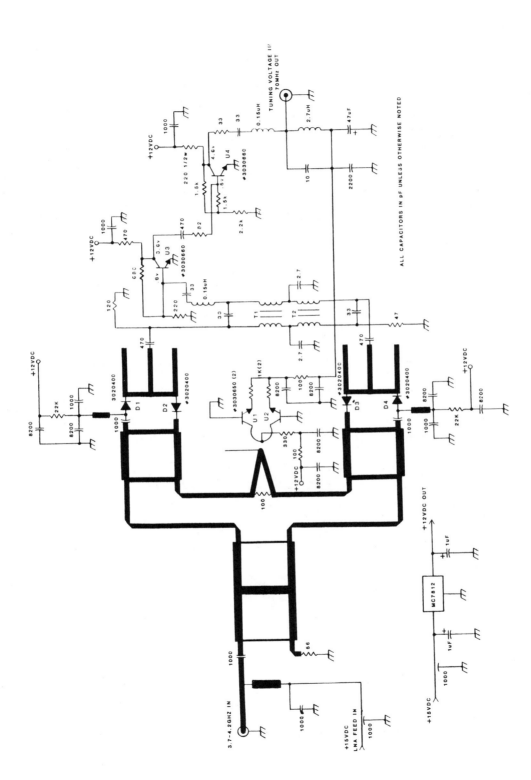

Fig. 18-25. Drake-made down converter. Courtesy of Conifer.

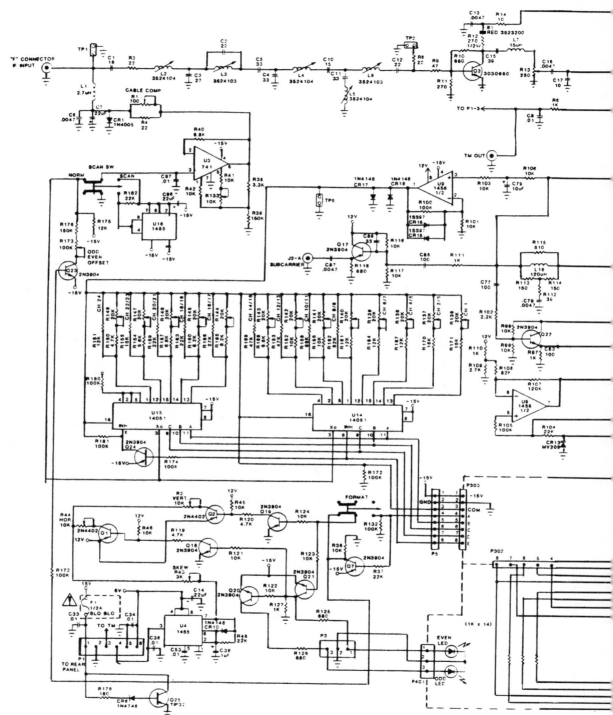

Fig. 18-26. The ESR 324 schematic. Courtesy of R.L. Drake.

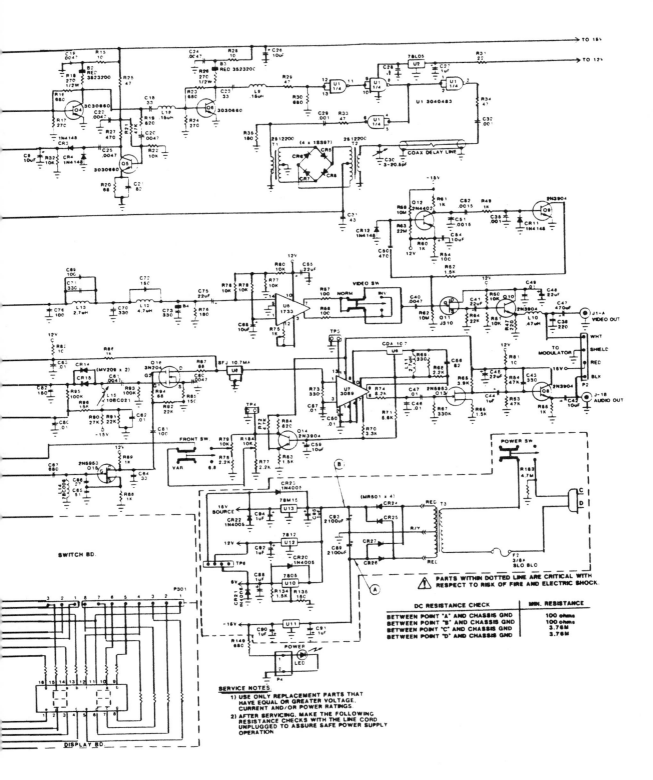

275

center tapped secondary. The delayed signal is also sent to a similar transformer, with the center tap being the detected video signal. The top and bottom leads from each transformer are applied to a double-balanced mixer, which consists of four matched diodes (CR5 to CR8). The output voltage from the center tap follows the incoming frequency variations. As the frequency goes above 70 MHz, the voltage goes positive. If the voltage goes below 70 MHz, the voltage goes negative. Thus the output is equivalent to the original video signal without the carrier.

This raw or baseband video signal is sent to a subcarrier output buffer (Q17), an afc amplifier (U9), the video processing circuitry (from L16 to U5 to Q11 and Q10) and the audio detector circuit (Q22, U9, Q15, CR14, CR15 and Q16). The afc circuit consists of an amplifier (typically a 741 or 1458 op-amp) and voltage limiter (CR17 and CR19) which feeds the tuning circuit (U14 and U15).

The video circuit consists of a deemphasis network (L16 and the attendant resistors and capacitors), a low-pass filter (L12, L13 and the accompanying capacitors), a balanced output video amplifier (U5, a 733) with a switchable output (normal/inverted video), a clamping circuit (Q12, Q9, CR11 and CR12), and a final buffer (Q10, a 2N3904), which drives the video output and the rf modulator.

The input of the audio circuit consists of a high-pass filter (C77 and R102) and an input buffer (Q22, a 2N3904). To get a 10.7 MHz audio i-f, an LO (2N5953), mixer (3N204), and varactor diode (MV209) is used. The i-f is filtered by using a standard FM radio-type ceramic filter (U8). This signal is sent to U7, a 3089 IC, which is an FM radio i-f amp and detector chip. The detected audio is deemphasized, amplified (by Q13, a 2N5953), and then buffered by Q8 (a 2N3904). It is then sent to the rf modulator and to the audio output.

The biggest differences between the various models is in the way that the video and audio channels are selected. Some models also have a polarizer control circuit and fancier channel number displays.

All tuning is done by using a negative tuning voltage applied up the 70 MHz coax cable. It typically goes from − 4 to − 10 volts as the channels are tuned from 24 to 1.

The following instructions apply to virtually all Drake made receivers. The biggest difference is in part ID numbers, which differ from model to model. Table 18-1

Table 18-1. Alignment Points for Various Drake Receivers.

Alignment Points	ESR-24	ESR-224	ESR-240	ESR-324
afc test point	TP-1	unlabeled TP	TP-7	TP-5
Even channel offset	—-	R132	R156 *1	R173
Channel 1	R15	CH1	R188	R136
Channel 24	R29	CH24	R164	R148
Audio gen. input	Q18 base	Q18 base	Q15 base	Q22 base
Audio afc test point	Q15 emitter	Q15 coll.	TP-6	TP-4
6.8 MHz test point	L4	L7	L14	L14
Quieting adjust	L3	L8	L16	L15
Detector trim cap	C628 *3	C21	C27	C30
Video level	R94	R57	R45	R75
Center-tune meter adj.	R50 *2	—-	R191	—-

*1 Is an odd channel offset control.
*2 on receivers below 20,000 only. R901 is used for all others.
*3 on receivers above 20,000 only.

is a Drake model-number to control-number cross reference.

Sweeping the i-f Filters. To sweep the i-f, connect a 70 MHz sweeper with markers at 55 MHz and 85 MHz to the i-f input jack. Adjust the input gain pot to halfway. Adjust the sweeper level for a – 60 dBm signal. The signal strength meter should be at 1/2 to 3/4 deflection. Connect a spectrum analyzer directly to the output of the first filter, which is TP2 in the ESR 240 and ESR 324. Connect it through a 22 ohm resistor on the rest of the models. The resistor should be connected to pin 11 of U5 in the ESR 224, pin 11 of U1 on later ESR 24s (after serial #20,000) or to pin 2 of the MWA120 transistor right after the filter in earlier ESR 24s (see Fig. 9-15). Alternately, a demodulator probe could be used with an oscilloscope to view the response.

Adjust the five coils on the filter board so that the two markers are about halfway up the skirts, at the – 6 dB points. The final response should look close to that in Fig. 18-27.

On receivers with two filters (later ESR 24s and all ESR 224s) repeat the adjustment for the second filter. Its output can be taken off of R801 on ESR 24s after #20,000 or R20 on the ESR 224s. Remember that you are going through the first filter, so that any anomalies in its response will show up in the second filter as well. In some cases, to get the ideal response requires that the first filter be tweaked again while looking at the output of the second filter.

Matching Receiver to DC. Put the dish onto F3-R and turn on the receiver for at least 20 to 30 minutes. This ensures that all components are at their full operating temperature. Set the cable compensation trimpot to 1:00 position. Set the channel fine tune control to 12:00 posi-

tion and do not adjust until after channel alignment is complete.

Connect an analog voltmeter (preferably one with a center scale tuning setting) or a DVM to the main afc test point. Select channel 1. Tune the internal channel 1 trimpot so that there is zero volts on the afc test point. Repeat this same procedure for all other odd channels. Change to horizontal polarization and channel 14. Adjust the even offset trimpot for zero volts. Switch to channel 24 and adjust its internal trimpot for zero volts. *Note*: The ESR 24 and ESR 240 use the even channels instead of the odd channels, so reverse the even/odd indications in the above instructions for those models. Use channel 13 to set the odd channel offset and channel 1 to set the low-end trimpot.

Audio Adjustments and Checks. There are five different ceramic filters used in the Drake-made receivers. These are identified by the coloring on the filter. Table 18-2 gives the color code.

To check and adjust the audio, first attach a signal generator to the input of the ceramic filter. It should be a 10 mV, 1 kHz tone, deviated 100 kHz. The generator should be set to the proper frequency according to the color of the filter as listed in Table 18-2.

Observe the waveform at the audio output jack on an oscilloscope. It should be an undistorted 0 dBm signal. Use the Audio A output on the ESR 24.

Measure the dc voltage at the audio afc test point. Reconnect the generator to the audio generator input as specified in Table 18-1. Set the output to 6.8 MHz. Adjust the front panel AUDIO TUNE control to the 6.8 MHz mark. Use the VARIABLE tuning setting on those receivers so equipped. Adjust the oscillator adjustment coil (labeled 6.8 MHz adjust in Table 18-1) to yield the same afc voltage as above.

Reduce the generator's output level until the 1 kHz tone becomes noisy. Adjust the Max Quieting control, listed in Table 18-1, for the least noise.

Video Adjustments. Connect a signal generator set at exactly 70 MHz at – 25 dBm to the 70 MHz input connector. Adjust the Detector Trim Cap, listed in Table 18-1, for exactly zero volts at the afc test point.

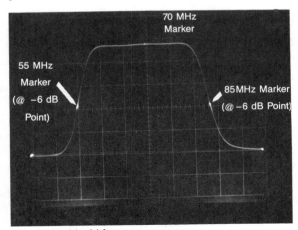

Fig. 18-27. Ideal i-f sweep response.

Table 18-2. Color Code for the Center Frequencies of the Ceramic Audio Filters.

Red	=	10.700 MHz
Blue	=	10.670 MHz
Orange	=	10.730 MHz
Black	=	10.640 MHz
White	=	10.760 MHz

Set the receiver on ESPN or on MTV and view the VITS with a waveform monitor. Adjust the Video Level control for 1 volts peak-to-peak into a 75 ohm load.

ESR-240 Remote Detector Adjustment. Attach an oscilloscope or frequency counter to the test point on the front panel board. There should be a 100 mV, 590 kHz (± 20 kHz) signal present. If it is off-frequency then trimpot R349 (at the top of the board) should be adjusted accordingly. One cycle would be 1.7 μs on the scope.

DEXCEL

Dexcel was one of the first manufacturers to offer an LNA specifically designed for the home TVRO market. They also came up with the first LNC. They have been involved in the TVRO marketplace since the first Oklahoma SPTS show in 1979. In 1983, they were acquired by Gould. And then in 1985, the TVRO section of Dexcel was purchased by Hyteck International. They are the only company to offer a two year warranty on all their products.

DXR 1000 and DXR 1100

Their receiver line has expanded over the years to include four different models. Their first receiver was the DXR 1000, it was built by Cybernet, a well known late 1970's CB radio manufacturer. It uses an NE564 PLL chip running at 17.5 MHz to detect the video and a 12124 FM radio chip for detecting the audio. It is switch selectable between six different audio subcarriers (via plug-in crystals). It has a built-in multiplex stereo decoder. It uses a crystal-controlled channel 2/3 rf modulator. The schematic for the DXR 1000 is shown in Fig. 18-28.

The initial design was improved upon in their DXR 1100. It has been their largest selling receiver to date. It improved upon the DXR 1000 demod by only dividing the i-f frequency in half (to 35 MHz) and by using a faster IC (a 74S74) to divide the frequency. It features variable audio tuning with a built-in matrix stereo decoder. It uses the same modulator as the DXR 1000. The DXR 1100 was built by Kyocera, who also manufactured the DXR 1300 and the STS MBA receiver and MBS-AA actuator controller. The DXR 1100 schematic is shown in Fig. 18-29.

The procedure on matching the receiver and LNC together is given in Fig. 18-30. Another Technical Bulletin on the DXR 1100 is given in Fig. 18-31.

DXR 1200

The DXR 1200 was a radical departure from the DXR 1000 and DXR 1100. It uses a delay-line discriminator

for the demod circuit and an MC10115 as the limiter. It uses two NE564s for dual-channel audio with a built-in matrix stereo decoder. It has an LED channel display and a built-in Polarotor I control circuit (which was described in Chapter 6). It uses an Astec LC-type rf modulator. It was designed to be used on either 110 Vac or $+12$ Vdc, and it has a provision on the back panel for hooking up the receiver to a $+12$ Vdc source. It also uses an LNC designed for $+12$ Vdc operation and thus with the addition of a $+12$ Vdc TV, it can be made mobile.

To match the receiver's tuning voltage to the LNC requires that the three FREQUENCY ADJUSTMENTS labeled LOW, MID, and HIGH (on the back panel) be adjusted by watching the picture and the center-tune meter. The proper procedure is given in Dexcel Technical Bulletin #S-003 (Fig. 18-32).

The DXR 1200 schematic is broken down into four parts. Three of these are shown in Fig. 18-33, Fig. 18-34 and Fig. 18-35. The fourth section, the polarity control schematic, has already been shown in Fig. 6-3. Figure 18-33 is the main part of the receiver. It contains the power supply, i-f and video sections. Figure 18-34 contains the audio circuitry, and Fig. 18-35 is the channel selector board schematic.

DXR 900

The DXR 900 is a stripped down low-cost receiver, without the compromises typically found in low-end receivers as to their video and audio performance. If uses the MC10115 for both the i-f amplification and as the limiter. It uses the MC1357 quadrature detector for video demod and an NE564 for audio detection. It has a built-in Polarotor I circuit and the same Astec rf modulator as the DXR 1200. The schematic is shown in Fig. 18-36.

The DXR 900 was sold in two packages. It was available with either an LNC or with a separate down converter. Because the LNC only needed $+12$ Vdc, the receivers that were manufactured to use the DC had their output dc voltage boosted to $+18$ to $+20$ Vdc. This allows any LNA to work with the DXR 900 and its matching DC. Unfortunately, this means that the receiver will not work with an LNC anymore because the voltage is too high. To determine which type of DXR 900 receiver you have and also to prevent LNC damage, the circuit shown in Fig. 18-37 should be constructed. If the receiver is to be changed between being used with an LNC and being used with a DC, then Technical Bulletin T-006 (Fig. 18-38) should be followed. It shows how to change the $+12$ Vdc to $+18$ Vdc.

How To Read Large Schematics

Each connection is labeled with a number and a letter (e.g. 2A,4b). The number refers to the schematics sheet page that the connection goes to or comes from. The letter refers to the connection. Capital letters are used for signal lines moving left and right. Lower case letters are used for signals moving up and down between pages.

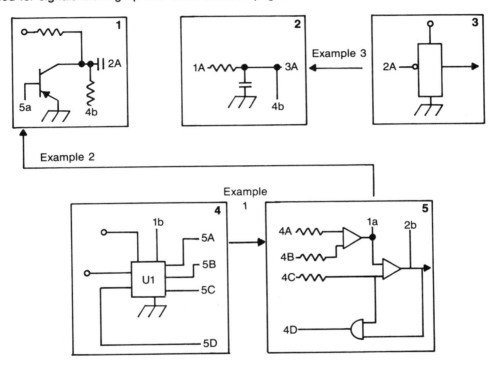

Example 1. This example shows that connection 5A on sheet four goes to sheet five connection A. Connection A on sheet five is labeled 4A. This means the connection comes from sheet four, connection A.

Example 2. This example shows that connection 1a goes to sheet one connection a. Connection a on sheet one is labeled 5a, which means that it comes from sheet five, connection a.

Example 3. This example shows that connection 2A on page three connects to connection A on page two. Page two has 2 connections labeled A: 1A and 3A. Since the connection you are looking for originated on page 3, Connection 3 A is the correct one.

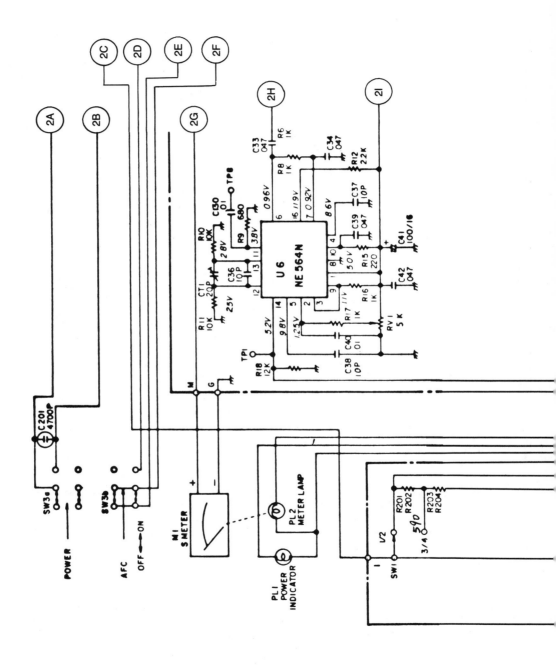

Fig. 18-28. The DXR 1000. Courtesy of Gould/Dexcel.

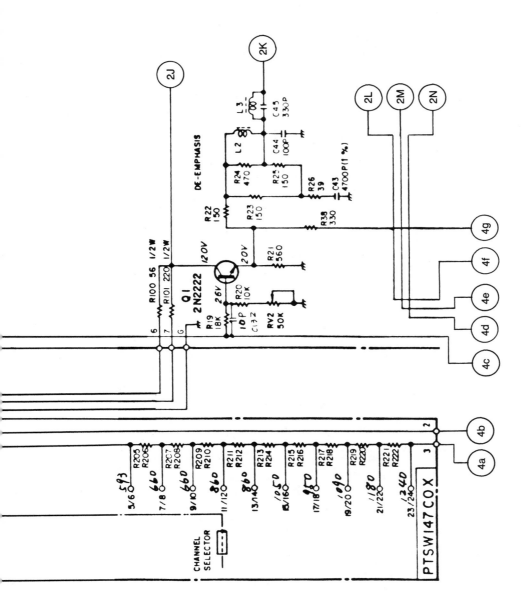

2

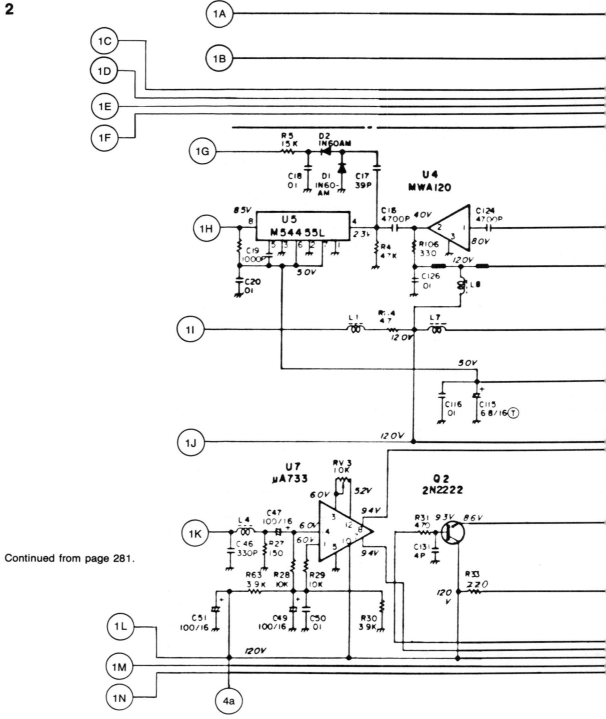

Continued from page 281.

282

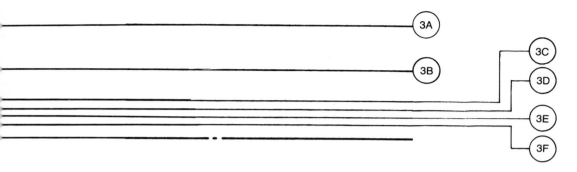

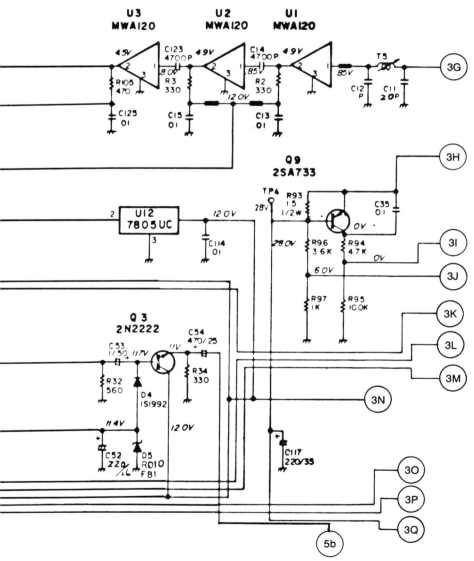

3

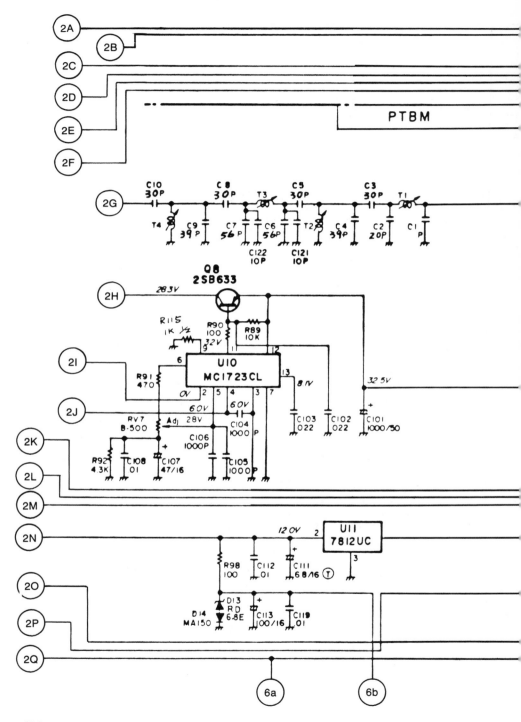

PTBM

Continued from page 283.

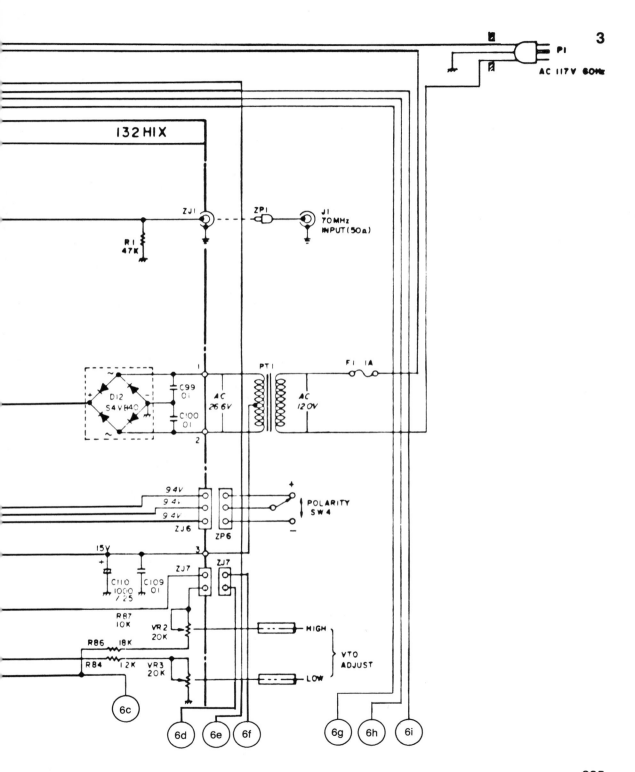

4

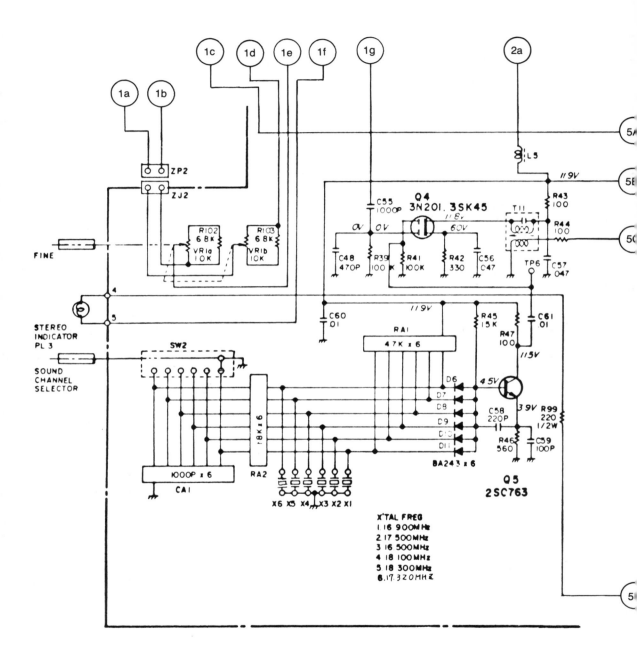

Continued from page 285.

286

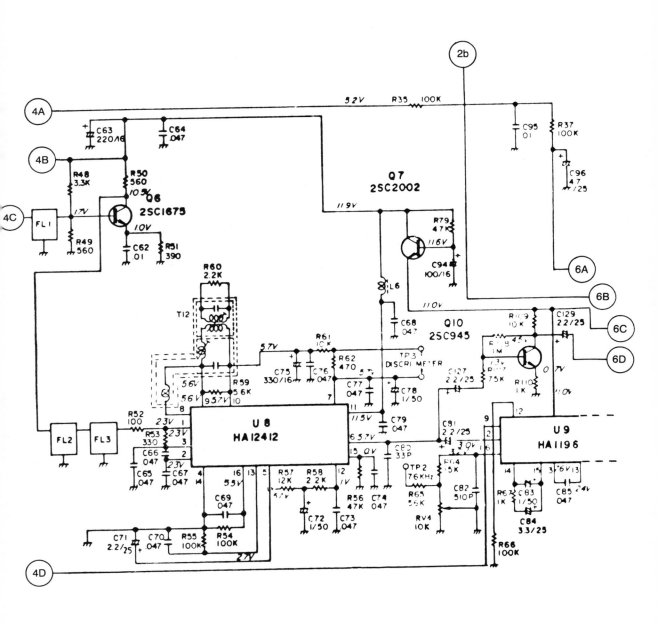

6

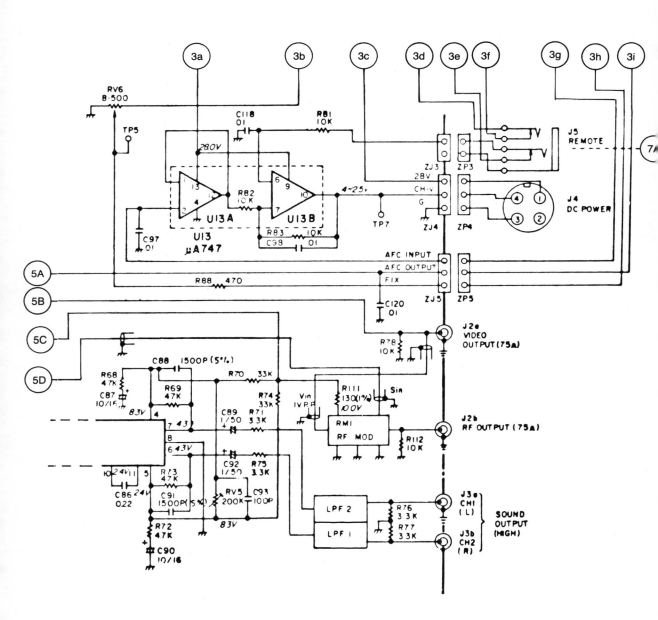

Continued from page 287.

1

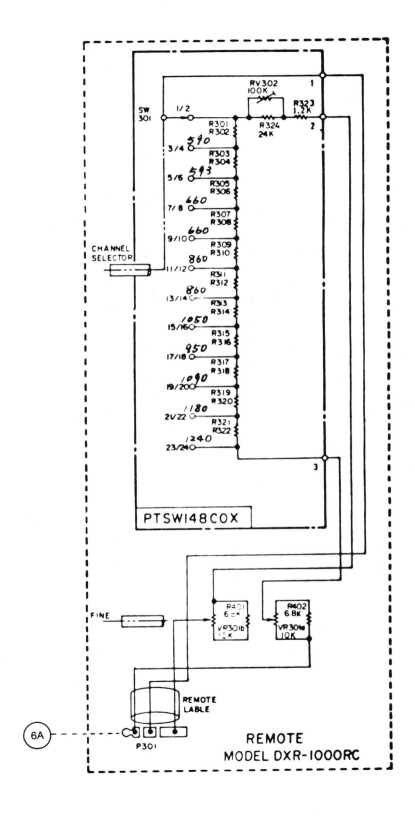

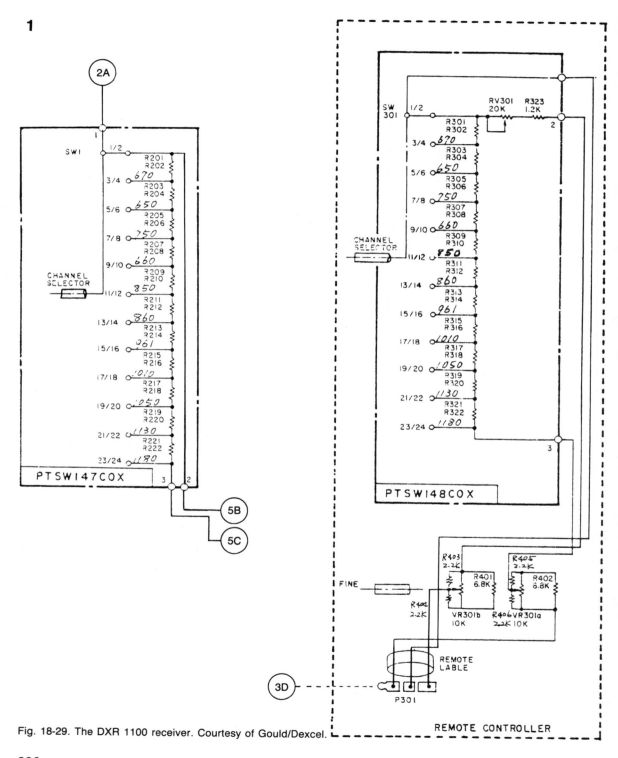

Fig. 18-29. The DXR 1100 receiver. Courtesy of Gould/Dexcel.

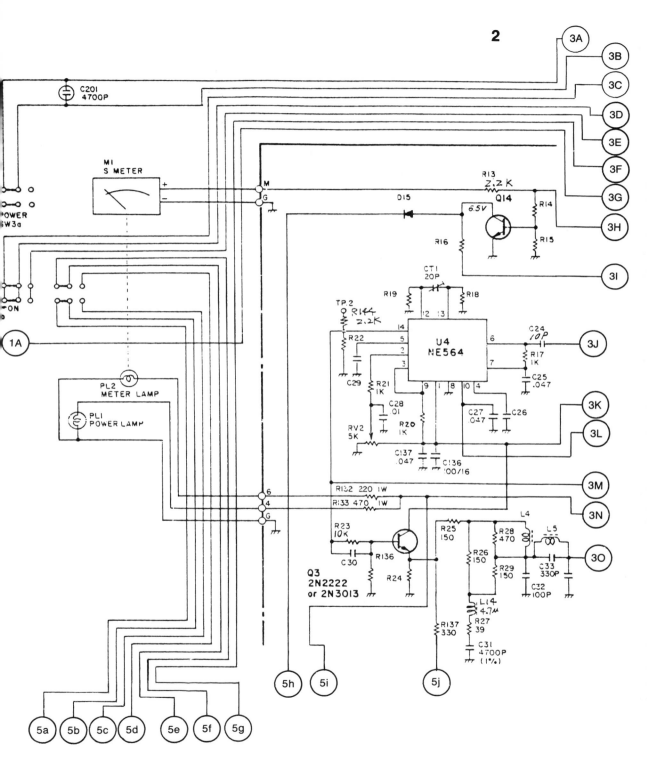

3

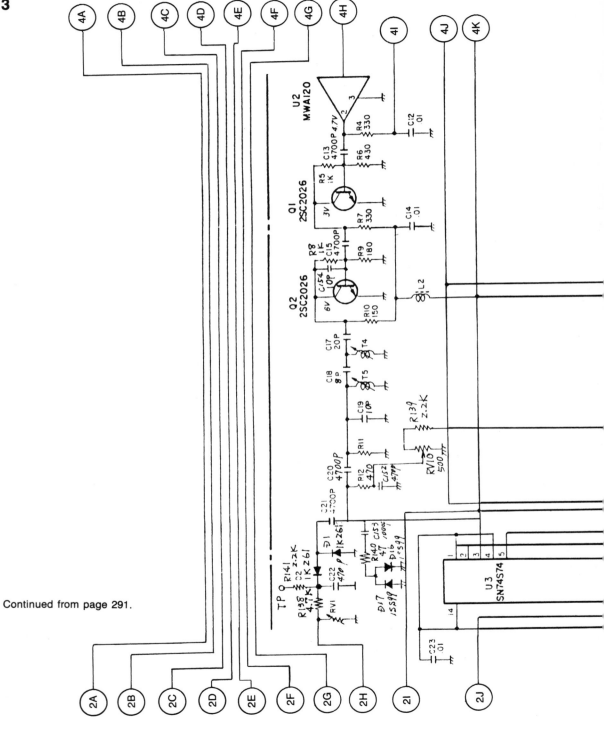

Continued from page 291.

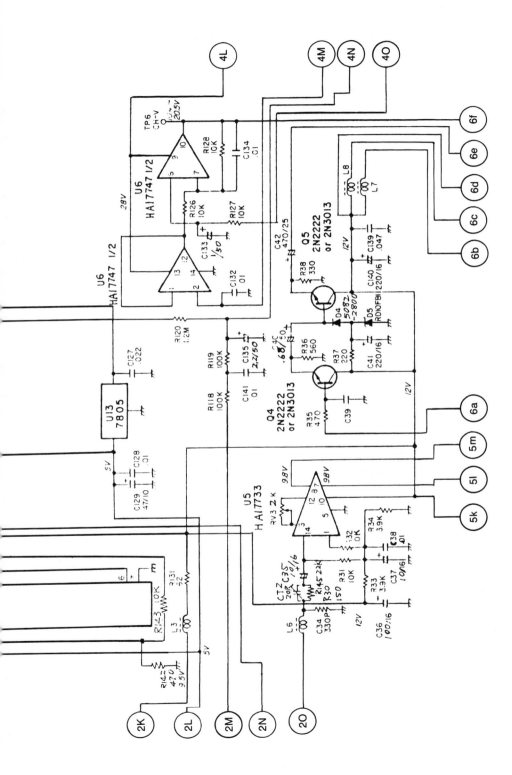

4

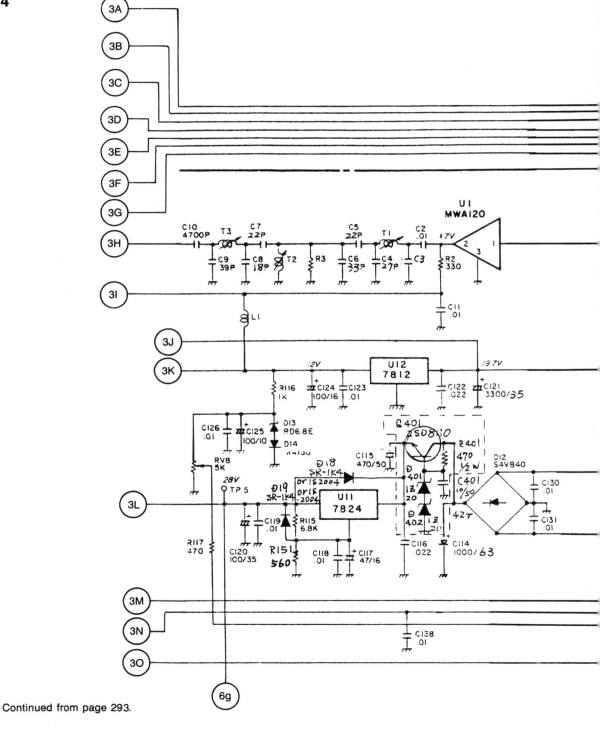

Continued from page 293.

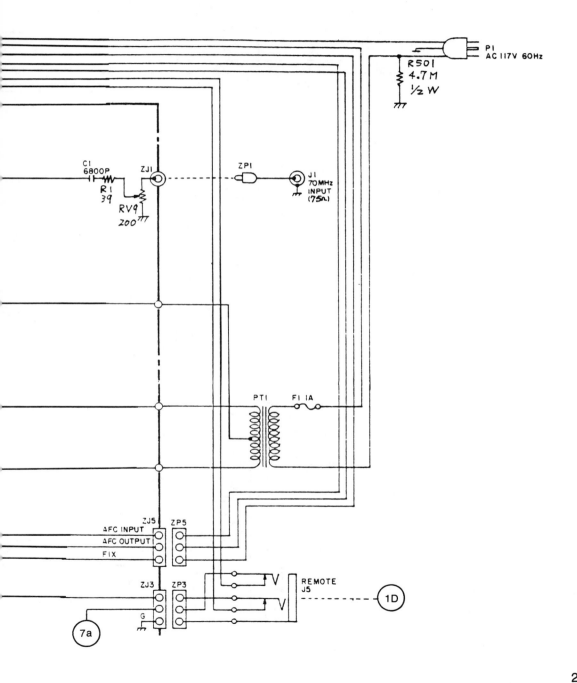

4

295

5

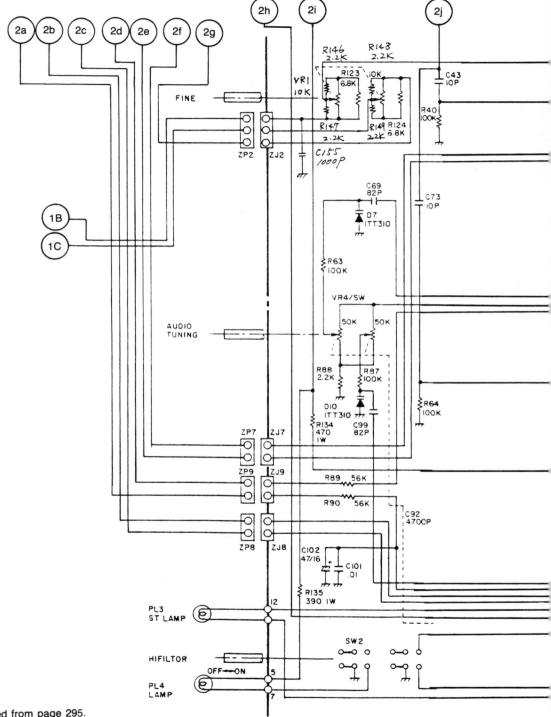

Continued from page 295.

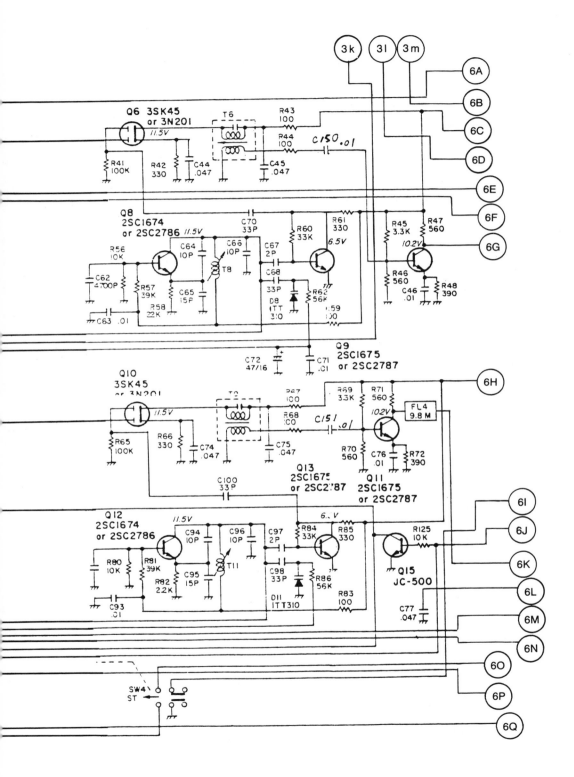

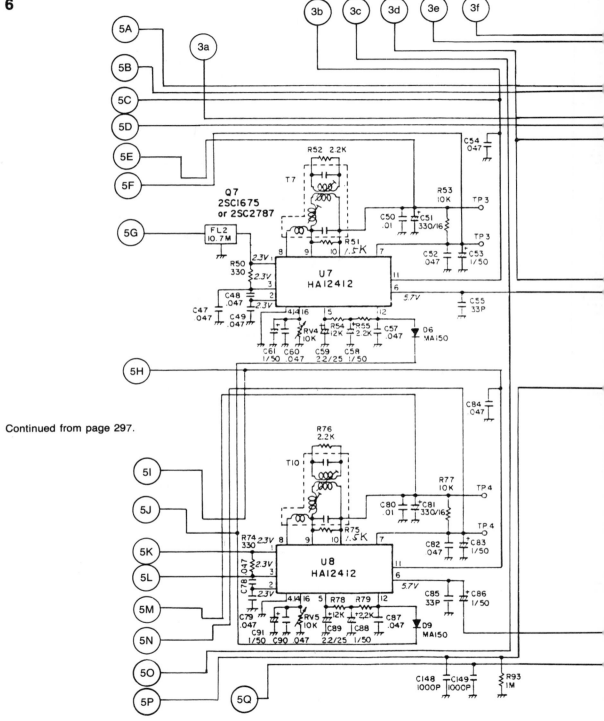

Continued from page 297.

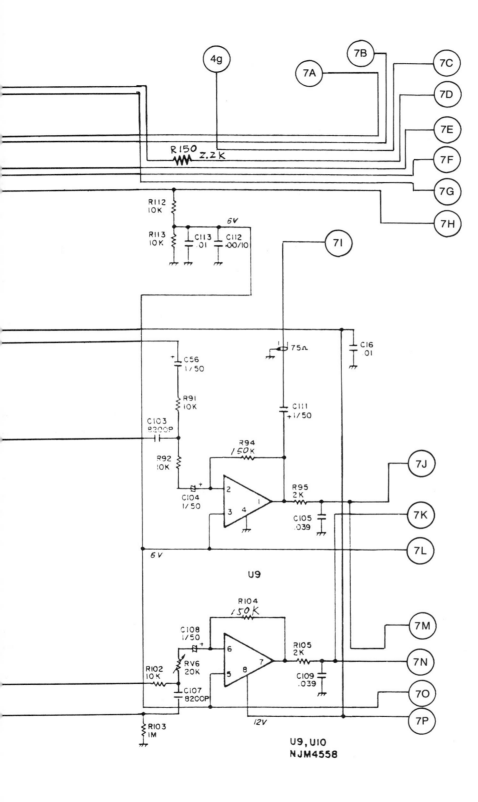

U9, U10
NJM4558

7

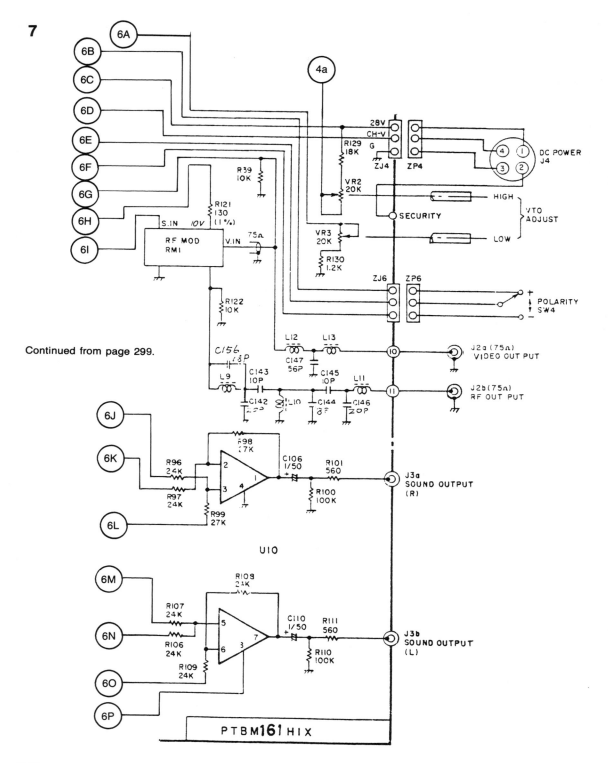

Continued from page 299.

PTBM**161**HIX

Gould Inc., Dexcel Division
2580 Junction Avenue
San Jose, California 95134
Telephone (408) 943-9055
TWX 910-338-0180

TECHNICAL BULLETIN

To: Innovision Distributors   From: Customer Service   Date: 11/19/84

Model #: DXR-1100   Bulletin #: S-002

Subject: DXR-1100 Matching Procedure

This procedure explains how to match any Gould Innovision DXR-1100 Satellite Receiver to any DXS-1000 LNC or DXD-1000 Downconverter. The dish should be pointed at F3R or Telstar 301 and a TV or video monitor should be connected. Perform the following steps:

1. Turn on receiver for a minimum of 25 minutes to insure temperature stability. Remove the top cover.

2. Set the CHANNEL FINE TUNE control to the ODD or VERTical mark. Turn to channel 1. DO NOT change the setting of the channel fine tune control until the matching is complete.

3. Turn the AFC off. Adjust RV8 (AFC offset trim pot) to bring in channel 1. Adjust for best picture.

4. Turn the AFC on. Adjust the VTO LOW ADJUST trim pot (using a small flatblade screwdriver through back panel) until the best picture on channel 1 is obtained.

   CAUTION: Make sure that there is no Terrestrial Interference (TI) entering the LNC at the low or high end of the band. If there is, use channels 3 and/or 21 to set low & high pots.

5. Turn the AFC on and off. If the picture tears or is different for any reason, repeat steps 3 & 4 until the picture continuity doesn't change.

6. Turn the CHANNEL SELECTOR to channel 23. DO NOT ADJUST THE FINE TUNING CONTROL.

7. Turn the AFC off. Adjust the VTO HIGH ADJUST control (also located through the rear panel) until the best picture is obtained.

8. Leaving the AFC off, change to channel 1. Check the picture. If it is not as good as it was in step 5, repeat steps 4 through 8 until picture continuity is maintained on both channel 1 and channel 23 as the AFC is switched on and off.

9. Adjust the RF INPUT LEVEL control (RV9) so that the SIGNAL METER peaks at 9.

10. Replace the cover. DO NOT adjust any other internal controls unless the proper test equipment and alignment information is available.

Fig. 18-30. Technical Bulletin S-002. Courtesy of Gould/Dexcel.

TECHNICAL BULLETIN

To: Innovision Distributors   From: Customer Service   Date: 11/29/84

Model #: DXR-1100   Bulletin #: S-006

Subject: CT-1 Trimmer Cap Failures

This procedure explains how to fix CT-1 Trimmer cap failures.

DXR-1100 receivers have been shipped using two types of trimmer caps for CT-1. These are identified by color as either red or white.

The red Trimmer Capacitor is the main cause of most of the following problems:

* Video scrambling (Unstable picture)
* No picture or No Sound
* Black and white horizontal lines throughout the picture

Most technicians assume that the NE 564 PLL chip is the cause and replace it, then adjust CT-1 to lock the picture back in, when in fact, CT-1 is causing the problem.

When receiving a unit for repair which show the above symptons and it has a red CT-1, try lightly pressing on CT-1 with your finger. Most of the time the picture will clear up. Always replace the red CT-1 with a white one in all DXR-1100 receivers, as the red CT-1 could again change capacitance.

All Gould/Dexcel specifications are subject to change without prior notice.

Fig. 18-31. Technical Bulletin S-002. Courtesy of Gould/Dexcel.

The LNC and DC matching procedure for the DXR 900 is shown in Fig. 18-39, which is Technical Bulletin #S-005.

DXR 1300

The DXR 1300 was the top of the Gould/Dexcel 1984 line. It features a built-in 35 watt/channel stereo amp, along with infrared remote control, matrix or discrete stereo, and random access channel tuning. It also has a built in Polarotor I controller. The audio and video sections are similar in design to the DXR 1200 but because of radically different manufacturing techniques used between the two, they do not outwardly appear so. The DXR 1300 uses a lot of chip components, caps, resistors, diodes,

and transistors. It also has a crystal-controlled channel 3/4 rf modulator built-in.

The DXR 1300 is an infrared remote-controlled receiver. The schematic for the receiver's remote control and built-in detector are shown in Fig. 18-40 and Fig. 18-41. The remote uses the SAA1350 chip to decode the selected button and to output the appropriate pulse sequence. Q703 and Q705 along with transformer (T701) drive the two infrared LEDs. The third LED (LED703) is used for operator feedback. It signals the operator that the remote is transmitting. The infrared detector (LED851) drives the TEA 1009 chip, which amplifies the signals. The output is a series of pulses which exit via connector ZJ854. Notice that the same board also

Gould Inc., Dexcel Division
2580 Junction Avenue
San Jose, California 95134
Telephone (408) 943-9055
TWX 910-338-0180

→ GOULD
Electronics

TECHNICAL BULLETIN

To: Innovision Distributors	From: **Customer Service**	Date: 11/29/84

Bulletin #: S-003

Model #: DXR-1200

Subject: DXR-1200 Matching Procedure

This procedure explains how to match any Gould Innovision DXR-1200 Satellite Receiver to any DXS-1200 LNC or GSD-012 Downconverter. Perform the following steps:

1. Install the receiver per the installation procedures in the DXR-1200 Owner's Manual. Remove the top cover.

2. Set the CHANNEL FINE TUNE control to 12:00 o'clock. DO NOT change the setting until the matching is complete. Make sure the AFC switch, S1 on the PCB, is on. (This switch is set toward the rear panel.) Set the CHANNEL SELECTOR knob to channel 2.

3. Check that the ANTENNA POLARITY SWITCH on the rear panel is in the correct position. The receiver is set for NORMAL POLARITY (Satcom, Comstar, Telestar satellites) when the switch is set toward the horizontal and vertical adjustment controls.

4. Check the AFC offset. Without any satellite signals reaching the receiver (with only snow for a picture) the FINE TUNE METER should be perfectly centered. To receive only noise, adjust the rear panel LOW FREQUENCY ADJUSTMENT control so that the receiver is tuned into the noise below channel 1.

   CAUTION: Make sure that there is no Terrestrial Interference (TI) entering the LNC at the low end of the band. If there is, use the high end of the band (above channel 24) or cover the LNC waveguide opening with some metal such as aluminum foil.

5. If the FINE TUNE METER is offset during step 4, adjust the trim pot RV3 (AFC offset) to center the FINE TUNE METER.

6. Set the CHANNEL SELECTOR knob to channel 2. Adjust the LOW FREQUENCY ADJUSTMENT trim pot (accessed through the rear panel using a flatblade screwdriver) to center the FINE TUNE METER on channel 2.

7. Turn to a mid-band channel like 12 or 14. Adjust the MID FREQUENCY ADJUSTMENT trim pot (also through the rear panel) to center the FINE TUNE METER.

8. Turn to channel 24. Adjust the HIGH FREQUENCY ADJUSTMENT trim pot (through the back panel) to center the FINE TUNE METER on channel 24.

9. Adjustments done during steps 6,7 and 8 interact. It is necessary to repeat those steps until proper tracking (centering) is maintained when switching through the even channels.

Fig. 18-32. Technical Bulletin S-003. Courtesy of Gould/Dexcel.

Continued from page 303.

contains the front panel pushbuttons for channel selection, odd/even switching, video fine tuning, audio level and wide/narrow audio.

The tuning and function-selector schematic is shown in Fig. 18-42. The SAA1351 is the decoder chip that detects both the front-panel function switches as well as the serial data-pulses from the remote detector-module. The remote-control signals are input to the chip at pin 17. They are output on pins 11 through 15, which are also tied into the front-panel switches through blocking diodes. Thus the front panel and the remote buttons are active at all times.

If the remote control is used to change channels, then the output from pin 8 resets the channel selection to channel 1, just before pin 10 outputs a series of pulses which cause the 1360 IC to count up to the proper channel. If the front panel buttons are used, then they directly control which line is latched to ground by the 1360. When the line goes low, then one of the diodes in D304 or D305 is turned on, allowing a preset tuning voltage to go out line A to the rest of the circuit.

The five transistors (Q306 through Q310) are used to drive the four LEDs on the front panel. They in turn are driven by the decoding circuitry consisting of U310 to U317. This circuit decodes the output of the 1351 or the output of the various front-panel pushbuttons directly.

Figure 18-43 is the display decoder and driver circuit for the front panel LED readout. Whenever a channel is selected, not only does a tuning voltage get output through D304 or D305, but a number is specified by the various gates made up of U303 to U307. This is input to U308 and transistors Q304 and Q305. The two transistors are used to decide whether the tens digit is blank (turned off), a 1 or a 2. The 4008 decodes the input signals and outputs a BCD number to the 4511, which decodes the input number to correctly drive the LED readout.

The audio amplifier is shown in Figure 18-44. It has a 35 watts RMS/channel output capability into 4 ohms. The amplifier is built around a Sanyo hybrid module called the STK465. There are two internal fuses (F601 and F602) to protect the amplifier from damage in case the speaker wires get shorted. There is also a loudness compensation switch (SW601) which adds a boost at the low end to compensate for the poor low end response of small speakers. The sound is muted by Q604 and Q605 during turn on and during channel selection.

The full schematic is shown in Fig. 18-45. The i-f strip (in the upper left part of the schematic) consists of Q101, U101, Q102 and U102. The discriminator circuit is composed of D103 through D106. *Note*: D104 is shown backwards and the two delay coax pieces are not shown at all. Refer to the DXR 1200 detector circuit for their positioning.

If a swept waveform is put into the DXR 1300, then the i-f response should look like Fig. 18-46. Before sweeping the 1300, a 100 pF capacitor must be unsoldered from ground. This capacitor is soldered to T104's metal can and is in series with a 47 ohm resistor which is connected to the base of Q102. They improve the at-threshold performance of the 1300, but only after the i-f filter is aligned properly without them in the circuit.

After the discriminator circuit, the signal makes a U-

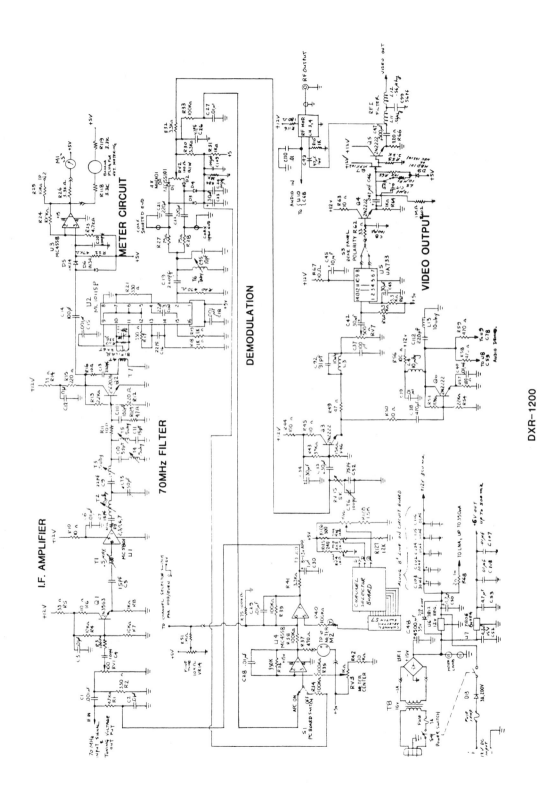

Fig. 18-33. Main board of the DXR 1200. Courtesy of Gould/Dexcel.

305

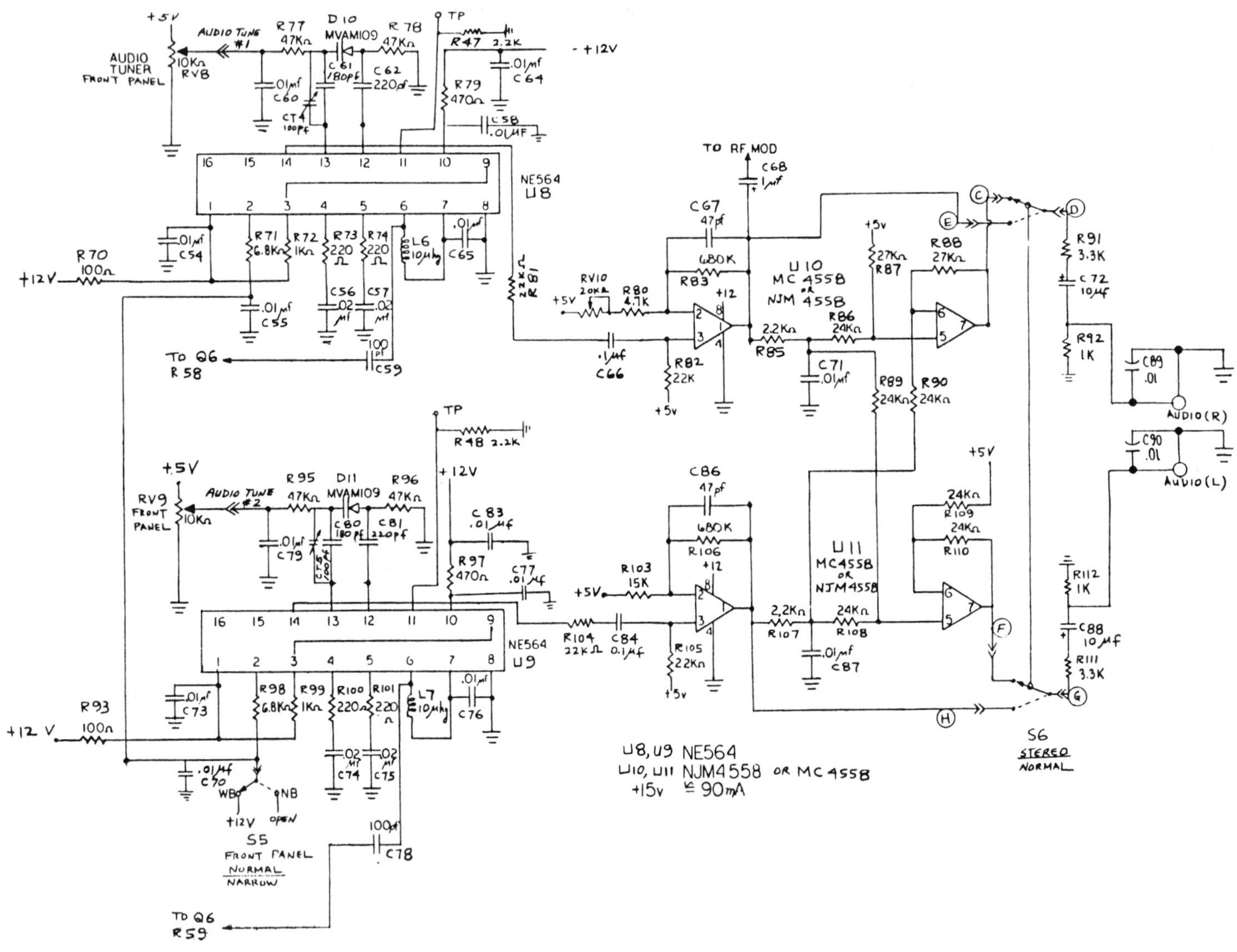

Fig. 18-34. Audio detector and matrix circuit from a DXR 1200. Courtesy of Gould/Dexcel.

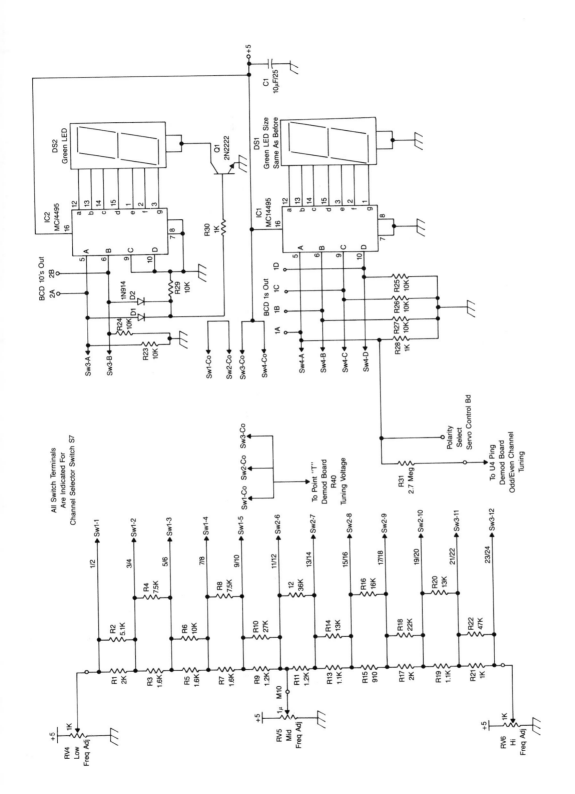

Fig. 18-35. Channel selector board with LED readout from DXR 1200. Courtesy of Gould/Dexcel.

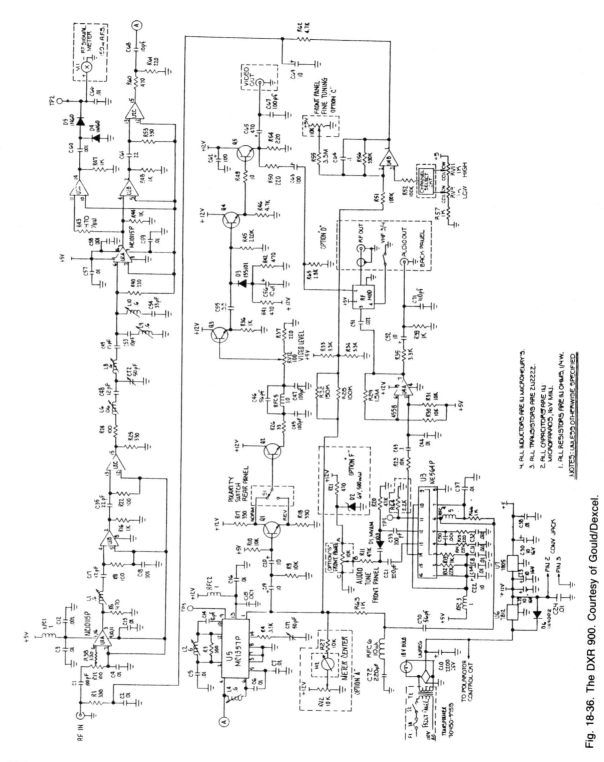

Fig. 18-36. The DXR 900. Courtesy of Gould/Dexcel.

Gould Inc., Dexcel Division
2580 Junction Avenue
San Jose. California 95134
Telephone (408) 943-9055
TWX 910-338-0180

GOULD
Electronics

TECHNICAL BULLETIN

To: Innovision Distributors   From: Customer Service   Date: 12/17/85

Bulletin #: T-007

Model #: DXR 900

Subject: DXR 900 Test Interface

Modified DXR 900 receivers supply unregulated +20 vdc to the LNA and downconverter. These are indentified by a stamp on the back panel (a circled CH). Modified receivers must not be used with LNC's, as the voltage is too high for the VTO to function properly.

This interface monitors the B+ voltage from the receiver, and if it is above +15 vdc it automatically switches in the 12 volt regulator, thus protecting the LNC from overvoltage.

It can also check to see if the receiver has been modified, if the LED lights up then the receiver is modified and has +20 vdc output, thus the receiver can only be used with a downconverter.

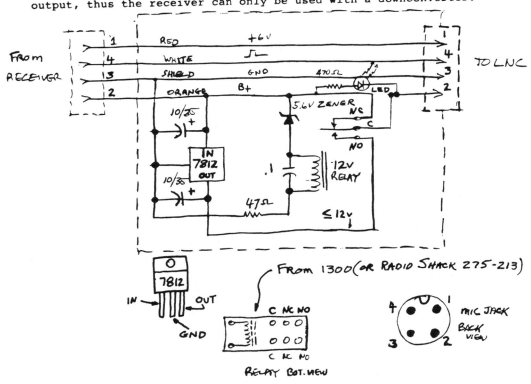

Fig. 18-37. Technical Bulletin T-007. Courtesy of Gould/Dexcel.

Gould Inc., Dexcel Division
2580 Junction Avenue
San Jose, California 95134
Telephone (408) 943-9055
TWX 910-338-0180

GOULD

Electronics

TECHNICAL BULLETIN

To: ALL INNOVISION DISTRIBUTORS   From: RICHARD MADDOX   Date: 11-15-84

Model #: XP-900   Bulletin #: T-006

Subject: INCREASING THE D/C AND LNA SUPPLY VOLTAGE

The XP-900 receiver is designed to work with the GT-series of LNA's which use a
+12 vdc power supply voltage. To insure XP-900 compatibility with all LNA's on
the market, the power supply voltage that feeds the D/C and LNA can be increased
to +20 vdc from the original +12 vdc.

This change has been implemented on all XP-900 receivers shipped since mid-
November 1984. Receivers that have the +20 vdc output are marked with a stamp,
a circled CH, on the back panel next to the HIGH FREQUENCY ADJUSTMENT access hole.

Modification of the XP-900 receivers without the stamp consists of cutting a trace
and installing a short jumper. Proceed as follows:

1. Unplug the receiver and remove the bottom cover.

2. Locate the +12 vdc trace that extends from the three pin plug to the voltage
   regulator. See Figure 1, (unmodified board).

3. As per Figure 2 (modified board), cut the +12 vdc trace as indicated, and,
   after scraping off the solder resist, jumper the trace to the +20 vdc output
   pin of the bridge rectifier.

4. Without plugging in the mic plug, turn on the receiver and check for +20 vdc
   on pin 2 of mic jack.

5. Replace the bottom cover.

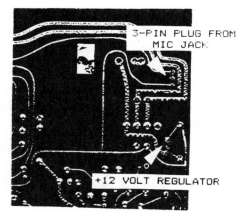

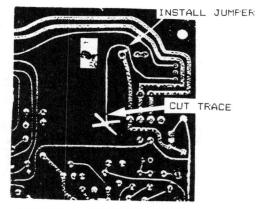

FIGURE 1. UNMODIFIED BOARD   FIGURE 2. MODIFIED BOARD

Fig. 18-38. Technical Bulletin T-006. Courtesy of Gould/Dexcel.

310

Gould Inc., Dexcel Division
2580 Junction Avenue
San Jose, California 95134
Telephone (408) 943-9055
TWX 910-338-0180

GOULD

Electronics

TECHNICAL BULLETIN

To: Innovision Distributors	From: Customer Service		Date: 11/29/84
Model #: DXR-900			Bulletin #: S-005

Subject: DXR-900 Matching Procedure

This procedure explains how to match any DXS-1200 LNC or GSD-012 Downconverter to a Gould Innovision DXR-900 Satellite Receiver.

Normally the CHANNEL FINE TUNE control is positioned within 45 degrees of the ODD or EVEN marks and should not require any major adjustments between channels except when Terrestrial Interference (TI) is encountered.

If more than a slight adjustment of the CHANNEL FINE TUNE control is required as channels are changed, it is an indication that the LNC needs to be rematched to the receiver.

NOTE: It could also mean that TI is entering the satellite receiver.

If the cable length increases or decreases from the standard 125 feet, the LNC or receiver is changed, or the matching controls are incorrectly adjusted, a rematching of the LNC and receiver may be required.

To rematch a receiver to an LNC requires that the dish be properly pointed at a satellite and the ANTENNA POLARITY be properly adjusted to receive ODD CHANNELS. Leave the receiver and LNC on for 15 minutes to allow for system stabilization before attempting any adjustments. Perform the following steps:

1. Set the CHANNEL FINE TUNE to the ODD mark.

2. On the back panel of the DXR-900 are four small holes labeled S, L, M and H (frequency controls). Behind the holes are three trim pots (TP's). M is not used on this model. DO NOT use excessive pressure when adjusting the trim pots since they are easily damaged.

    S is the 70 MHz signal level adjustment. It also affects the RF SIGNAL METER on the front panel.

    The L and H TP's affect the tuning voltage going from the receiver to the LNC. The L TP is the low channel (#1) adjustment. The H TP is the high channel (#23) adjustment.

    Using satellite F3R as an example, set the ANTENNA POLARITY control to VERTICAL. Set the CHANNEL FINE TUNE control to the ODD mark. Turn the CHANNEL SELECT knob to number one.

3. Adjust the L control, using a small flatblade screwdriver until channel one comes in with equal salt and pepper (black and white sparklies) in the picture.

Fig. 18-39. Technical Bulletin S-005. Courtesy of Gould/Dexcel.

Continued from page 311.

turn and heads to the right to Q103. Here it is split between the audio and video circuits. The video circuit consists of U103, and Q105 to Q107. The clamp diode is D107. The clamp can be turned off by changing switch SW101's position.

The audio circuit is almost identical to the DXR 1200's. The biggest difference is in the channel selection and volume control circuitry. The channel selection is done by U107 and U110, which are 4066 analog switches. The LA2600 (U108) is a dual channel VCA (Voltage Controlled Amplifier), which is controlled by the audio level UP and DOWN buttons. It determines the audio level going to the audio output jacks and the stereo amplifier. Early production units had the tap for the rf modulator audio after the VCA chip so that the volume would also change on the rf modulator. But feedback from the field made it clear that it was better to have the modulator audio at a fixed level. Most receivers had a modification done to them to change this.

The modification involved moving the plus lead of C186, which is the rf modulator blocking capacitor, from its original position (connected to pin 8 of U110 and the − lead of C185), to the + lead of C179. This changed the signal from after the VCA to before the VCA. Thus

a fixed level is obtained. In most receivers, the change involved cutting a trace and installing a jumper. By moving the jumper so that it jumps the cut trace again, then the sound level can be controlled going to the TV set. If speakers are not used with the receiver, then it would be advantageous to have the audio level controls affect the TV level. *Note*: Later units have a new board which has this change incorporated in it, and thus to change them involves cutting a trace on the top side of the board and jumpering across a gap put into the original trace. Figure 18-47 is a close-up of the audio section schematic. It shows the two steps involved in making the change.

Figure 18-48 is a circuit that can be used to drive a coaxial relay from the odd/even logic signal that drives the Polarotor I control circuit in the DXR 1200 or DXR 1300 (or any other receiver that uses a logic level signal to switch from horizontal to vertical). It can be put inside the receiver if built up on a small perf board. Be sure to insulate it from the internal circuitry of the receiver.

In the DXR 1200, the 10 K input resistor would connect to the satellite polarity switch on the back panel. It would connect to the center terminal (purple wire). In the DXR 1300 the 10 K resistor would be connected to pin 1 of the polarity control board. It is the small board that

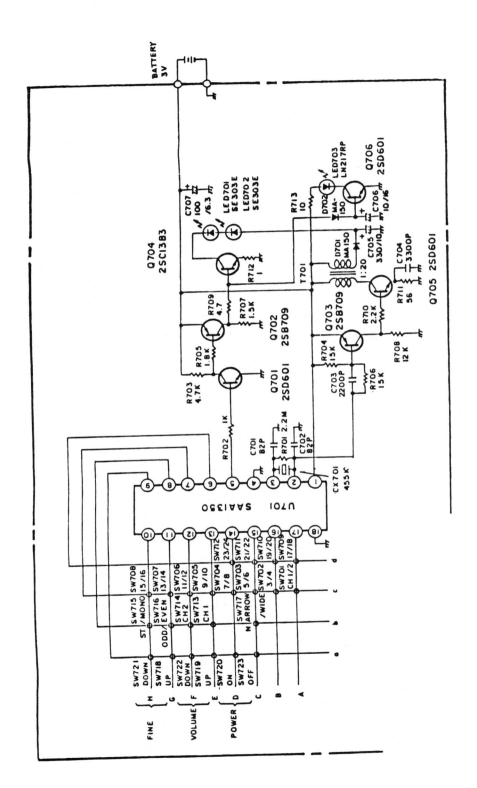

Fig. 18-40. The DXR 1300 remote control schematic. Courtesy of Gould/Dexcel.

313

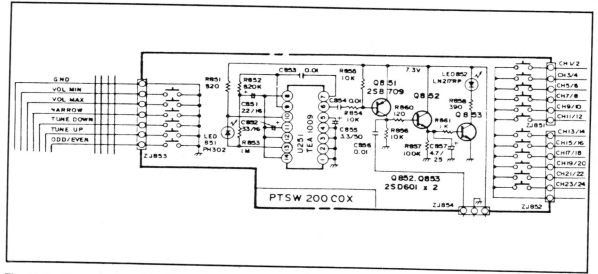

Fig. 18-41. Close-up of the infrared receiver circuit. Courtesy of Gould/Dexcel.

is plugged into the left side of the main board, and has a clamp that holds it in place. Pin 1 is the one toward the front of the receiver.

Figure 18-49 is the instructions for matching the DXR 1300 to an LNC.

LUXOR

One very important thing to remember whenever you are hooking up Luxor receiver is that there is +18 Vdc on the rf output. This will be shorted out and cause regulator failure if the output is directly attached to a TV set's 75 ohm input. This +18 Vdc is used to supply power to the 9536 remote sensor. If the 9536 is used, then it will block the dc from the TV set. If a 9536 is not used, then a dc block or a balun transformer must be used between the TV and the rf output to prevent regulator damage.

Beware of the shorting out the coax cable that runs from the receiver's +18 V jack to the down converter, as this will also cause regulator failure. The +18 volts must be checked with the down converter connected. This can be done by putting a two way splitter at the DC's output. Measure the voltage on the centertap of the unused output. This is placed at the output of the down converter. Alternately, you could use a power inserter attached to the input of the down converter as the test point. The voltage must be above +17 Vdc or you will have problems with the down converter. If it is below +17 Vdc, then the coax cable's center conductor is too small. RG-6 is recommended for both coax cables on runs of over

200 feet. On shorter runs, RG-59 can be used if it has a 20-gauge center conductor.

If for some reason it is not possible to replace the cabling and the voltage is 17 volts or less, then the voltage can be raised to +23 Vdc by changing the dc pick-up point inside the receiver. This is accomplished by unsoldering the red wire (that runs from the board to the +18 volt F connector) at the board and moving it to the DIN connector's +23 Vdc contact. The proper pin can be identified by the two red wires that are already attached to it. This change will also help stabilize the DC in cold weather because it increases the voltage drop in the internal regulator and thereby warm up the interior slightly. Alternately this change is not recommended if the DC is exposed to high temperatures, as in desert climates during the summer.

LNC and DC Matching

Before matching the LNC or DC's tuning range to the receiver and cable length, the receiver and DC must be turned on for at least 15 minutes to make sure that the system is temperature stabilized. If the weather is extremely cold outside, then allow about 20 to 30 minutes for the system to stabilize.

Once it has stabilized, turn off the afc. Select channel 1. If it is not tuned properly, then it can be fine tuned by pressing the up (right facing arrow) or down (left facing arrow) tuning buttons on the front panel. The speed at which the tuning voltage changes is set by the sound

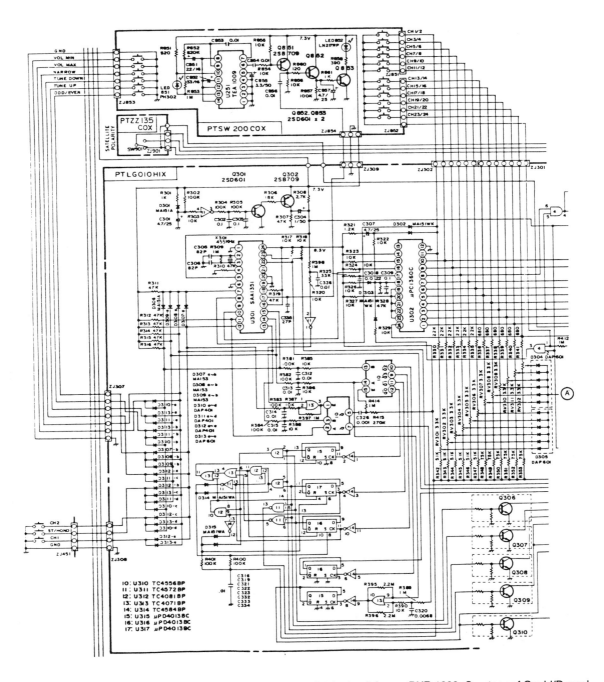

Fig. 18-42. Close-up of the tuning and front panel push button latch circuit from a DXR 1300. Courtesy of Gould/Dexcel.

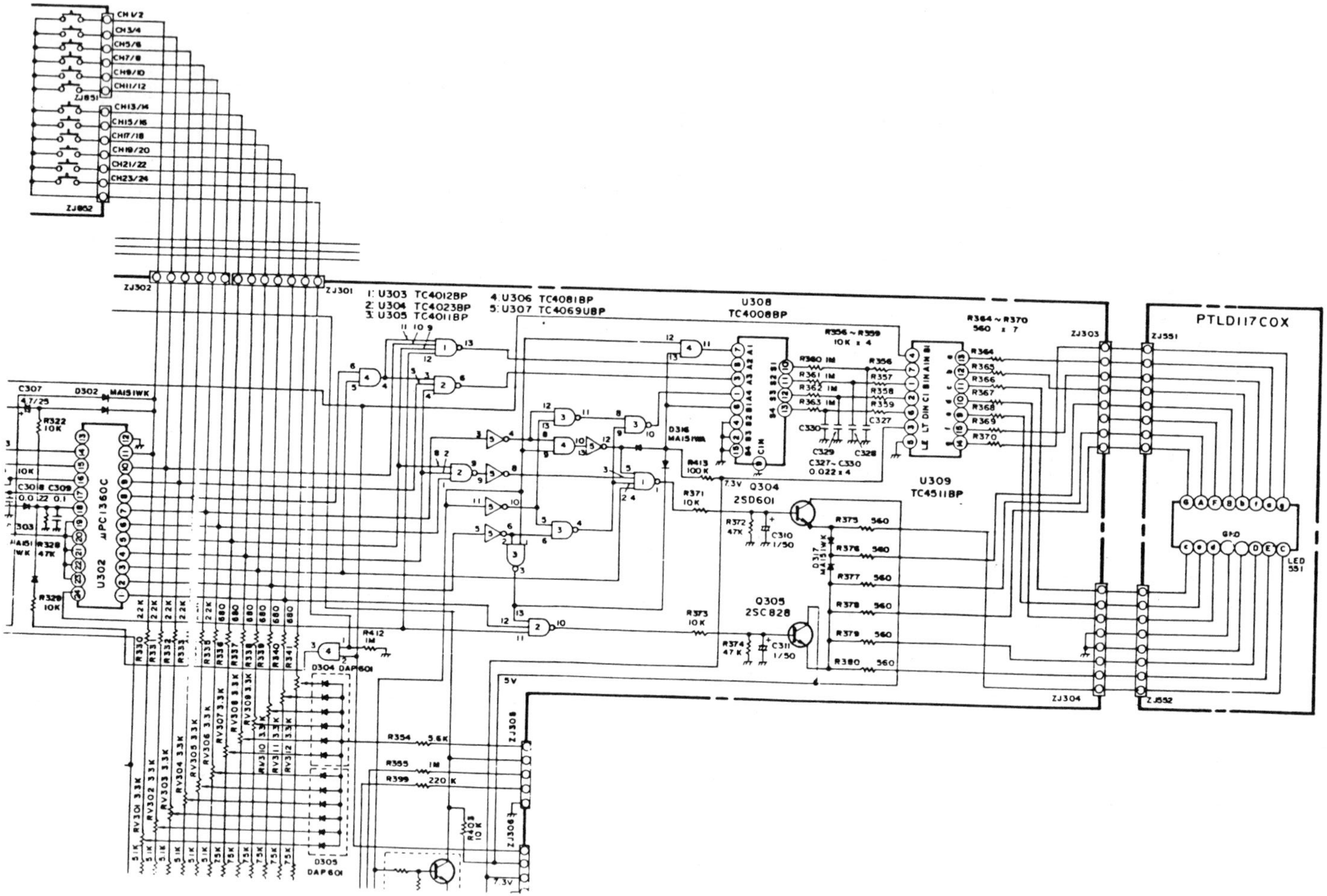

Fig. 18-43. LED display decoding and driving circuit. Courtesy of Gould/Dexcel.

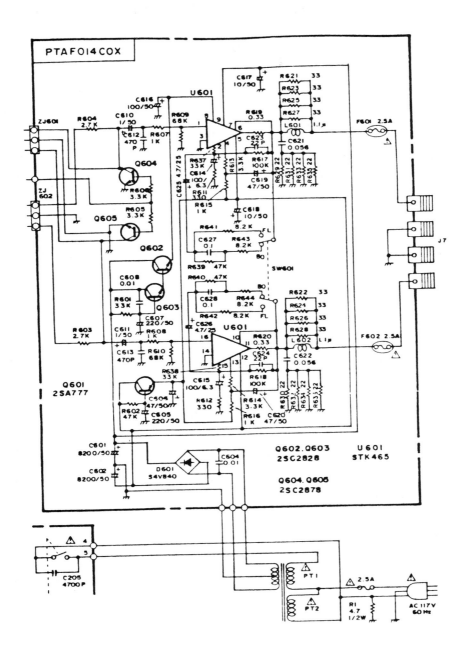

Fig. 18-44. 35 watt/channel stereo amplifier, from a DXR 1300. Courtesy of Gould/Dexcel.

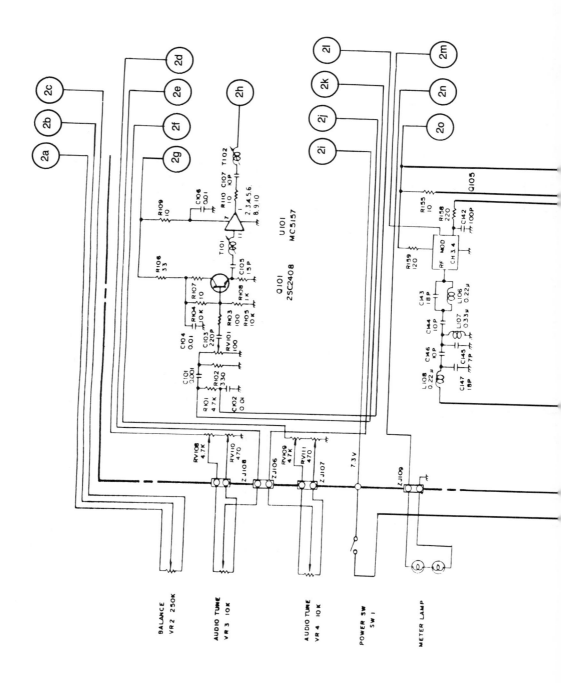

Fig. 18-45. The DXR 1300, full schematic. Courtesy of Gould/Dexcel.

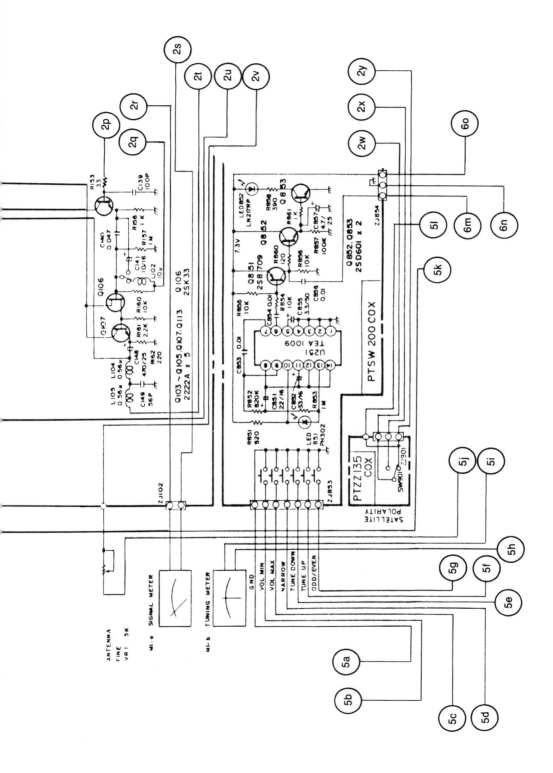

2

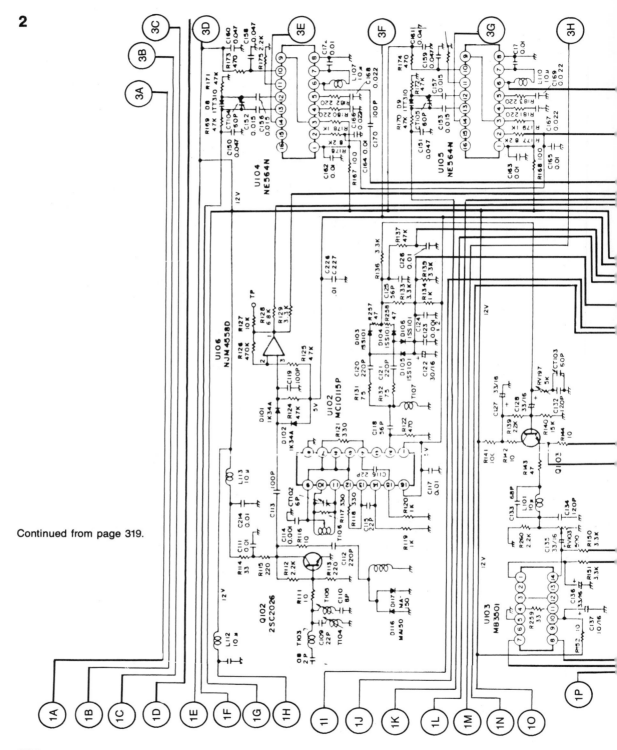

Continued from page 319.

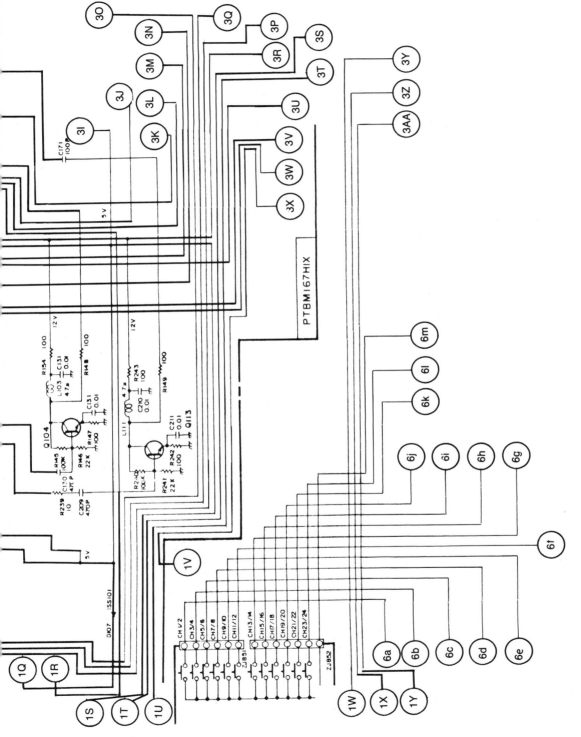

3

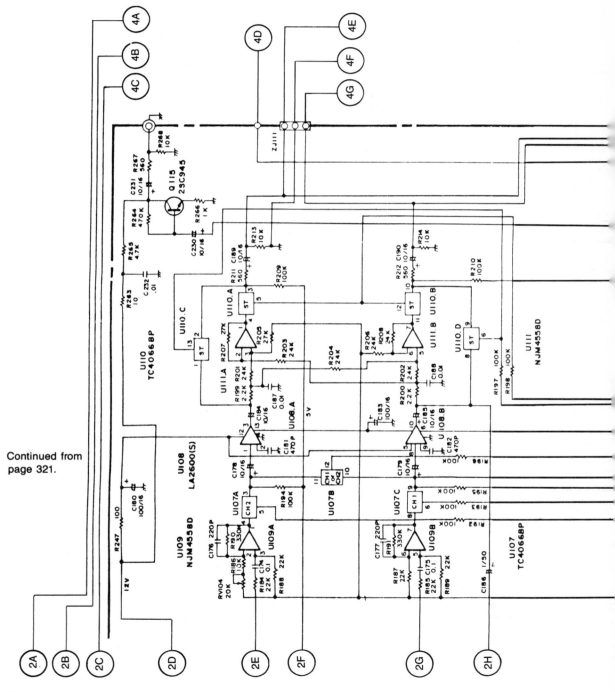

Continued from page 321.

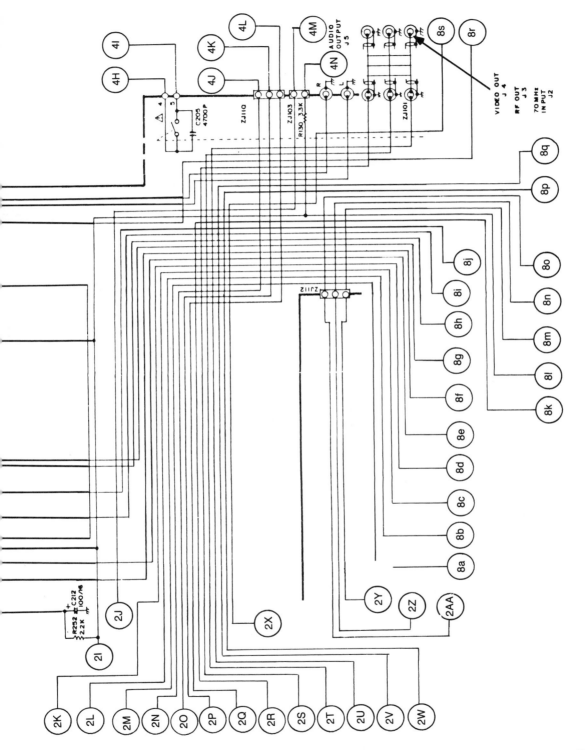

4

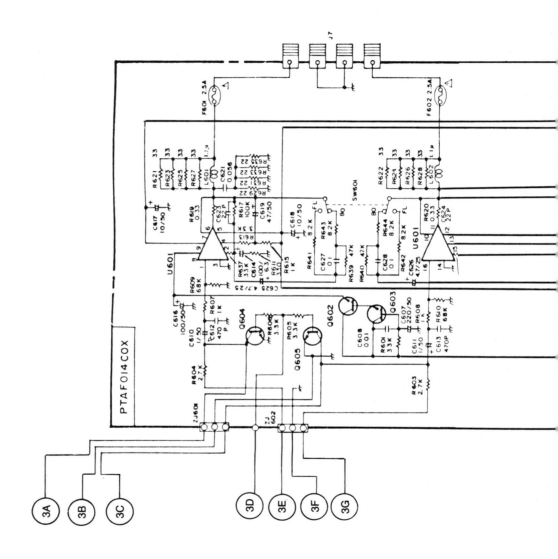

Continued from page 323.

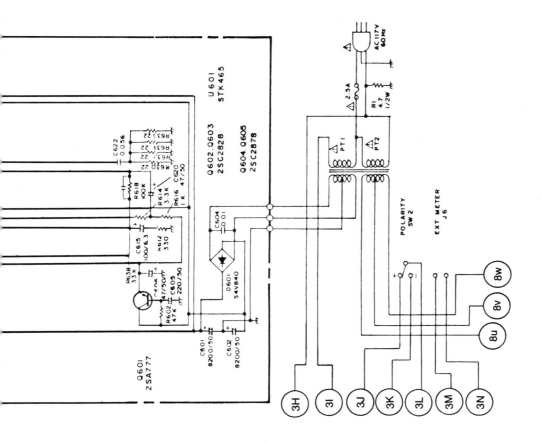

5

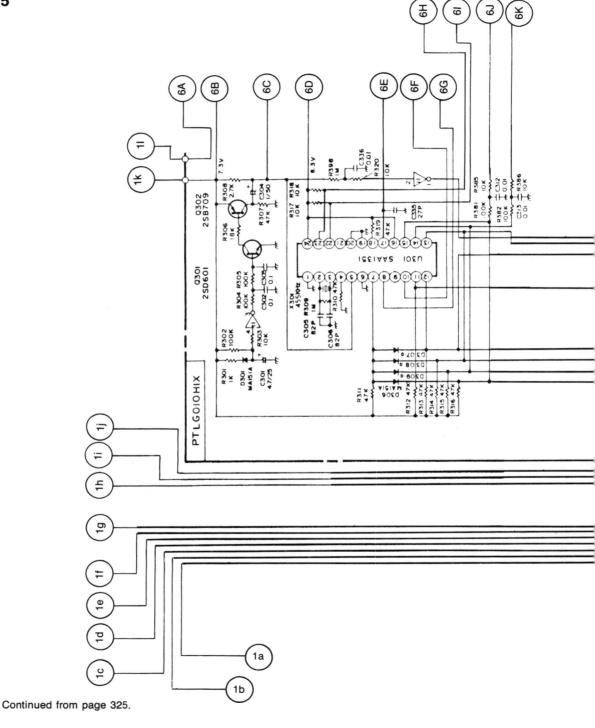

Continued from page 325.

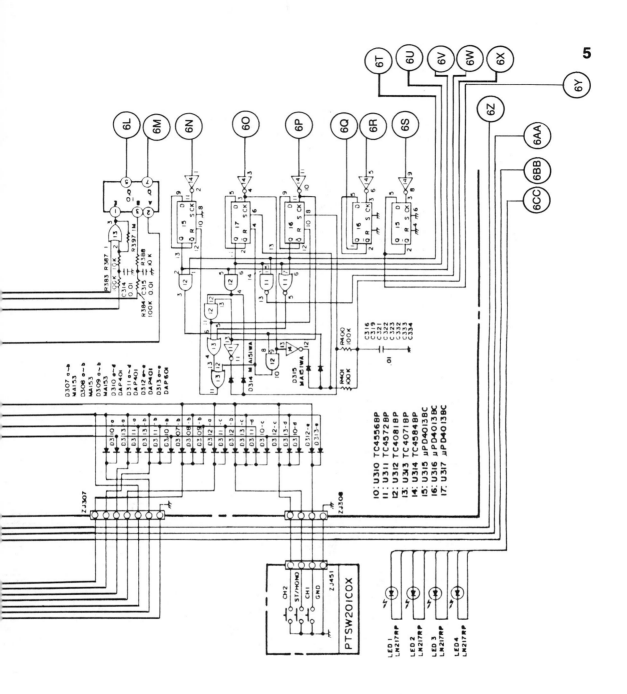

6

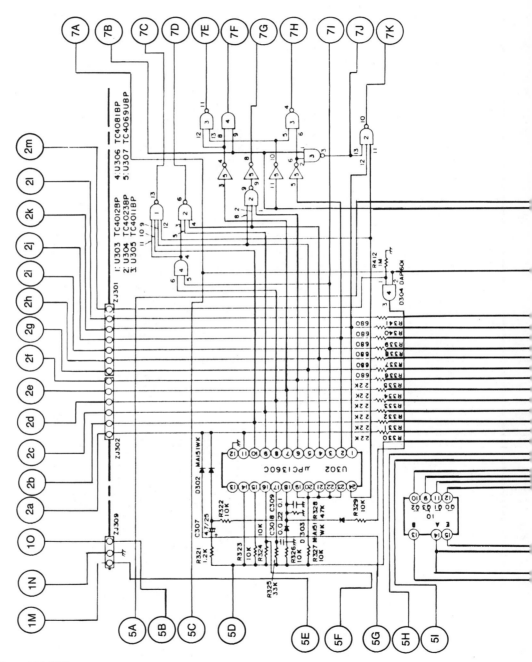

Continued from page 327.

328

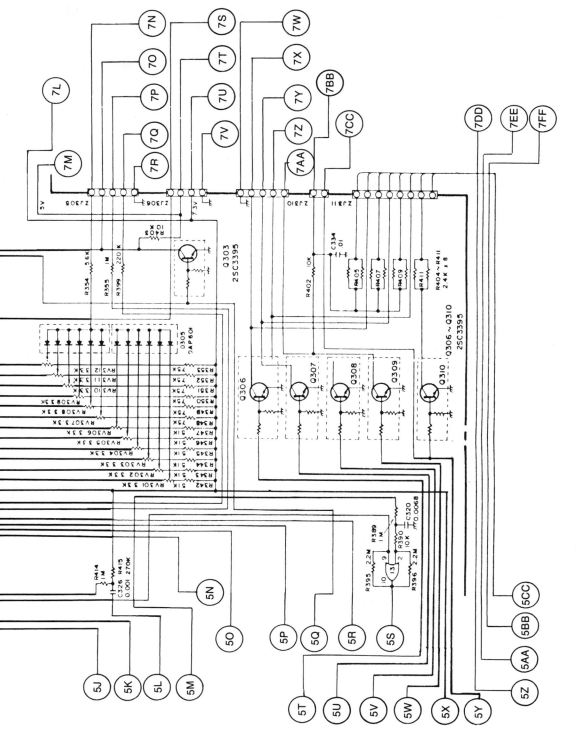

7

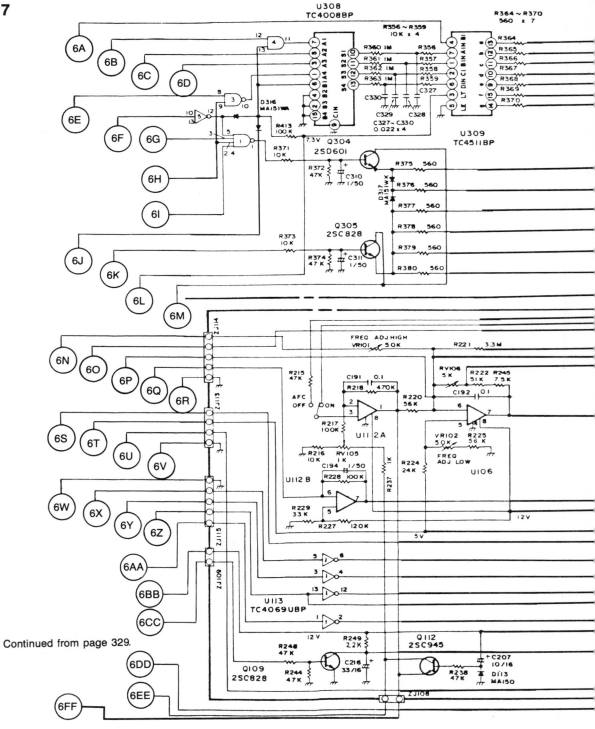

Continued from page 329.

330

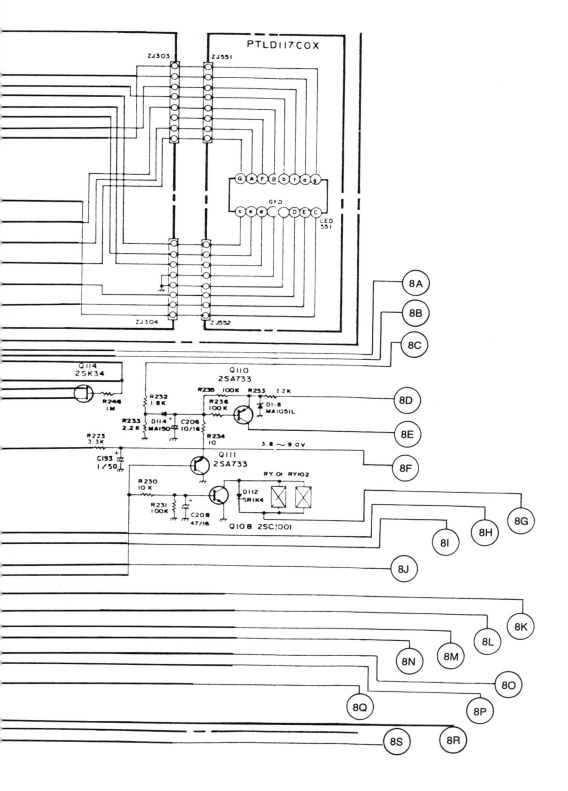

PTLDII7COX

ZJ303   ZJ551

G A F B b f g

GND

c g d   D E C
LED 551

ZJ304   ZJ552

8A

8B

8C

Q114
2SK34

Q110
2SA733

R246
1M

R232
1.8K

R238 100K R253 2.2K

R236
100K

D118
MA1051L

8D

R233
2.2K

D114
MA150

C206
10/16

8E

R223
3.3K

R234
1Ω

3.8 ~ 9.0v

8F

C193
1/50

Q111
2SA733

R230
10K

RY101 RY102

D112
SR1K4

8G

R231
100K

C208
47/16

8H

8I

Q108 2SC1001

8J

8K

8L

8M

8N

8O

8P

8Q

8R

8S

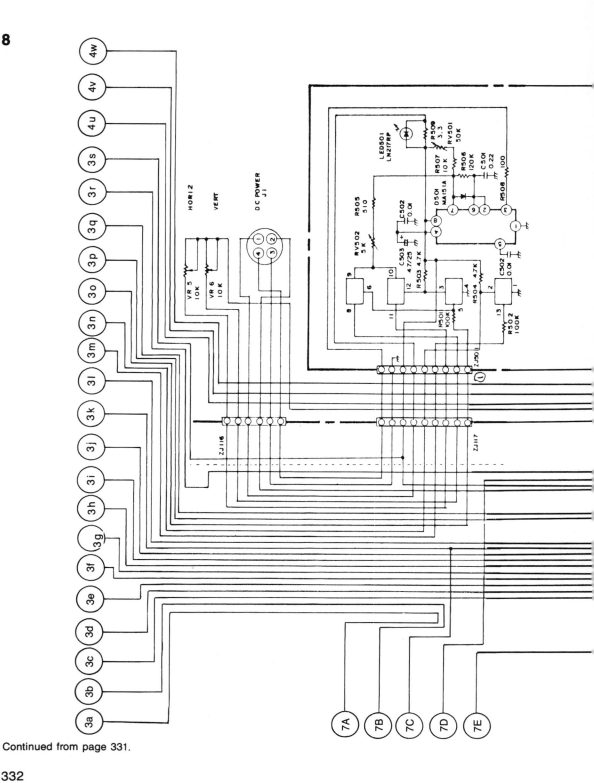

8

Continued from page 331.

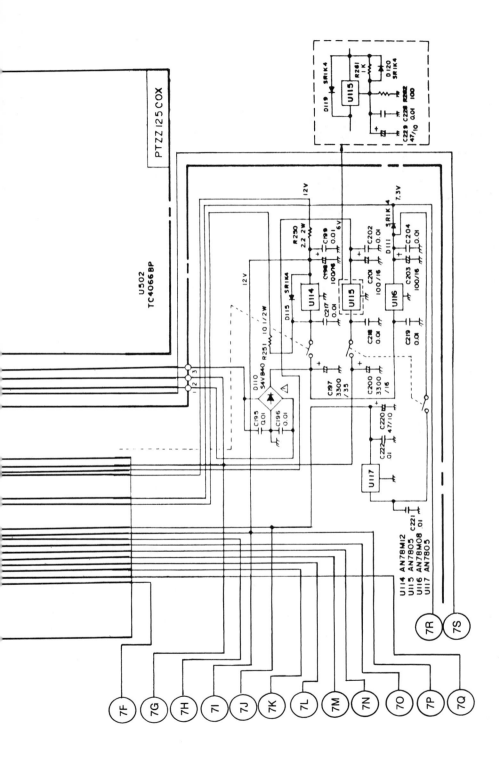

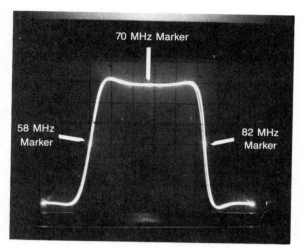

Fig. 18-46. Ideal i-f response of a DXR 1300.

select setting (tunes fastest in mono 1) and on how long the button has been pressed. Once the video is fine tuned, then the audio mode can be chosen and stored by pressing the store button. This procedure is repeated for all 24 channels. Be sure to change the audio mode to either direct or mono 1 for the fastest tuning.

Once all 24 channels are tuned in, then the afc should be switched on. The picture quality should not change at all when the afc is turned on. If it does, then check to see if it is only on one or two channels or if it is on all channels. If all channels get slightly worse, then the afc centering trim pot should be readjusted. If switching the afc on causes only certain channels to fill with sparklies or to be blacked (or greyed) out, then strong TI is entering the system and causing the afc to try to lock onto it. A switchable 60/80 MHz TI filter can be tried (make sure it passes dc voltage) to see if it will decrease the interference. If the TI carriers are wideband, then the notch filters will not fully trap out the interference. In such cases, the afc should be switched off and the fine tuning buttons used to minimize the TI by tuning away from it (i.e., by tuning above the center frequency of the channel or below the center frequency of the channel).

Drifting Channels

Luxor advises that the receiver be left on continuously, especially during the winter, to prevent turn-on drift. The receiver only draws about 50 watts, so this shouldn't be a problem. If the receiver is turned off and the DC allowed to get very cold, then it is possible that the oscillator will take a few minutes to start up when

first turned on. Sometimes 5 to 10 minutes of warm-up is required before the channels will tune in.

There are several mechanisms that will cause the channels to be in need of continual fine tuning. The foremost is the temperature with moisture second. The first thing to try to prevent drift is to leave the receiver continuously turned on. Make sure the afc is turned on and that it is aligned. If the DC is not mounted behind the dish (rather it is attached directly to the LNA), then move it behind the dish and put it in a weatherproof box. It should be wrapped in insulation for maximum stability. If an LNC is used, it should be covered with a breathable cover. Wrapping the LNC with insulation will also help stabilize the LNC.

If rapid channel changing or jumping is occurring, then it is most likely a sign of moisture entrance into the connectors and hence into the cable's dielectric material. It could also signal oscillator failure in the DC (or LNC). To check for moisture intrusion, disconnect both ends of both coax cables. Measure the resistance from the center conductor to the shield. There should be infinite resistance. If there is any resistance at all, then there is definitely moisture in the cable and it will have to be steamed out by heating up the end of the cable with a hair dryer of other heat source. It the resistance is measured while doing this, then the meter will show a drop in resistance as it is first heated, and after a couple minutes, it will begin to rise until it is infinite again. An alternate method is to cut off the last 5 or 6 inches and put new ends on them. Be sure to check the resistance on the new ends before connecting them. Usually water only gets a couple inches into the cable, but it's better to be safe than sorry.

If there is water in the cable, there's a good possibility it has also entered the DC or LNC. It is not recommended that the units be opened in the field. Instead try putting them in a warm dry place, like an oven set on low heat (about 150 degrees) overnight. If the cable has been dried out, and the weather is dry, then reconnect the system. If after 20 to 30 minutes of warm-up there is still noticeable drift, then the DC or LNC has probably been damaged by the water entrance. Substitution of the DC or LNC will definitively prove this.

If the drift is stopped and everything appears to be working normally, then make sure the cable ends and connectors are sealed. The best way to do this is by using self adhering tape (or even regular electrical tape) and a sealant. Wrap the tape around the connector on the DC or LNC's body and continue up past the coax connector and onto the coax cable for about 1 inch. Cover the tape with your sealant. Chapter 4 lists several types of seal-

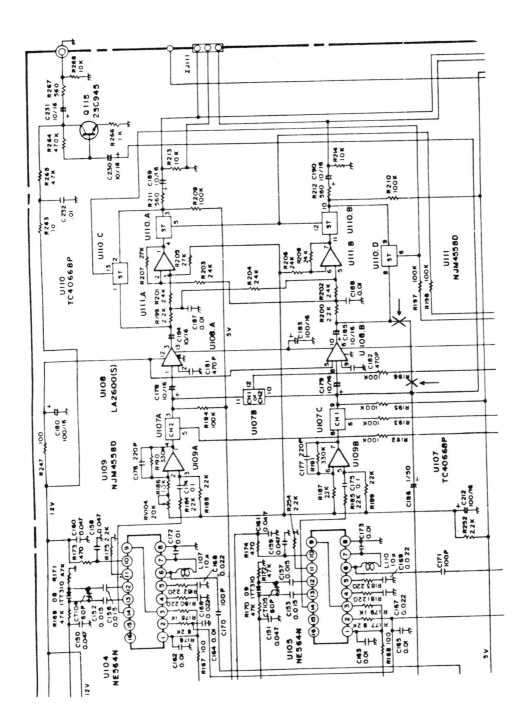

Fig. 18-47. Modification that was done to output a fixed audio level to the modulator. Cut trace and jumper as shown by the arrows +lead of C186 to + lead of C179. Courtesy of Gould/Dexcel.

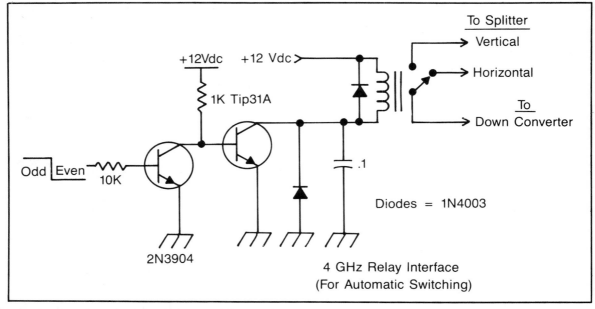

Fig. 18-48. Interface circuit for driving a 4 GHz relay from a logic pulse.

ant that can be used. This is one situation where the sealant should be applied to the body of the LNC or DC.

It should be noted that some of the down converters that were sold with Luxors are very temperature sensitive, and will require retuning as the weather and temperature changes. If you appear to have one of these models, you'll have to either retune the channels as the seasons change, or place the DC in a weatherproof box and wrap it with insulation. If the temperature is normally cold for most of the year, then heat tape could be used to keep the DC warm. The receiver should also be left on continuously.

Other Tips

Whenever you are installing a system that is using a Polarotor I, always mount the feedhorn so that the LNA (or LNB or LNC) is mounted at a 45 degree angle from the dish. This ensures that the probe travel will be sufficient to get both polarizations. If only one polarity can be received, then try physically rotating the feedhorn 20 degrees clockwise. If this does not improve the situation then rotate it another 20 degrees. See Chapter 6 for more information on polarity controllers and polarization.

There are two pots on the bottom of the receiver that are used to set the shortest and longest Polarotor pulse width that can be sent to the PD. These should only be adjusted by using an oscilloscope. Attach the scope to the

polarizer pulse output on the back panel of the receiver. The pulse width should be between .8 and 2.2 ms. To find the minimum and maximum settings, hold down the POL FINE buttons on the remote control until the length does not change. If it is within spec, then press the other POL FINE button to find the other end point. If either direction yields a length of less than .8 ms or more than 2.2 ms then adjust the corresponding trim pot.

During extremely cold weather the PD may freeze up, although more likely the voltage drop to the motor is too much. Check to see that there is 5 volts between the red and black Polarotor wires. If it is below 5 volts then larger diameter wire will be required. In some cases, putting a 1,000 μF electrolytic capacitor between the red (+) and black (–) wires will help. This will also cure the hum bars that are sometimes caused by the Polarotor's current draw when the motor is running.

All audio tuning is done by the 6 trim pots extending from the back of the receiver. These are labeled as #10 to #13 in the photo on page 6 of the owner's manual. They are tuned typically so that MONO 1 is set for 6.8 MHz, MONO 2 for 6.2 MHz, DIRECT (or discrete stereo) is set for L equal to 5.8 MHz and R equal to 6.8 MHz, and MATRIX is set using the + for 6.62 or 6.8 MHz and the – for 5.8 MHz. Once these are stored for the various transponders, then they can be recalled using the audio step button on the remote. If the preprogrammed mode

Gould Inc., Dexcel Division
2580 Junction Avenue
San Jose, California 95134
Telephone (408) 943-9055
TWX 910-338-0180

 **GOULD**
Electronics

TECHNICAL BULLETIN

To: Innovision Distributors	From: Customer Service	Date: 11/29/84
		Bulletin #: S-004

Model #: DXR 1300

Subject: DXR 1300 Matching Procedure

This procedure explains how to match the Gould Innovision DXR 1300 Satellite Receiver to the LNC. Observe the following steps:

1. Install the DXR-1300 Satellite Receiver per the installation instructions found in the DXR-1300 Owner's Manual. Be sure the receiver has been turned on for a minimum of 30 minutes.

2. Unplug the receiver from the wall plug for 30 seconds. Plug it back in and turn it on. This procedure ensures that the VIDEO FINE TUNE offset is centered. DO NOT touch either VIDEO FINE TUNE button until matching is complete.

3. Select channel 1. Be sure the SATELLITE POLARITY button is in the proper position for the satellite being used. Adjust VR102 through the back panel (low frequency adjustment) to get channel 1. Adjust for perfect centering of the CENTER TUNING METER.

4. Select channel 23. Adjust VR101 (high frequency adjustment) to get channel 23. Adjust for perfect centering of the CENTER TUNING METER.

5. Switch back and forth between channels 1 and 23. If there is any movement in the CENTER TUNING METER or if either channel fails to come in, repeat steps 3 and 4.

    NOTE: There are individual channel tuning trim pots inside the receiver. If all channels do not center tune, then these trim pots may need to be adjusted. The trim pots are located on the CHANNEL SELECTOR BOARD behind the CHANNEL SELECTOR BUTTONS. They are labled RV1 through RV12. RV12 affects channel 23, RV11 affects channel 21 and so on. It will be necssary to remove the top cover to reach these adjustments.

    Once all 12 odd channels are perfectly centered in the CENTER TUNING METER, switch to an even channel. Adjust RV106 (even channel offset trim pot) to center the channel using the CENTER TUNING METER. Check the other even channels. TI may cause some deflection of the CENTER TUNING METER, so perfect tuning of all channels may not be possible. Adjust RV101 (RF INPUT level) trim pot to get a reading of between 4 and 5 when receiving signals from Galaxy 1.

    Replace top cover. DO NOT adjust any other internal controls without the proper equipment and alignment information.

Fig. 18-49. Technical Bulletin S-004. Courtesy of Gould/Dexcel.

or frequency does not match when changing satellites, then the AUDIO TUNE buttons on the remote can be used to tune in the desired channel. First press the TUNABLE AUDIO button. Then select the desired mode by pressing the AUDIO STEP button. Once the mode is found, then the channels can be tuned by pressing the AUDIO TUNE up and down buttons.

It should be noted that full ccw on all the back panel controls is approximately 5.5 MHz, while full CW is approximately 8.0 MHz. On most matrix and discrete broadcasts the second carrier is almost always at 5.8 MHz. It is typically the first carrier that is tuned as the control is moved cw from a full ccw position. In a matrix stereo broadcast, the difference channel is located at 5.8 MHz and will be very low in level or even totally quiet depending on the programming. It is typically found at about the 9:00 position on the DIRECT L and MATRIX – trim pots. If switching to matrix stereo causes one audio channel to quit, then the two trim pots are tuned to the same subcarrier. See Chapter 11 for more information on audio tuning and audio transmissions in general.

There are internal video and audio level adjustments, as shown in Fig. 18-50 (for the 9550) and 18-51 (for the Skantic receiver). The video should be adjusted for 1 volt peak to peak on ESPN or a similar transponder with consistent video levels. If the TV sound has a buzz t it when lettering appears on the screen, then the video level is probably set too high and should be turned down slightly. The video level is similar to the contrast control on your TV. If it is turned down the contrast control will have to be turned up slightly to compensate. Note that there are trim pots for both WB (wideband) and NB (narrowband), so check the AUDIO switch setting on the front panel before adjusting the trim pots.

Skantic Receiver

The 9539 receiver is basically the same a the 9550 without the interfacing for the motor drive or remote control. The biggest difference is in the polarization adjustment procedure. There is a trim pot on the front panel (POL FINE) versus the POL FINE buttons on the remote control for the 9540 and 9550.

The easiest way to set polarity is to hook the PD directly to the receiver. Set the POL FINE control at 12:00. Press the EVEN/ODD button and mark the probe's position. Press the button again and notice the probe's position. It should be 90 degrees from the first mark. If it is not, then adjust the trim pots on the bottom of the receiver for a 90 degree swing as the EVEN/ODD button is pressed. Make sure that there is skew (POL FINE)

adjustment in both settings and in both directions. If the probe will not move any farther in one direction as the POL FINE control is turned, then the probe is against the stop and both settings must be readjusted to properly center it. Refer back to the skew polarization chart in Chapter 6. It shows the typical pulse widths to achieve proper polarization. These can be measured on the pulse output terminal on the back panel.

Once it is set at the receiver, the PD should be placed on the feed such that the LNA is at a 45 degree angle to the dish or so that the two marks line up in a horizontal and vertical plane. If polarization is not exact (with the POL FINE control set at 12:00), then the PD should be rotated for proper polarization. *Note*: F-3R is offset from all the other satellites and so should not be used as a polarization reference. It is better to use G-1.

There are no separate baseband or unclamped outputs. Baseband video is obtained from the collector of TA01. To get prefiltered but unclamped video use the collector of TA05. The circuitry used in the Skantic is basically the same as the 9550, but parts designation and board location is different.

9534 Antenna Actuator Controller

The 9534 is covered in some detail in Chapter 14, so you should refer to that chapter. The most important points to remember when installing or troubleshooting the controller is to be sure that shielded wire is used to carry the feedback and polarization signals and that the bypass capacitors and resistors as noted in Chapter 14 are added. Both the controller and the receiver should be plugged into a line filter to protect them from power line transients adversely affecting the circuitry. Another important point, especially in areas prone to lightning storms, is to make sure that the dish is properly grounded. Doing all of the above will cut down tremendously on drive problems.

If the microprocessor does go into an interrupt mode and it will not respond to the front panel buttons or the remote control, then powering off and then back on should reset it. As a safety precaution, it is a good idea to either put a heavy duty switch in line with one of the motor leads or to unplug the MOTOR connector whenever the unit will not be used for an extended period of time. It could also be unplugged from the ac line or turned off, but this will cause the batteries to wear down and may cause loss of memory.

The best place to install the motor protection switch is right above the MOTOR connector. Use a single-pole,

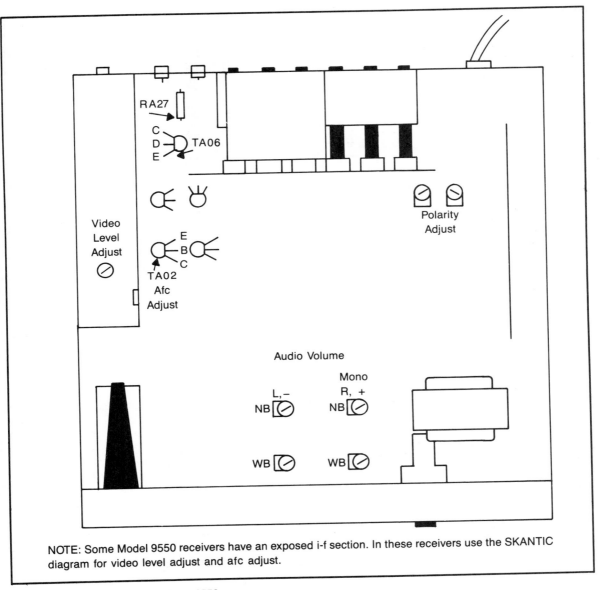

NOTE: Some Model 9550 receivers have an exposed i-f section. In these receivers use the SKANTIC diagram for video level adjust and afc adjust.

Fig. 18-50. Adjustment location for a 9550.

single-throw switch capable of handling 6 amps at 35 volts. Connect it to the green wire from the connector and to the corresponding 1 ohm, 10-watt resistor. When the switch is open, then power cannot be applied to the motor. In this position, there is no possibility of accidental dish movement due to line spikes, lightning, or other electronic failure. If a keyswitch is used, then it will double as a parental lockout. To use it as such, move the dish to the desired satellite or even between satellites are remove the key.

SAT-TEC

The R-5000 by SAT-TEC is one of the best selling low-end receivers. It has undergone several performance im-

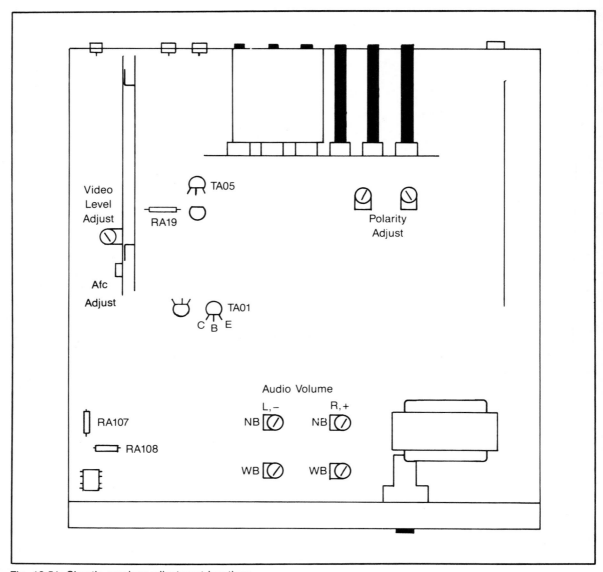

Fig. 18-51. Skantic receiver adjustment locations.

provements over the last four years that it has been available, but overall, it is still basically the same circuit it started out as. The receiver consists of the D-2000 down converter (DC), the R-5000 receiver and a wall socket ac transformer. The receiver is all-on-one board design with a plug-in rf modulator board. All connections are through RCA jacks on the back panel. The receiver does not offer any polarization control, although there is sufficient power available internally to power a Polarotor II.

Circuit Description

The down converter is a single-conversion, image-reject mixer design with the tuning voltage being supplied up the coax. The tuning voltage is typically +2 to +8 Vdc. A separate shielded cable is used to supply the power supply voltage. Any voltage from +15 to +24 Vdc can be used to power the DC as it has an integral 12 volt regulator. The unregulated voltage is passed on to the input connector (type-N) to supply voltage to the LNA. If the down converter is used in multiple receiver installations, the diode that is in series with the supply line to the N connector can be removed so that an external dc block is not required. The LO leakage out the input connector is very low for a single conversion DC (-60 dBm) because a half-frequency LO and mixer are used. Thus an isolator or splitter with 20 to 30 dB isolation should be sufficient to keep crosstalk to a minimum.

The receiver consists of an i-f amplifier that uses two NE021 UHF transistors before and two more after the 70 MHz filter. The filter is a tunable LC type using five coils for adjusting the low, mid, and high sections of the filter, as well as the upper and lower skirts. There is no input attenuation or level control. The only internal control in the i-f is a meter sensitivity control labeled METER CAL (R18). It does not affect the i-f signal level.

After the i-f amplifier/filter section there is an ECL chip (MC10116) that serves as the limiter. A 1496 balanced demodulator is used for recreating the video signal from the carrier. Video amplification and clamping is done by using standard npn transistors. The audio detector is an NE564 PLL chip. The only adjustment in the video or audio circuits is the front panel AUDIO TUNE control. There is no video level control or frequency adjustment.

Matching DC to Receiver

The R-5000 is a continuously tuning receiver (meaning a potentiometer is used), thus a separate fine tune control is not necessary. DC tuning voltage matching is set by two pots, a LO SET (R35) and a HIGH SET (R39). These actually are only used to match the received channels to the channel numbers printed on the front panel. The trim pots will need to be adjusted if the quadrature adjustment setting (C63) is changed, the afc centering setting (R34) is changed, the DC is changed, or if the coax cable length or RG-type is changed.

Quadrature Detection Set-up

To adjust the 1496 chip for proper quadrature detection, tune in a weak channel for best reception. Adjust C63 for the best picture. If necessary, retune the front panel channel control. Retuning C63 may offset the channel selector's position, if this happens the LO and HI set pots can be readjusted to properly match up the numbers on the front panel.

Checking the AFC Centering

When a good clear picture is obtained on any chan-

nel, there should be no voltage differential on the two test points called out as TP2. If there is any voltage, adjust R34 for zero volts. Again recalibration of the LO and HI pots may be necessary.

Troubleshooting

Always verify that both $+12$ Vdc and $+20$ Vdc are present before suspecting anything else. The $+12$ Vdc is measured on the front leg of the 7812 voltage regulator. If no voltage is present, check the rear leg for $+24$ volts. If it is present, then the regulator is defective or there is a short on the $+12$ volt line. The output leg of the regulator could be carefully unsoldered and checked again. The $+20$ Vdc regulator is checked by measuring the center pin of J2 (DWN CNVTR PWR). If it is damaged, check for a short in the DC power supply cable, the DC, and the LNA.

If there is no input voltage to the regulators, check the input connector for voltage. If it's present try cleaning the jack and reseating the plug. The most likely failure points are the wall mounted transformer, the interconnecting cable or plug, the four diodes, and the voltage regulators.

No Audio, Video, or Meter Reading. Check the DC by substitution or by running the 70 MHz line into another receiver. Check the LNA and DC cables for moisture. Check for a tuning voltage of $+2$ to $+8$ Vdc on the center conductor of the 70 MHz cable. If it's not present, check U3 (LM358) pin 7 for the same voltage. Check that pin 6 changes as the front panel pot is changed. Check the LO, HI and channel tuning pots for continuity.

Check for water in the coax cable. This is most easily done by disconnecting both ends and measuring from the center conductor to the shield. There should be no resistance at all. If there is any resistance, the tuning voltage can be shunted to ground. To repair, cut off the last six inches of cable at the D/C and put on a new connector.

No Audio, Video OK. Check for audio at the RCA jack and out of the TV. If there is no sound coming from either place, substitute U4 (NE564). If the problem is still present, check Q12 and Q13 for proper bias and signals.

If the sound is going to the rf modulator but not the back panel RCA jack, check C53 (a 10 μF cap next to J3) and the RCA jack itself for a cold solder joint.

If the sound is not going to the TV but is present at the RCA jack on the back panel, try fine tuning the TV. If that doesn't get it, adjust the 4.5 MHz can on the rf modulator board (it is the metal can coil on the rf mod board). Do not adjust beyond one full turn in either direction. Mark your starting point so that the control can be left in that position if it doesn't solve the problem. If there is still no rf audio check for audio on pin 1 of the rf mod connector (it is the one closest to the back panel). If it is present, check that CR-1, the diode right above the metal can, is passing the audio. If it is try changing the LM1889 IC.

No Audio, or Video, but Good Signal Indication. Check for raw video output. If present check Q15 for a base/emitter or base/collector short. If there is no raw video output either, check Q6, U2, and U1. U1's output is on pin 15. There should be about 650 mV of rf present at that point. U2's output is pin 6. There should be 350 mV of video present there. Q6 should be checked for proper bias if the correct signal is obtained from pin 6 of U2.

No Meter Indication but Good Video and Audio. Check the setting of R18. Check R18 continuity. Check the meter movement by momentarily placing a voltage across the meter. This is most easily done by applying a low voltage to Q5, which is the transistor right next to R18. Hook-up a 9-volt battery to the chassis and to a 3.3K resistor. Momentarily touch the resistor to Q5's collector (which is the leg toward the rear of the receiver) while viewing the meter. If it moves at all, then Q5 is probably defective. If it does not move, then the meter is probably defective.

Jumping Channels, Tuning Instability. Check for moisture in the DC and the interconnecting cables (especially the 70 MHz cable). Check for continuity in the tuning potentiometer. Check for continuity in the HI and LO set pots.

STS

The MBS-AA by STS was covered in Chapter 14 on motor drives. Since it is virtually identical to the Luxor 9534, the section on troubleshooting the Luxor model applies to it as well. However it does use a different actuator than the Luxor does. The arm attachment and cabling drawings are shown in Fig. 18-52.

Figure 18-53 is a drawing of the MBS-SR receiver's internal adjustment locations. SW102 is the afc on/off switch. It should be left in the on position. RV109 is the signal-strength meter level control. It is turned cw to increase and ccw to decrease meter indication. It is a very touchy adjustment. To adjust the afc (R103), first turn off SW102, and then adjust the video fine tuning for the best picture. Turn SW102 on. If there is any picture quality change, then adjust RV103 for the best picture quality.

RV104 is the audio balance control. It is adjusted by tuning to MTV and turning on the matrix mode. Adjust both audio channels to the L + R carrier (6.62 MHz). Monitor the left audio output. It should be blank or very low in volume. RV104 is adjusted to null the sound. CT101 is adjusted to allow channel 1's tuning range to go from 5.5 to 8.0 MHz. CT 102 does the same for channel 2.

STS has manufactured three different down-converters, all of which have been called model DC-102. The first one was made to match the Luxor receiver with a tuning voltage of + 1 to + 16 volts. The second version had a + 12 volt regulator added into it. It was also compatible with the Luxor receivers. The third version is designed around the MBS-SR and its + 1 to + 10.5 Vdc tuning range. Only the third version will work properly with the MBS-SR receiver. It is identified by the # sign before the serial number. A version one or two DC-102 will not have the # sign in front of their serial number.

WINEGARD

The Winegard SC-7032 schematic is shown in Fig. 18-54. The signal enters through P17, which is an F connector. It passes through C1 (.0047 μF capacitor), which blocks the tuning voltage. It goes through two transistors (Q1 and Q2) which provide the initial i-f gain to drive the first filter module. R163 is the i-f gain setting, which can be accessed through the rear panel. The output of the first i-f filter is sent to the signal-strength meter via an amplifier (Q3) and detector circuit (D2 and D3). It is also sent to U1 which is the main i-f amplifier. It drives another filter module and the MWA130, which is the limiter for the circuit. The output of the MWA130 is split, both going to the delay line and directly to the DBM circuit.

C19 is the delay-line phase adjustment. It should be set so that there is a symmetrical voltage to the output waveform. The output of T2's center tap is a baseband video signal that is about 100 to 200 mV in level. It is sent to the afc circuit, (comprised of U3 and the four diodes D8 through D11), the subcarrier out jack (through emitter follower Q4, 2N3904), and to the audio and video circuits.

The video and audio circuitry is virtually identical to the Drake circuit. The video is amplified by a 733 IC, whose output goes to the normal/invert switch. From there, the video signal is sent to the clamping circuit and after being buffered by Q6, out to the modulator and the back panel.

The audio is sent to emitter follower Q14, which buffers the signals. C23 and R91 comprise a high-pass filter, which only allows the upper baseband signals through. The desired channel is tuned and up converted to 10.7 MHz by Q13. The signal is then filtered before being sent to U6, which is the i-f amplifier, limiter, and detector. Q10 and Q9 boost the audio level to drive the modulator and the audio output jack.

R2 is the cable compensation trim pot, which is ac-cessed through the rear panel. It is used to raise or lower the overall tuning voltage to compensate for cable losses. The tuning voltage comes from U18 (741 op-amp), the input of which consists of the output of the fine tune pot (R116), the output of the even channel offset, and the out-put of the channel selector switches.

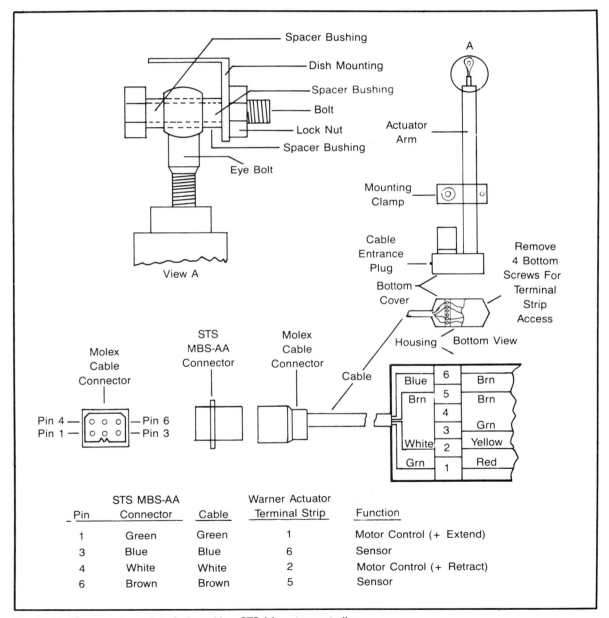

	STS MBS-AA		Warner Actuator	
Pin	Connector	Cable	Terminal Strip	Function
1	Green	Green	1	Motor Control (+ Extend)
3	Blue	Blue	6	Sensor
4	White	White	2	Motor Control (+ Retract)
6	Brown	Brown	5	Sensor

Fig. 18-52. Warner actuator interfacing with a STS-AA motor controller.

Winegard models SC-7032 and SC-7035 use a momentary double-pole switch to change channels. By holding it one way or the other, it will sequentially step through the channels. The same type of switch is used to move the dish. Holding it in a cw direction will move the dish east, while holding it in a ccw direction moves the dish to the west. A single audio channel is tuned by a potentiometer. There is also a scan tuning and a format switch on the front panel.

Actual channel selection is done through three 4051 analog demultiplexer chips. These chips select which tuning pot voltage will be sent to the DC. The channel is actually selected by the 4510 chips which are BCD up/down counters. As the channel selector switch is turned, the 555 IC (U14), which is set in a one-shot mode, puts out one pulse. If the switch is held closed, then the 555 continuously resets and puts out a series of pulses for as long as the switch is turned. If the SCAN switch is on, then the 555 is put in an astable oscillator mode and outputs a steady stream of pulses.

These pulses clock the 4510 up/down counters that, in turn, turn on the various analog switches (U19 to U21)

that output the various tuning voltages on pin 3 of each chip. This voltage is summed together with the fine-tuning voltage and the even-channel offset voltage (if an even channel is selected). The odd/even channel selection is determined by the output of one section of U12 at pin 10. A low output turns on Q22, which adds in an extra voltage boost to offset the tuning for even channels. It also drives Q20 which drives the polarity-control circuit.

Figures 18-55 and 18-56 are the two satellite select control circuits used in the 7032 and 7035 receivers. Figure 18-55 is the ACT-832. It consists of an analog-to-digital IC (the ADC0804), the digital decoder and latch IC (the MCM7641), and the 4511 LED readout driver IC. The two readouts are selected by the two 2N4401 transistors which supply the ground for the readouts.

The circuit in Fig. 18-56 consists of a 7217 up/down counter and latch IC, and a 4011 quad NAND gate IC. The 7217 drives the same display as the other circuit. The blue wire on pin 4 is the positional feedback from the motor pot. The two wires on pins 5 and 6 supply the ground to two relays which selectively turn on a motor to drive the actuator either in and out.

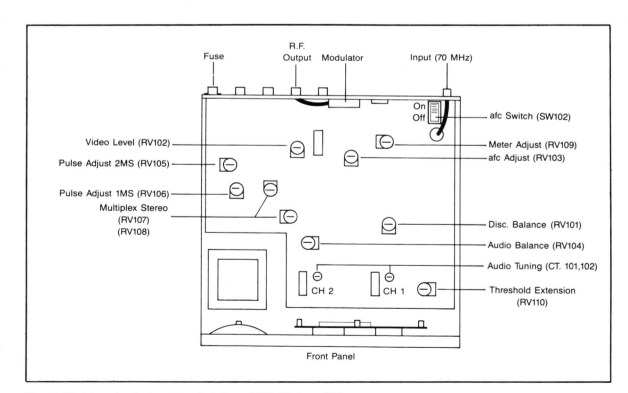

Fig. 18-53. Internal adjustment controls for a MBS-SR from STS.

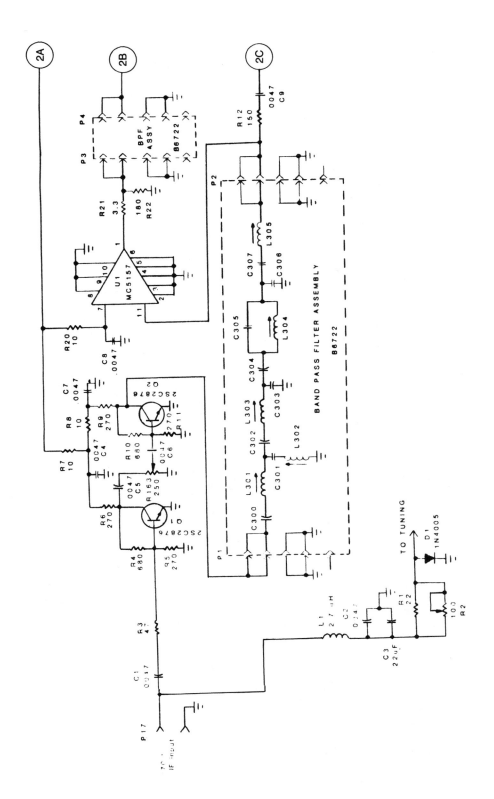

Fig. 18-54. Winegard main board schematic. Courtesy Winegard.

2

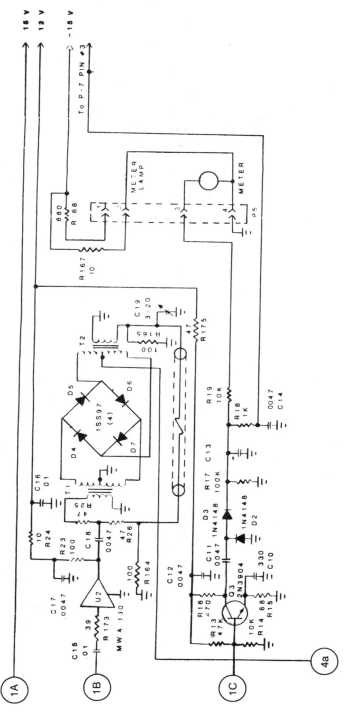

Fig. 18-54. Continued from page 345.

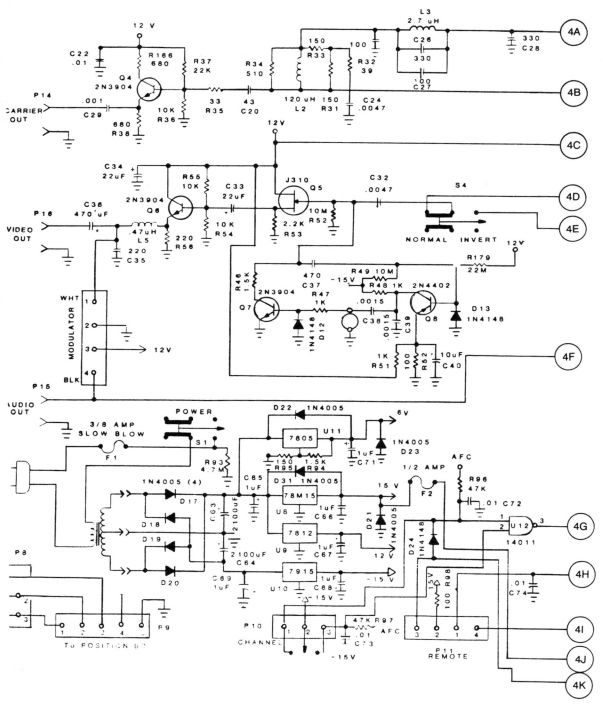

4

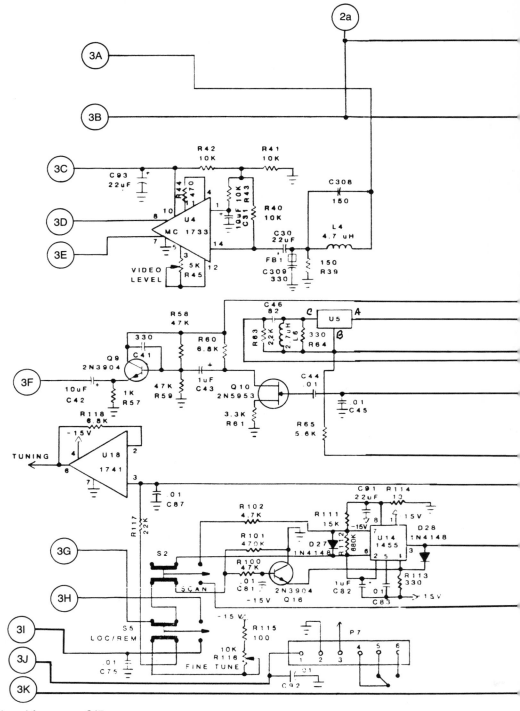

Fig. 18-54. Continued from page 347.

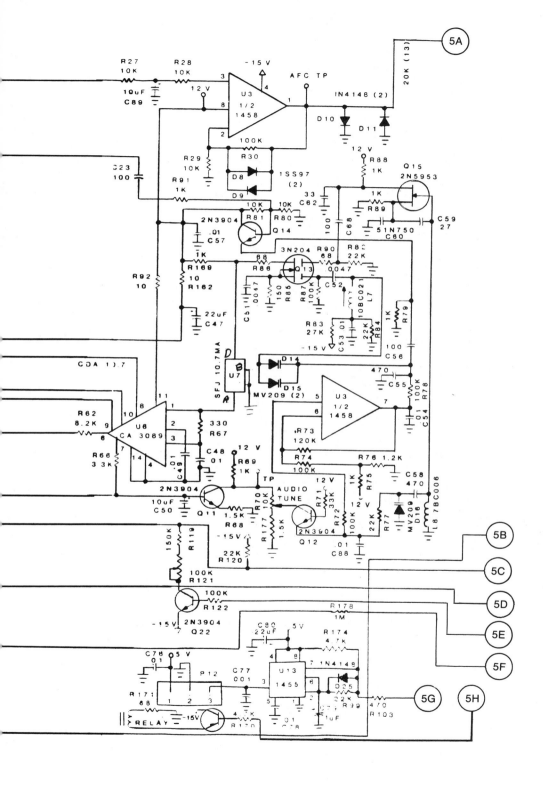

5

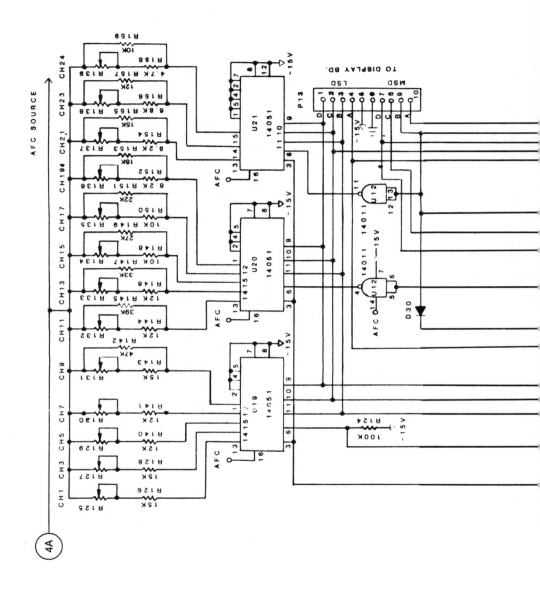

Fig. 18-54. Continued from page 349.

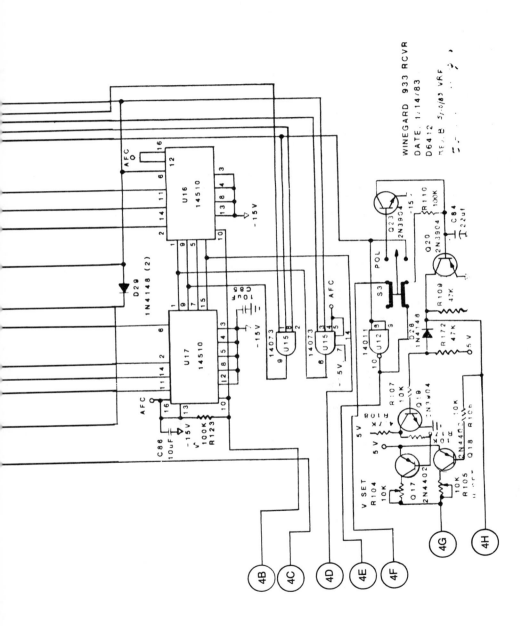

WINEGARD 933 RCVR
DATE: 1/14/83
D6412

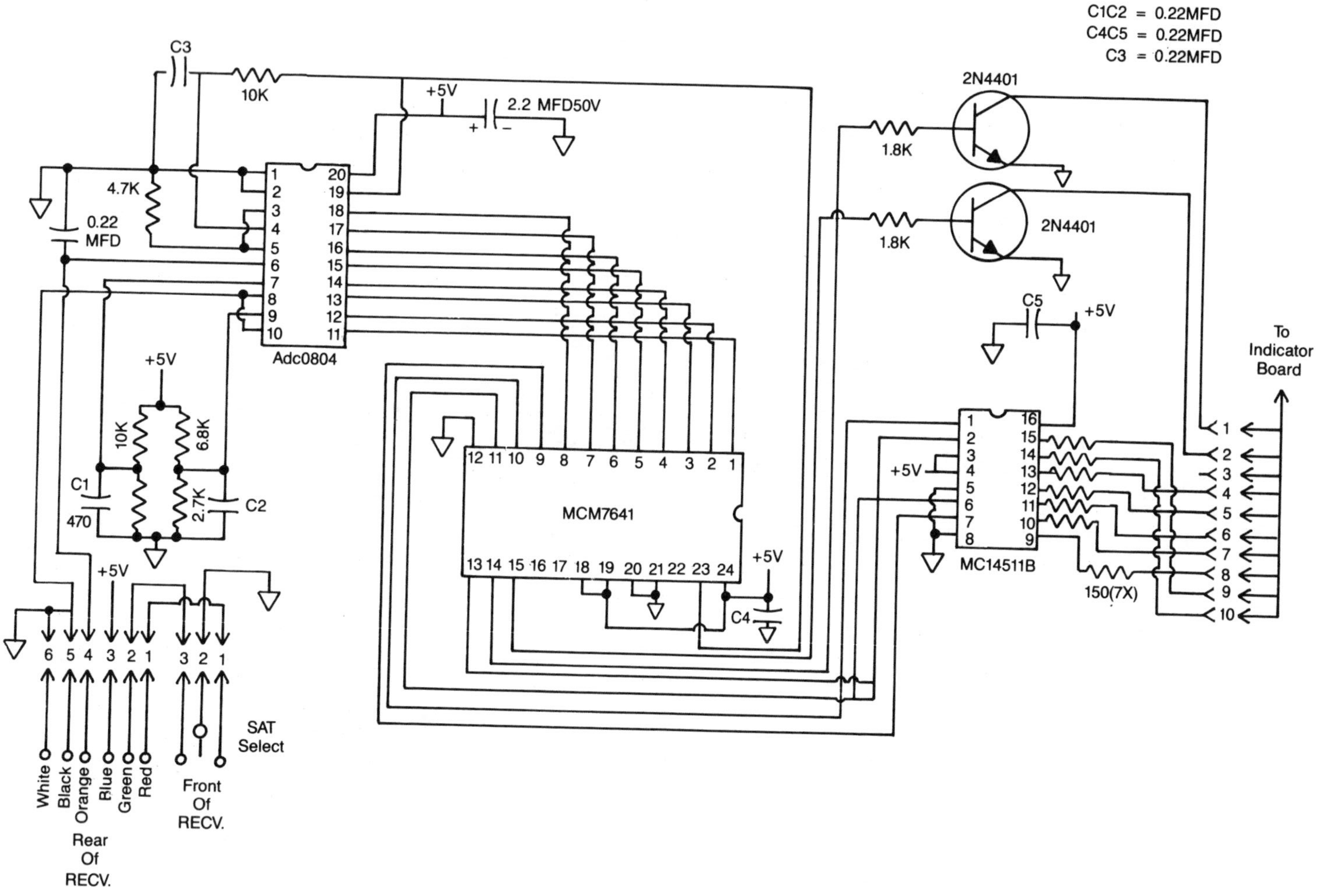

Fig. 18-55. Motor controller circuit. Courtesy Winegard.

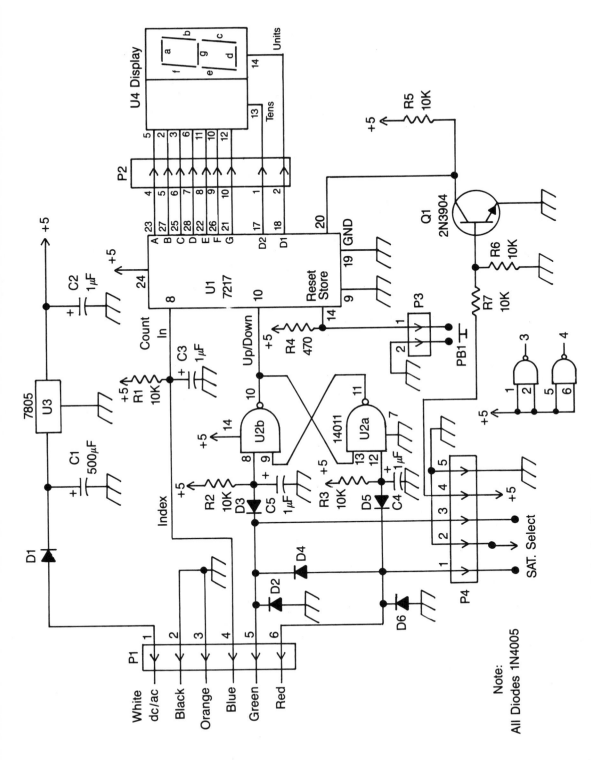

Fig. 18-56. Motor controller circuit. Courtesy Winegard.

353

Figure 8-5 (back in Chapter 8) is the power supply circuit for the motor drive. It is also the + 5 Vdc supply for the ACT-832 circuit. It consists of two transformers, one for the + 5 Vdc and the other for the ± 36 Vdc for the motor. The voltage is applied to the motor through the two relays K1A and K2A.

Figure 18-57 is the readout for the channel selection. It consists of two 4511 ICs that are driven by – 15 Vdc. This is backwards from most uses. Notice that pin 16, which is normally at + 15 Vdc, is at ground, and pin 8 which is normally at ground is at – 15 Vdc. As long as the relationship between the two remains the same, (i.e., pin 16 is more positive than pin 8), then the chip will function properly.

Adding a Hard Limiter

To improve upon the picture, a hard limiter can be inserted between the MWA130 and C18 (see Fig. 18-58). It consists of three sections of a 74S00 IC. This ensures that a clean square wave is output to the detector. This is also shown schematically in Fig. 18-58.

To add it into the receiver, unsolder the input side of the .0047μF capacitor. It will be soldered to pin 8 of the limiter IC. Take the IC and solder pins 1 and 2 together, pins 3, 4, and 5 together, pins 9 and 10 together, and pins 12 and 13 together. Do not apply heat for too long a time or the IC may be damaged.

Use an insulated wire to connect pin 7 to pins 12 and 13. Use another wire to connect pin 6 to pins 9 and 10. Solder a 4 inch wire to pin 7, another 4 inch wire to pins 1 and 2 and a 47 ohm resistor to pin 14. Mount the IC upside down, onto the MWA130 using double-sided tape. Position it so that the .0047μF capacitor can easily reach pin 8. Solder the cap to pin 8, the wire from pins 1 and 2 to the output of the MWA130 and the wire from pin

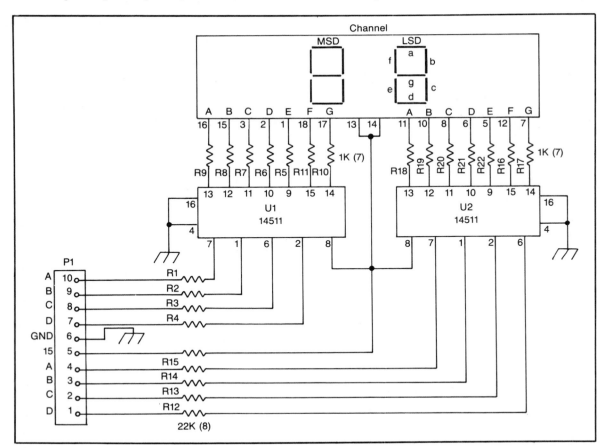

Fig. 18-57. LED readout driver and display. Courtesy Winegard.

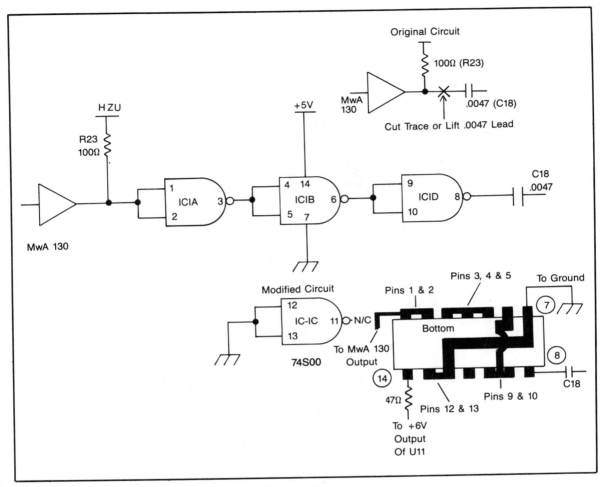

Fig. 18-58. Adding a hard limiter into a receiver with soft limiting.

7 to a convenient ground pad. Run a wire to the output of the 7805 regulator (+6 Vdc) from the 47 ohm resistor. Solder this wire at both ends. Be sure to use shrink tubing on the resistor. This modification should result in cleaner pictures with less solid color streakiness to them.

Winegard Technical Bulletins

Figure 18-59 is the adjustment locations for the SC-7032 and SC-7035 receivers. Winegard has supplied the following technical information for the SC-7035.

Stabilizing the Digital Readout. If the satellite location readout flickers or changes between numbers on the SC-7035S, then add a .22 μF/50 volt capacitor between pins 2 and 6 of the A-to-D chip (ADC0804). Pin 6 is the positional feedback voltage from the motor drive pot. It can get noise pulses on it which will disrupt the functioning of the motor drive controller circuit. The capacitor swamps out these spikes.

Polarization Device Problems. The PR-I probe can become unstable under certain conditions. This will lead to probe oscillations or possibly binding of the probe. Typically this can be cured by placing a 1,000 μF/15 volt electrolytic capacitor between the red and black wires on the PR-I. Follow the diagram in Fig. 18-60. If the problem is not resolved, then try putting a 100 ohm/1/4-watt resistor in series with the control pulse (usually white) wire at terminal #8 of the receiver's interface terminal strip. See Fig. 18-61.

Notes on Actuator Controllers. There are six

1. i-f Gain Potentiometer

2. Cable Compensation Potentiometer

3. Horizontal Potentiometer (Polarotor)

4. Vertical Potentiometer (Polarotor)

5. Modulator (Selectable Ch. 3 or 4)

6. Audio Level Adjustment - If audio has a "buzz" or too low of level, remove "RF out" nut on "F" fitting out back of receiver to allow access to tuning slug. May also require adjustment of Item 8 video level adjustment.

7. 15 VDC - 1/2 Amp. Slo-Blow Fuse - Check if no +15VDC supply out back terminal of receiver. Must check with ohmmeter for open circuit.

8. Video Level Adjustment Potentiometer Adjust - in connection with Item 6 Audio Level Adjustment if audio problem present. If picture "jitters," reduce video level slightly.

9. Local Oscillator Adjust - If audio is "raspy" and cannot be cleared with audio tune knob, center audio tune knob and adjust coil for best audio.

10. Sound IF Adjust - Tune receiver to WGN Channel 3 on Satcom 3R. Turn audio tune knob counterclockwise to second sub-carrier. Tune adjust coil for best audio.

11. AFC TP (Test Point)

12. Channel Tuning Potentiometers

13. Odd - Even Potentiometer (R121) - Whenever cable type or length between downconverter and receiver is changed from 150 ft. standard, Item 12 channel tuning potentiometers must be retuned. To do this, set fine tune knob on receiver front to center position. Attach positive probe of voltmeter to Item 11 AFC TP. Attach negative probe of voltmeter to chassis ground. Set voltmeter to lowest DC voltage scale. Tune receiver to Channel 1 and varify proper program is viewed. Find Ch. 1 potentiometer in Item 12 group and tune to 0 VDC. Repeat this procedure for all odd number channels. Next, tune receiver to Channel 12 and set Item 13 Odd-Even Potentiometer to 0 VDC. Last step is to tune receiver to Channel 24 and returning to Item 12 group, tune Ch. 24 potentiometer for 0 VDC.

NOTE:

1. No items should be touched or adjustments made to anything not mentioned above.

2. Use non-metallic tuning tool to make all of the above adjustments.

Fig. 18-59. Internal adjustment locations on an SC-7032 and 7035 receiver. Courtesy Winegard.

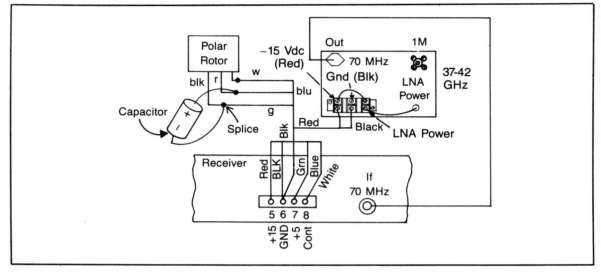

Fig. 18-60. Curing Polarotor I problems by adding a capacitor.

wires that run between the receiver and the PS-8032 controller power supply. These are called out as white, black, orange, green, blue, and red. On the PS-8032 end, which is marked TO REC, the white wire is the ground or shield. The black wire is the return wire for thepositional feedback. The orange wire is the center tap of thepositional feedback pot. It will read between + 0.2 and + 4.6 Vdc. It should read the lowest voltage when the inner tube is retracted within the actuator. The blue wires is + 5 Vdc. The green wire is the actuator out return line. It will

measure + 5 Vdc until the SAT SELECT switch is turned cw. The red wire is the actuator in return line. It will read + 5 Vdc until the SAT SELECT switch is held ccw.

On the TO MOTOR connector, the white wire is the return ground for the motor. The black and red wires are the motor drive controls. They will be at zero volts until either the actuator out switch is closed (in which case the red wire will go to + 38 Vdc) or the actuator in switch is closed (which will put – 38 Vdc on the motor). The actuator switch is the SAT SELECT switch on the SC-7032

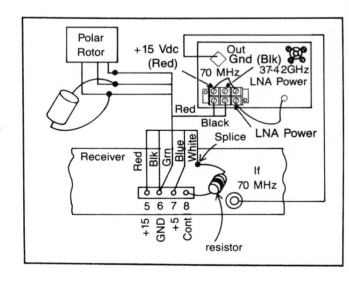

Fig. 18-61. Curing Polarotor I problems by adding a resistor.

358

or 7035 receiver. The blue wire reads +5 Vdc, and the green wire will measure between +0.2 and +4.6 Vdc, depending on the actuator position.

Actuator Problems. If there is no +5 Vdc on the two blue wires, then the 7805 regulator in the PS-8032 is open. If the voltage is less than +4.5 Vdc, then it is being pulled down. This may be caused by a bad IC, thus disconnect both blue wires and recheck the voltage. If it is still low, then the 100 µF cap may be shorted or the 7805 may be bad. If the voltage is above +6 Vdc, then the regulator has shorted and has most likely damaged the ADC0804, the MCM7641, and the 4511 ICs. These are located on the ACT-832 board inside the receiver.

If the positional feedback output does not change as the motor is moved in or out, then either a broken wire

or burned out pot has occurred. The pot is located inside the motor housing. It is a 10 K ten-turn pot.

If the motor will go in one direction but not the other, check the voltages on the red and black TO MOTOR wires. If both voltages (±38 Vdc) are present as the SAT SELECT switch is turned, then one of the end limit switches in the motor drive is open. This can be confirmed by measuring the resistance between the two motor control terminals. There should be very low resistance between the red and black wires. In addition, there should be a few hundred ohms resistance between the white and red or the white and black wires.

If the above voltages are not present, then try grounding the TO REC red terminal to ground (white terminal), and then the green terminal to ground. If the motor drives

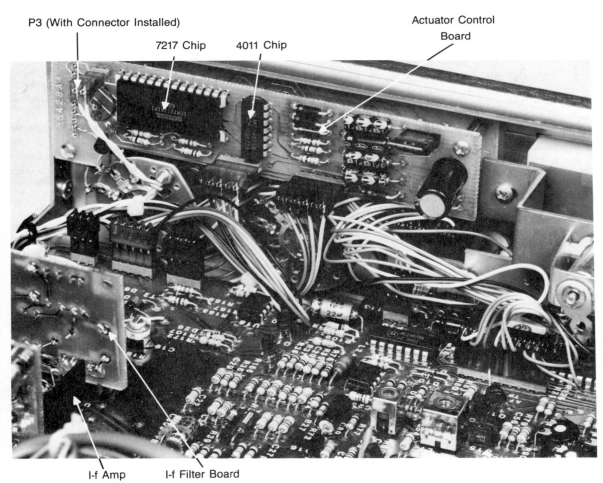

P3 (With Connector Installed)

7217 Chip   4011 Chip

Actuator Control Board

I-f Amp   I-f Filter Board

Fig. 18-62. Location of P3, for adding in a reset switch.

in and out, then the problem is in the controller. If the voltage is still only present on one wire then the relay is bad or the contacts are frozen open.

If the circuit breaker in the PS-8032 opens almost immediately upon being reset, then the transformer, one of the rectifier diodes, or one of the capacitors is shorted. If the motor drives in one direction without control, then one of the two relays is shorted. Sometimes the contacts can be pried apart and smoothed out with fine sandpaper, but it is better to replace the relay.

Receiver Tips. Whenever multiple receivers are used, isolators are required between the DC and the splitter. When an isolator is used, do not jumper the + 15 Vdc terminal to the LNA POWER terminal on the DC. Use a dc power inserter between the splitter and the LNA. Connect a diode in series with each + 15 Vdc terminal from each DC. From each of the diode's cathode lead (the one with the band) run a wire to the power inserter. This allows any receiver to power the LNA, without causing interference to the others.

If an audio buzz is heard in the TV, then the rf modulator may need to be aligned or the video level may be too high. The sound carrier on the rf modulator should be set at 4.5 MHz. This frequency is adjusted by the coil in the modulator. It can only be accessed by removing the modulator from the back panel. R45 is the video level control. Turning the level down will darken the picture slightly. Turn the contrast up on the TV to compensate.

If the tuning is off and if channels below about 10 cannot be received, then the 741 op-amp is probably bad. Check the voltage on the center conductor of the coax cable. It should be between − 4 and − 15 Vdc. If it only goes to − 8 volts or so, then the op-amp should be replaced.

Unclamping the Video. To unclamp the video requires that a jumper be added onto the main board in the receiver. This jumper goes from the junction of R47 and C38 to ground. There are two pads on the board about an inch behind U4 (733 IC) to which this jumper can be soldered. If necessary, a clamp/unclamp switch could be installed on the back panel. A hole will have to be drilled or punched, unless it is a 7035, in which case, the RESET hole could be used. A single-pole, single-throw toggle switch is all that is necessary. Connect the two terminals to the two pads with insulated 22 gauge wires.

Adding Reset Switch to S Models. If there is no actuator reset switch, then one can be added by installing a single-pole, double-throw toggle switch to the back panel in the hole labeled RESET. Solder two 22 inch wires to the switch and run these to the actuator control board. There is a two pin connector labeled P3. Connect the wires to the pins by soldering or by using a two-pin female molex plug. See Fig. 18-62 for the connector's location.

Sound and Video Problems. If the receiver has audio popping noises, heavy herringbones in the video and poor color, then try turning the receiver off, and then back on. If the problem disappears for a while but then returns, then the most probable cause is the 2100 μF electrolytics (C-63 and C-64) in the main $\pm$ 23 Vdc supply. These should be replaced. Try replacing the + filter first. Once the problem is cured, check the various power supply voltages. They should be + 15 on the output of the 7815, + 12 on the 7812, − 15 on the 7915, and + 6 on the 7805. If any voltages are incorrect, replace the regulator. The regulators and filter capacitors are located in the left rear section of the receiver.

Ac Line Fluctuation. At no time can the incoming power-line voltage rise above 130 volts. If it does for any extended length of time, then power supply failure will probably result. Conversely, operating the receiver on a power line that is below 108 volts will cause hum bars to appear in the picture.

In some cases, if the ac line is poorly regulated, then constant fluctuations can result. This often occurs when heavy loads like motors are added to the line. This will cause the voltage to momentarily drop below the minimum level, and then to slowly rise as the motor load decreases. This can cause down converter drifting depending on the variance of the voltage. In some cases, an autotransformer may be necessary to regulate the incoming ac voltage.

Polarotor Supply Voltage Is not + 6 Vdc. If the supply voltage is + 5 volts, then the heatsink that is mounted on the regulator is shorted to ground. If the voltage is below + 5 Vdc even with the PR-I disconnected, then suspect either the 555 IC (U13) or the 7805 regulator (U11).

Appendix A

Active Component Guide

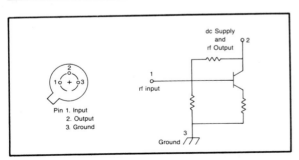

MWA110, MWA120 and MWA130 pinout and internal circuit. Single transistor amplifier with internal biasing resistors.

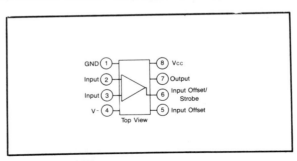

311 pinout. Voltage comparator.

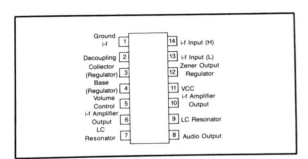

TBA 120 pinout. 10.7 MHz FM radio i-f amp and detector.

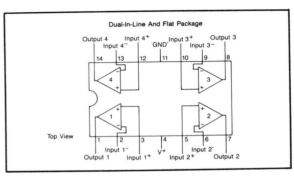

324 pinout. Low power quad op-amp.

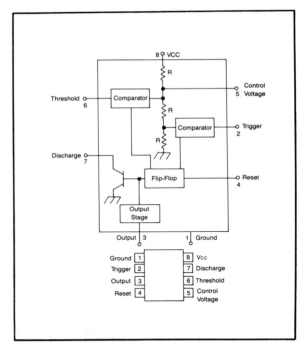

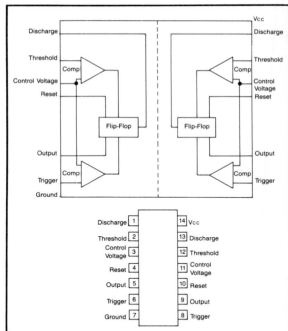

555 pinout and internal block diagram. Timer.

556 pinout and block diagram. Dual timers.

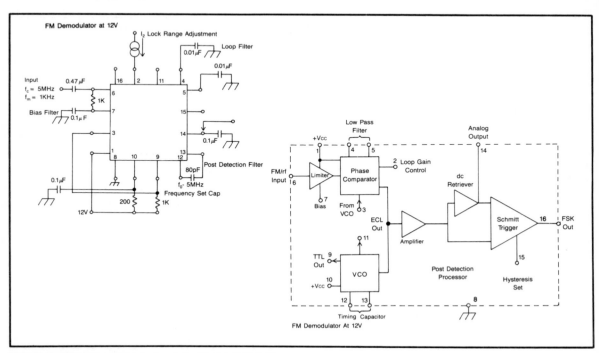

564 block diagram and typical usage. Phase lock loop.

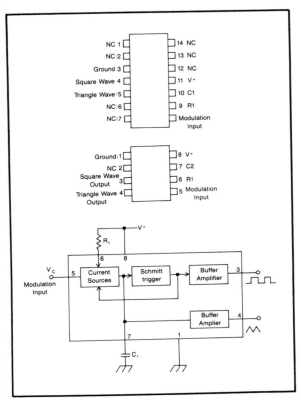

566 pinout and block diagram. Function generator.

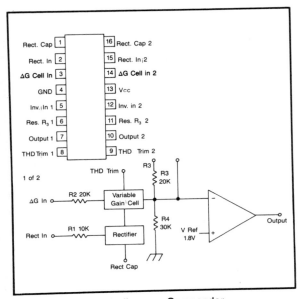

571 pinout and block diagram. Compander.

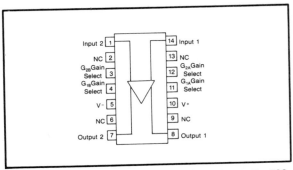

592 pinout. Video amplifier (pin-for-pin equivalent to the 733 chip).

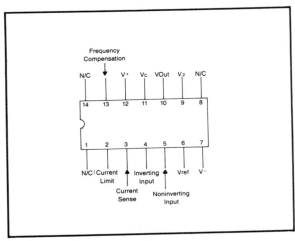

723 pinout. Adjustable regulator.

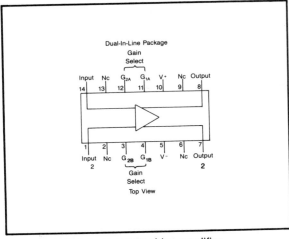

733 pinout. Balanced output video amplifier.

363

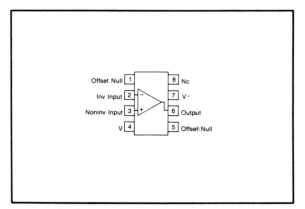

741 pinout. Operational amplifier.

1112 pinout. Dolby noise reduction chip.

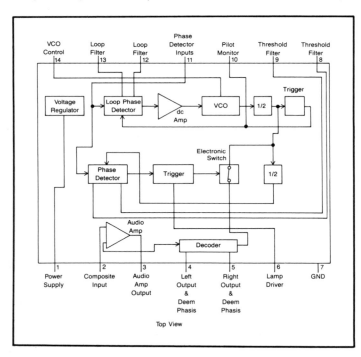

1310 pinout and block
diagram. An FM multiplex decoder.

1357 pinout. Quadrature FM detector.

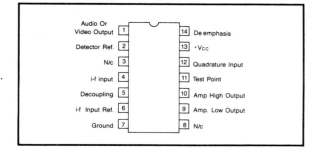

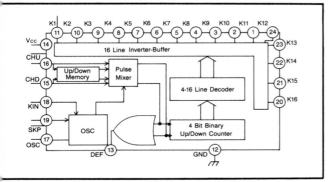

1360 pinout and block diagram.
Up/down counter/decoder.

1458 (or 4558) pinout. Dual op-amp IC.

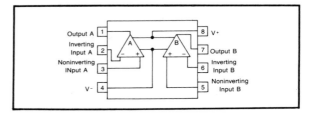

1496 pinout. Balanced demodulator.

1648 pinout with typical usage and schematic. ECL VCO chip.

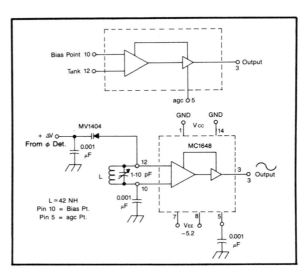

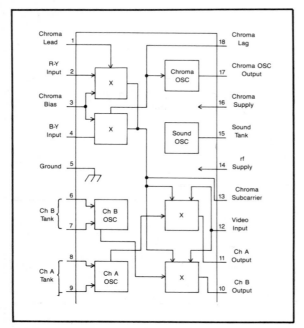

1889 pinout and block diagram. TV video modulator.

1894 pinout and block diagram. Dynamic noise reduction chip.

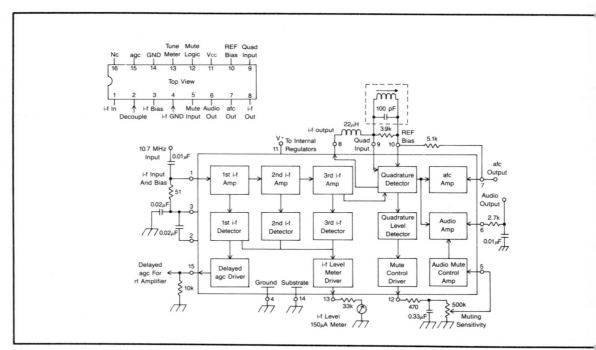

3089 pinout and block diagram. FM receiver i-f section.

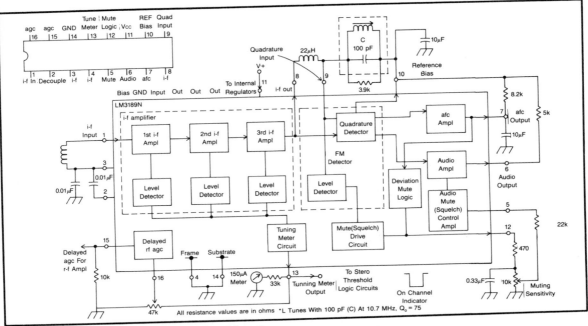

3189 pinout and block diagram. FM receiver i-f section.

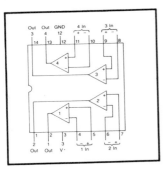

3302 pinout and block diagram. Quad op-amp.

$$V_{REF} = 1.25V \left(1 + \frac{R2}{R1}\right) + R2 \times 80\mu A$$

$$I_{LED} = \frac{12.5}{R1} + \frac{V_{REF}}{2.2}$$

Note 1: Capacitor C1 is required if leads to the Led supply are 6″ or longer.

Note 2: Circuit as shown is wired for dot mode. For bar mode, connect pin 9 to pin 3. V_{LED} must be kept below 7V or dropping resistor should be used to limit IC power dissipation.

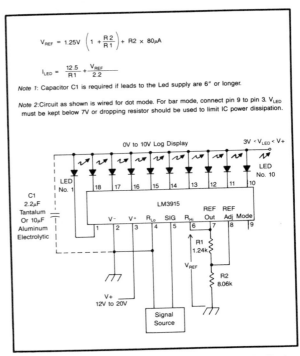

3915 pinout and typical usage. Logarithmic LED display driver.

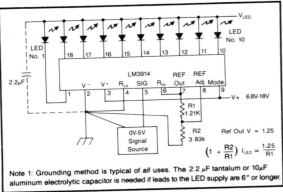

Note 1: Grounding method is typical of *all* uses. The 2.2 μF tantalum or 10μF aluminum electrolytic capacitor is needed if leads to the LED supply are 6″ or longer.

3914 pinout and typical usage. Linear LED bar graph driver.

7217 pinout and functional description. 4 digit CMOS up/down counter/display driver.

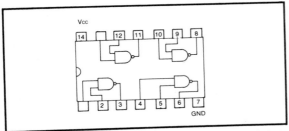

74S00 pinout and block diagram. A quad low-power Schottky 2-input NAND gate.

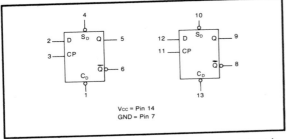

74S74 pinout. A dual low-power Schottky positive edge triggered flip-flop.

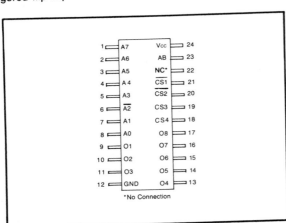

7641 pinout. 4K PROM (Programmable Read Only Memory).

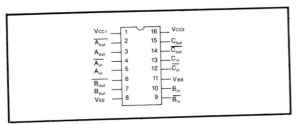

10114 pinout. Triple line receiver.

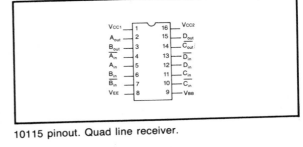

10115 pinout. Quad line receiver.

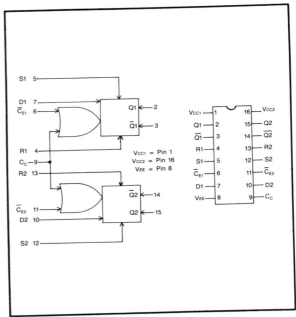

10131 pinout and block diagram. Dual D flip-flop.

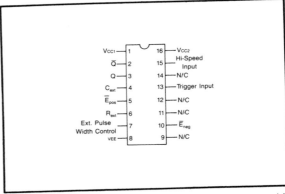

10198 pinout and block diagram. Retriggerable monostable multivibrator.

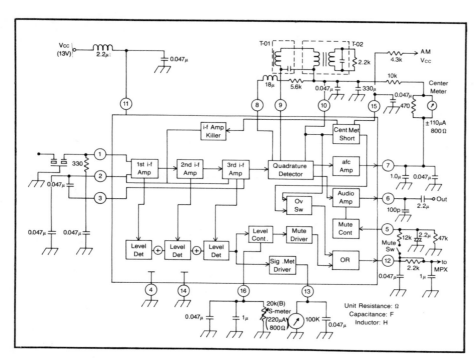

12124 pinout and block diagram. FM radio i-f amplifier and

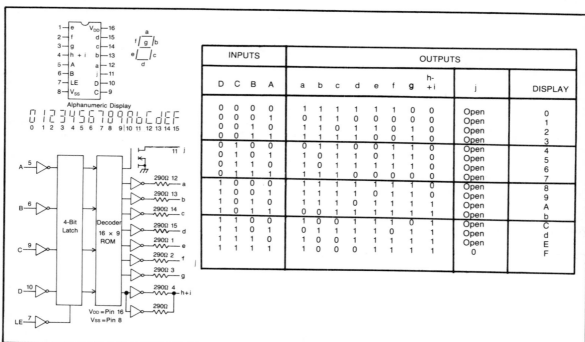

14495 pinout, block diagram and truth table. CMOS hexadecimal to 7 segment display latch/decoder/driver with integral current limiting resistors.

INPUTS				OUTPUTS									
D	C	B	A	a	b	c	d	e	f	g	h-+i	j	DISPLAY
0	0	0	0	1	1	1	1	1	1	0	0	Open	0
0	0	0	1	0	1	1	0	0	0	0	0	Open	1
0	0	1	0	1	1	0	1	1	0	1	0	Open	2
0	0	1	1	1	1	1	1	0	0	1	0	Open	3
0	1	0	0	0	1	1	0	0	1	1	0	Open	4
0	1	0	1	1	0	1	1	0	1	1	0	Open	5
0	1	1	0	1	0	1	1	1	1	1	0	Open	6
0	1	1	1	1	1	1	0	0	0	0	0	Open	7
1	0	0	0	1	1	1	1	1	1	1	0	Open	8
1	0	0	1	1	1	1	1	0	1	1	0	Open	9
1	0	1	0	1	1	1	0	1	1	1	1	Open	A
1	0	1	1	0	0	1	1	1	1	1	1	Open	b
1	1	0	0	1	0	0	1	1	1	0	1	Open	C
1	1	0	1	0	1	1	1	1	0	1	1	Open	d
1	1	1	0	1	0	0	1	1	1	1	1	Open	E
1	1	1	1	1	0	0	0	1	1	1	1	0	F

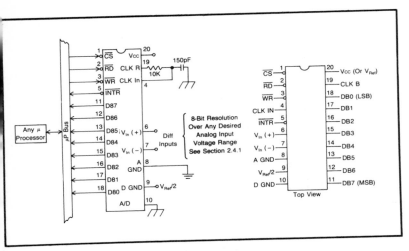

ADC0804 pinout and typical interface. Digital to analog converter.

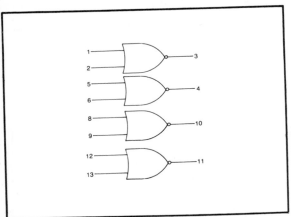

4001 pinout and logic diagram. Quad 2-input NOR gate.

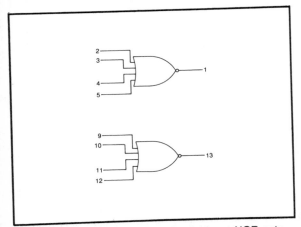

4002 pinout and logic diagram. Dual 4-input NOR gate.

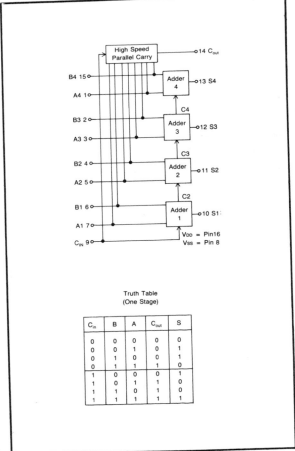

Truth Table
(One Stage)

C_{in}	B	A	C_{out}	S
0	0	0	0	0
0	0	1	0	1
0	1	0	0	1
0	1	1	1	0
1	0	0	0	1
1	0	1	1	0
1	1	0	1	0
1	1	1	1	1

4008 block diagram and logic diagram. 4 bit full adder.

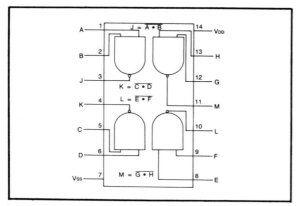

4011 pinout and logic diagram. Quad 2-input NAND gate.

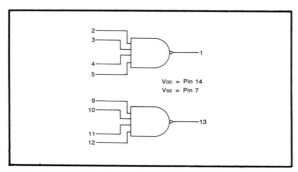

4012 pinout and logic diagram. Dual 4-input NAND gate.

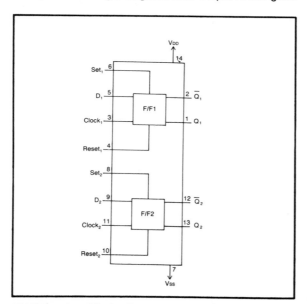

4013 pinout and logic diagram. Dual D flip-flop.

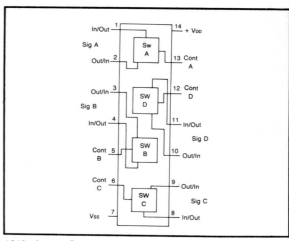

4016 pinout. Quad analog switch.

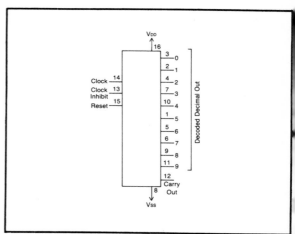

4017 pinout. Decade counter/divider.

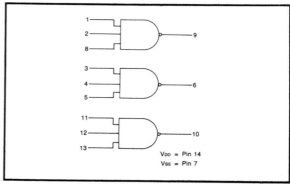

4023 pinout and logic diagram. Triple 3-input NAND gate.

372

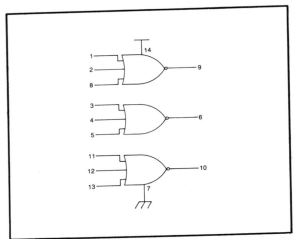

4025 pinout and logic diagram. Triple 3-input NOR gate.

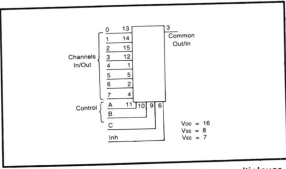

4051 pinout. 8-channel analog demultiplexer multiplexer.

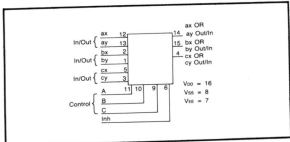

4053 pinout. Triple 2-channel analog demultiplexer multiplexer.

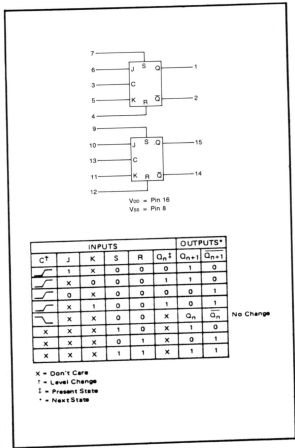

INPUTS						OUTPUTS*	
C†	J	K	S	R	Qn‡	Qn+1	Q̄n+1
⤴	1	X	0	0	0	1	0
⤴	X	0	0	0	1	1	0
⤴	0	X	0	0	0	0	1
⤴	X	1	0	0	1	0	1
⤵	X	X	0	0	X	Qn	Q̄n
X	X	X	1	0	X	1	0
X	X	X	0	1	X	0	1
X	X	X	1	1	X	1	1

No Change

X = Don't Care
† = Level Change
‡ = Present State
* = Next State

4027 pinout and truth table. Dual J-K flip-flop.

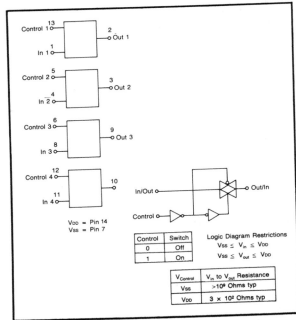

4066 pinout, logic diagram and truth table. Quad analog switch.

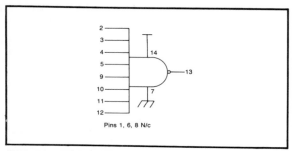

4068 pinout and logic diagram. 8-input NAND gate.

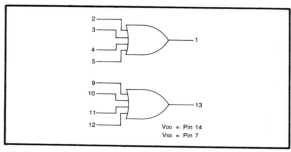

4072 pinout and logic diagram. Dual 4-input OR gate.

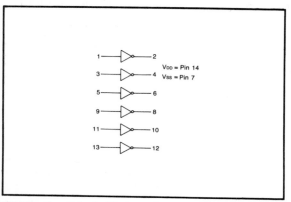

4069 input and logic diagram. Six inverters.

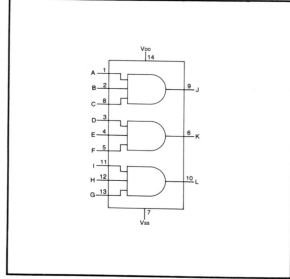

4073 pinout and logic diagram. Triple 3-input AND gate.

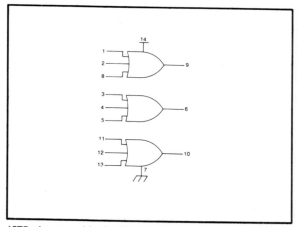

4071 pinout and logic diagram. Quad 2-input OR gate.

4075 pinout and logic diagram. Triple 3-input OR gate.

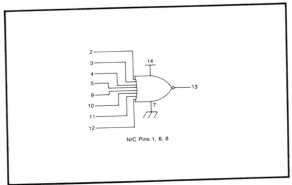

4078 pinout and logic diagram. 8-input NOR gate.

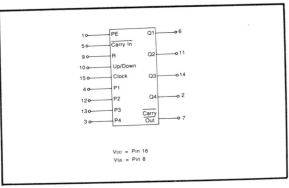

V_{DD} = Pin 16
V_{SS} = Pin 8

4510 pinout. BCD up/down counter.

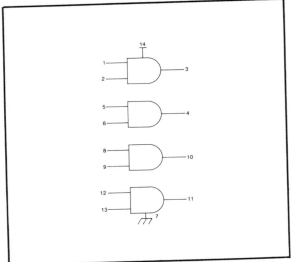

4081 pinout and logic diagram. Quad 2-input AND gate.

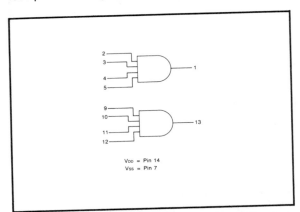

V_{DD} = Pin 14
V_{SS} = Pin 7

4082 pinout and logic diagram. Dual 4-input AND gate.

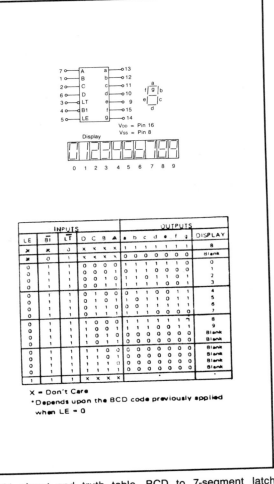

INPUTS							OUTPUTS							DISPLAY
LE	BI	LT	D	C	B	A	a	b	c	d	e	f	g	
X	X	0	X	X	X	X	1	1	1	1	1	1	1	8
X	0	1	X	X	X	X	0	0	0	0	0	0	0	Blank
0	1	1	0	0	0	0	1	1	1	1	1	1	0	0
0	1	1	0	0	0	1	0	1	1	0	0	0	0	1
0	1	1	0	0	1	0	1	1	0	1	1	0	1	2
0	1	1	0	0	1	1	1	1	1	1	0	0	1	3
0	1	1	0	1	0	0	0	1	1	0	0	1	1	4
0	1	1	0	1	0	1	1	0	1	1	0	1	1	5
0	1	1	0	1	1	0	0	0	1	1	1	1	1	6
0	1	1	0	1	1	1	1	1	1	0	0	0	0	7
0	1	1	1	0	0	0	1	1	1	1	1	1	1	8
0	1	1	1	0	0	1	1	1	1	0	0	1	1	9
0	1	1	1	0	1	0	0	0	0	0	0	0	0	Blank
0	1	1	1	0	1	1	0	0	0	0	0	0	0	Blank
0	1	1	1	1	0	0	0	0	0	0	0	0	0	Blank
0	1	1	1	1	0	1	0	0	0	0	0	0	0	Blank
0	1	1	1	1	1	0	0	0	0	0	0	0	0	Blank
0	1	1	1	1	1	1	0	0	0	0	0	0	0	Blank
1	1	1	X	X	X	X				*				*

X = Don't Care
*Depends upon the BCD code previously applied
 when LE = 0

4511 pinout and truth table. BCD to 7-segment latch decoder/driver.

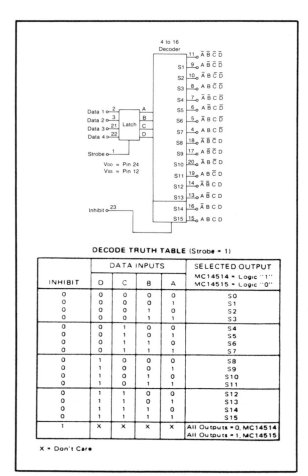

DECODE TRUTH TABLE (Strobe = 1)

INHIBIT	DATA INPUTS				SELECTED OUTPUT MC14514 = Logic "1" MC14515 = Logic "0"
	D	C	B	A	
0	0	0	0	0	S0
0	0	0	0	1	S1
0	0	0	1	0	S2
0	0	0	1	1	S3
0	0	1	0	0	S4
0	0	1	0	1	S5
0	0	1	1	0	S6
0	0	1	1	1	S7
0	1	0	0	0	S8
0	1	0	0	1	S9
0	1	0	1	0	S10
0	1	0	1	1	S11
0	1	1	0	0	S12
0	1	1	0	1	S13
0	1	1	1	0	S14
0	1	1	1	1	S15
1	X	X	X	X	All Outputs = 0, MC14514 All Outputs = 1, MC14515

X = Don't Care

4514 and 4515 pinout, block diagram and truth table. 4 bit latch/4 to 16 line decoder.

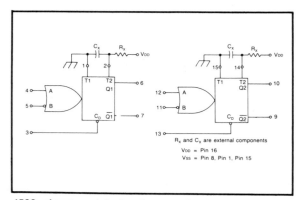

4538 pinout and logic diagram. Dual precision retriggerable/resettable monostable multivibrator.

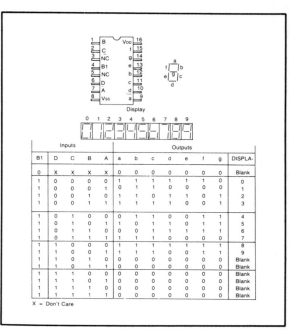

B1	D	C	B	A	a	b	c	d	e	f	g	DISPLAY
0	X	X	X	X	0	0	0	0	0	0	0	Blank
1	0	0	0	0	1	1	1	1	1	1	0	0
1	0	0	0	1	0	1	1	0	0	0	0	1
1	0	0	1	0	1	1	0	1	1	0	1	2
1	0	0	1	1	1	1	1	1	0	0	1	3
1	0	1	0	0	0	1	1	0	0	1	1	4
1	0	1	0	1	1	0	1	1	0	1	1	5
1	0	1	1	0	1	0	1	1	1	1	1	6
1	0	1	1	1	1	1	1	0	0	0	0	7
1	1	0	0	0	1	1	1	1	1	1	1	8
1	1	0	0	1	1	1	1	0	0	1	1	9
1	1	0	1	0	0	0	0	0	0	0	0	Blank
1	1	0	1	1	0	0	0	0	0	0	0	Blank
1	1	1	0	0	0	0	0	0	0	0	0	Blank
1	1	1	0	1	0	0	0	0	0	0	0	Blank
1	1	1	1	0	0	0	0	0	0	0	0	Blank
1	1	1	1	1	0	0	0	0	0	0	0	Blank

X = Don't Care

4547 pinout and truth table. High current BCD to 7-segment decoder/driver.

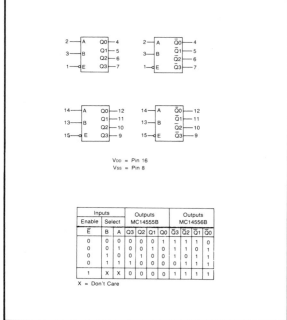

Inputs			Outputs MC14555B				Outputs MC14556B			
Enable	Select									
Ē	B	A	Q3	Q2	Q1	Q0	Q̄3	Q̄2	Q̄1	Q̄0
0	0	0	0	0	0	1	1	1	1	0
0	0	1	0	0	1	0	1	1	0	1
0	1	0	0	1	0	0	1	0	1	1
0	1	1	1	0	0	0	0	1	1	1
1	X	X	0	0	0	0	1	1	1	1

X = Don't Care

4555/4556 pinout and truth table. Dual binary to 1-of-4 decoder/demultiplexer.

376

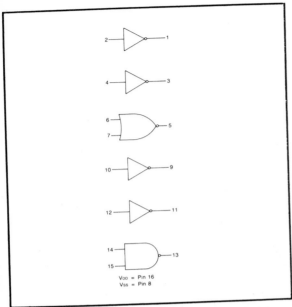

4572 pinout and logic diagram. Six logic gates in one package. 4 inverters, a 2-input NOR gate and a 2-input NAND gate.

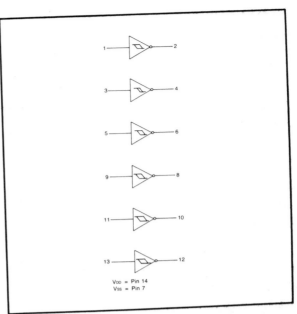

4584 pinout and logic diagram. Six Schmitt triggers in one package.

Appendix B

Equations, Calculations, and Charts

1. Wavelength = Velocity/Frequency
Velocity = 186,000 miles/second or 300×10^6 meters/second

2. One cycle of waveform (in time) = 1 / Frequency

3. Time constant of RC network = R (ohms) × C (farads)

4. Time constant in RL network = L (henries) / R (ohms)

5. Ohm's law for dc circuits

   E (voltage) = I (current) times R (resistance)

   or I = E/R

   or E = I × R

   or R = E/I

6. Ohm's law for ac circuits

   E (voltage) = I (current) × Z (impedance)

7. Circuit impedance Z = E/I

8. Power in watts = E^2/R

   or = E × I

9. Inductive reactance

   X_L = 6.28 × F (frequency) × L (inductance)

   Current lags voltage by 90 degrees in an inductive circuit.

10. Capacitive reactance

   X_L = 1/6.28 × F (frequency) × C (capacitance)

   Current leads voltage by 90 degrees in a capacitive circuit.

11. As frequency goes up the inductive reactance goes up and the capacitive reactance goes down. As the frequency goes down, the reverse is true.

12. The fr (resonant frequency) = 1/6.28 times LC

13. C/N formula

$$C/N = G/T + EIRP + 32 \text{ dB} - 10(\log B)$$

where B = receiver bandwidth in MHz

EIRP = satellite broadcasting power in dBw

G/T = figure of merit for antenna

14. G/T formula

G/T = antenna gain − 10 log (LNA temp. + antenna temp in dB/K)

15. focal point = D^2 (diameter squared)/(dish depth × 16)

16. Antenna Gain = k (antenna efficiency) × 6.28 × D^2 (diameter squared) / 9 inches

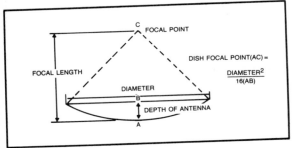

Finding the focal point of a Dish.

Table of Offset Angles to Latitude.

Lat.	Offset Angle	Lat.	Offset Angle	Lat.	Offset Angle	Lat.	Offset Angle	Lat.	Offset Angle
1	.01	16	2.8	31	5.1	46	7.0	61	8.3
2	.36	17	3.0	32	5.3	47	7.1	62	8.3
3	.54	18	3.1	33	5.4	48	7.2	63	8.4
4	.72	19	3.3	34	5.5	49	7.3	64	8.4
5	.90	20	3.5	35	5.7	50	7.4	65	8.5
6	1.07	21	3.6	36	5.8	51	7.5	66	8.5
7	1.25	22	3.8	37	6.0	52	7.6	67	8.5
8	1.43	23	4.0	38	6.1	53	7.7	68	8.6
9	1.60	24	4.1	39	6.2	54	7.8	69	8.6
10	1.78	25	4.3	40	6.3	55	7.8	70	8.6
11	1.95	26	4.4	41	6.5	56	7.9	71	8.7
12	2.13	27	4.6	42	6.6	57	8.0	72	8.7
13	2.3	28	4.7	43	6.7	58	8.0	73	8.7
14	2.5	29	4.8	44	6.8	59	8.1	74	8.8
15	2.6	30	5.0	45	6.9	60	8.2	75	8.8

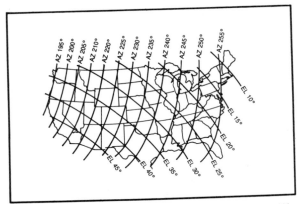

Azimuth-elevation measurements for F3 @135 degrees W.

Decibel Table.

dB	Voltage or Current ratio	Power Ratio
1	1.12	1.25
3	1.41	2
6	2.00	4
10	3.16	10
20	10	100
30	31.6	1,000
40	100	10,000
50	316	100,000
60	1,000	1,000,000

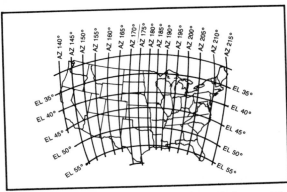

Azimuth-elevation measurements for F4 @83 degrees W.

Unit Value Table.

pico (micro-micro)	one-millionth of a million	1×10^{-12}
nano (milli-micro)	one-thousandth of a million	1×10^{-9}
micro	one-millionth	1×10^{-6}
kilo	one thousand	$1 \times 10^{+3}$
mega	one million	$1 \times 10^{+6}$
giga	one thousand million	$1 \times 10^{+9}$

Unit Conversion Table.

1 meter (m)	= 39.37 inches
1 centimeter (cm)	= 0.3937 inches
1 cm	= 10^4 microns
1 kilometer (km)	= 0.62137 mile
1 inch	= 2.54 cm
1 mile	= 1.6093 km
degrees Kelvin (K)	= degrees C + 273
degrees Centigrade (C)	= 5/9 × (degrees F − 32)
degrees Fahrenheit (F)	= (9/5 × degrees C) + 32
Pi	= 3.1416
Pi^2	= 6.28
1 pF	= .001 nF
1 nF	= 1,000 pF
1,000 pF	= .001 µF
1 µF	= 1,000 nF

Resistor Color Code Table.

Color	Digit	Muliplier	Tolerance
Black	0	1	20%
Brown	1	10	1%
Red	2	100	2%
Orange	3	1,000	
Yellow	4	10,000	
Green	5	100,000	.5%
Blue	6	1,000,000	.25%
Violet	7	10,000,000	.1%
Grey	8		.05%
White	9		
Silver		0.01	10%
Gold		0.1	5%
No color			20%

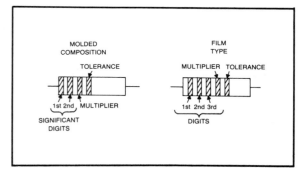

Resistor color code drawings.

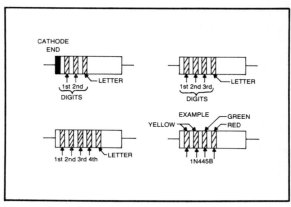

Diode color code drawings.

Diode Color Code Table.

Color	Digit	Letter
Black	0	
Brown	1	A
Red	2	B
Orange	3	C
Yellow	4	D
Green	5	E
Blue	6	F
Violet	7	G
Grey	8	H
White	9	J

Antenna Comparison Table.

Size	Temp	Midband Gain	3.0 dB Point	1st Null	1st Sidelobe
4	70	32.5	4.3	5.8	7.5
6	58	36.0	2.9	3.9	5.0
8	48	38.5	2.2	2.9	3.7
9	44	39.6	1.9	2.6	3.3
10	40	40.5	1.7	2.3	2.9
12	32	42.1	1.4	1.9	2.5
15	26	44.0	1.2	1.6	2.0
16	21	44.8	1.1	1.4	1.9

1. Antenna Noise Temperature at 30 degrees elevation.
2. At 70% efficiency.

On the average, there is about a .3 dB increase or decrease whenever the antenna efficiency increases or decreases by 5%.

C/N for Various Dish Sizes to EIRP Levels.

EIRP

Dish Size	33	34	35	36	37	38
6	2.6#	3.6#	4.6#	5.6#	6.6#	7.6#
8	5.3#	6.3#	7.3#	8.3	9.3	10.3
10	7.6#	8.6	9.6	10.6	11.6	12.6
12	9.4	10.4	11.4	12.4	13.4	14.4

55% efficient dish with 100 degree LNA
= below or right at threshold for most receivers

Wind Loading Table.

Torque per foot/pound

Antenna Diameter	25 mph	50 mph	75 mph	100 mph
4	10	50	120	200
6	40	165	400	700
8	100	400	900	1600
10	200	800	1800	3200
12	350	1400	3000	5500
14	550	2200	5000	9000
15	6754	S2600	6000	10500
16	800	3300	7000	130000

Wind pressure calculated at 32°F using a solid reflector with a f/D ratio of .37. At a temperature of 60°F, loading is 5% less, while at 0°F the loading is 7% more and at −40°F the loading is 17% more.

Noise Figure to Noise Temperature Chart.

Noise Figure in dB	Temperature in Kelvin	Noise Figure in dB	Temperature in Kelvin
2.0	170	.9	67
1.9	159	.8	59
1.8	149	.7	51
1.7	139	.6	43
1.6	129	.5	35
1.5	120	.4	28
1.4	110	.3	21
1.3	101	.2	14
1.2	92	.1	7
1.0	84	0	0
1.0	75		

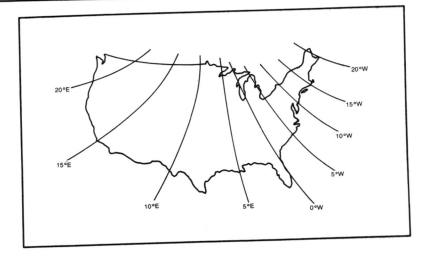

Magnetic deviation map.

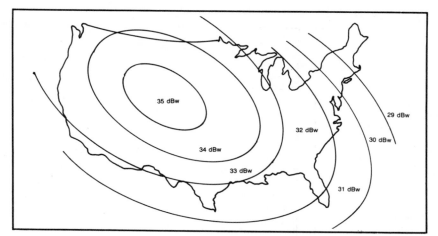

Footprint for F3-R.

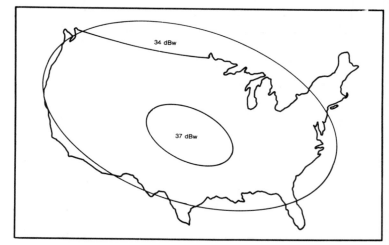

Footprint for D3/D4.

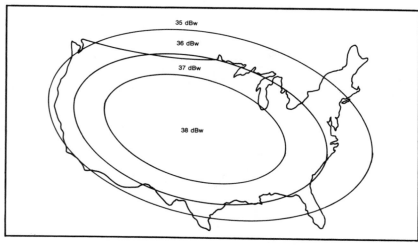

Footprint for G1.

Receiver i-f and Connector Type.

MANUFACTURER	MODEL #	DNCV./RECEIVER JUMPER CONNECT.	FINAL MHz IF FREQ.
Advanced Elect.	Star Tech 1000	F	70
	Star Tech 2000	F	70
Amplica Inc.	All Models	F***	70
Anderson Scientific	ST-1000	F	70
Arunta	DD-3000	F	70
	DD-3300	F	70
	Interceptor-416	F	70
Astron	AR-110	F	70
Automation Techniques	All Models	F	70
Avcom	COM-2	BNC*	70
	COM-2A	F	70
	COM-2B	F	70
	COM-3	BNC*	70
	COM-3R	BNC*	70
	COM-11	BNC*	70
	COM-12	BNC*	70
Basic Systems	#3250	F	70***
Bearcat	CSR-2001	F	70
Beddingfield	Falcon	F	70
Birdview	20/20	BNC*	70
Blonder-Tongue	#6008	F	230**
Bowman Industries	SR-800B	BNC*	70
	SR-2000	F	70
	SR-2500	F	70
Channel Master	All Models	BNC*	70
Collins-Rockwell	All Models	F	70
Comet	HSR-200H	SMA** *	70
Comtech	RCV-550	SMA* ***	70
	RCV-650	SMA* ***	70
Conifer Corp.	RC-2001	F	70
Consumer Sat. Syst.	Ranger-I	F	70
D.B. Electronics	Modusat MK-6	BNC*	70
Dexcell	DXP-1000	Type "M"	70
	DXR-1100	Type "M"	70
	DXR-1200	F/"M"	70
	DCR-4000	SMA*	510**
Drake, R.L. Co.	All Models	F	70
D.X. Engineering	DSA-642	F	130**
	DSA-643	F	130**
	DSA-606	F	70
Earth Terminals	Washburn	BNC*	70
Electron Consulting Assoc.	Video Four	F	70
El-Tech	DSV-804D	F	70
Equinox	#102	F	70
Farral Inst.	SR-200	N***	70
Freedom Systems	Freedom 7	BNC*	70
	Freedom X	F	70
Futurex Syst.	JFX-1000A	F	70
General Instrument (Interstat)	Baby "Q"	F	70
	CRHF I.Q.	F	70
	C4R	F	70
Gensat Comm. Corp.	CSR-1200	F	70
Gillaspie & Assoc.	All Models	F	70
Harris Corp.	6521	SMA*	70
	6522	SMA*	70
	6528	"N"* ***	520**
	6529	"N"* ***	520**
	6550	SMA/N*	70
Heath Corp.	SRA-8100-3	F	230**
Hopson Elect. Labs	RX-718	N***	70
Hughes	SVR-463	F	300**
Hustler	SVS-1000	RG-59*	70
I.C.M.	ICM-4400	F	70
	ICM	BNC*	70
	SR-4600P	F	70
Intersat	SRDC	F	70
	IQ-160	F	70***
	MBS-5R	F	70
Kaul-Tronics	Nova-MSR	F	70
Kem-Tron Industries	Interceptor 2400	F	70
	Interceptor 24	F	70
KLM	X-Pioneer	F	70
	7X	F	70
	SR-3	BNC*	70
	Sky Eye 4	F	70
	KLM-5	F	70
	Olympiad 1	F	70
	Kablevision 4	N*	610**
LOCOM	BR-220	F* * ***	70
Lowrance Elect.	System 7	F	70
Luxor (Skantic)	All Models	F	70
Maspro	SR-1	F	70
Maxum	"007"	F	70
McCullough	All Models	F	70
Microdesign (Interstat)	Communicator	N***	70
Micro-Dynamics	All Models	F	70
Microdyne	1100TVRM	SMA***	70
	1100CSR	SMC*	70
	1100X-24	SMA*	70
	1100BDC	SMA*	70
	1100DCR-12	SMC*	70
	1000TVRM	N* ***	70
M/A Comm. (Microw. Assoc.)	Series 4/MS	N* ***	70
	MA-2GU	N* ***	70
	100-H1	F	403**
	100-T1	F	140**
M.T.I. (Microtenna Assoc.)	MTI-2000	BNC*	70
	Starbaby		
National Microtech (Apollo Series)	Z-1	F	70
	Z-70	F	70
	XA	F	70
NEC Broadcast (Standard Comm.)	NEC-24	BNC*	70
	NEC-204	BNC*	70
Norstal	2040	F	70
Northwest Satlabs	Cosmos II	F	70
Phasecom	#3900	BNC*	300**
Pinzonne Comm.	8200	BNC*	70
	8250	BNC*	70
Prime Source	PS-24	N* ***	70
Regency	SR-3000	F	70
	SR-1000	F	70
Sales Inc.	Astro-20	F	70
Satelinc	SR-3000	F	70
Satellite America	SA-2000	F	70
Satellite Consultants Inc.	BR-200	F	70
Sat. Finder Syst.	Freedom I, II	F	70
Sat. Supplies Inc.	HR-100	BNC*	70
Sat-Tech (Div. of Ramsey Elec.)	The Entertainer R-5000	RCA/RCVR* F/DNC*	70
Scientific Atlanta	6601	BNC*	70
	6602	BNC*	70
	6603	BNC*	70
	6650	F***	230**
	7500	BNC*	70
S.E.D. Syst. Inc.	RX-04-22	F	70
	RX-04-44	F	70
SRI Sat. Receivers Inc.	All Models	F	70
Standard Comm. Corp.	Agile 24 M/S	BNC*	70
	24 Slave	BNC*	70
Telcom Industries	TX-1240	BNC*	70
	TX-2440	BNC*	70
	TX-4000	BNC*	70
T.L. Systems	7678-24	F	70
Toki	TR-110(All)	F	70
Tripple Crown Elect. Inc.	TSR-4000	F	70
Uniden Corp.	UST-1000	F	70
	UST-3000	F	70
Universal Comm.	DL-2000	F	70
Vedacom	SR-1200	F***	70*
Vista Electronics	XR-1000	F	70
	XR-750	F	70
Wilson	YM-400	F	70
	YM-1000	F	70
World-Tech	WSR-8401	F	70
	WSR-8402	F*	70
	WSR-8403	F*	70***

*Adapters required. Connectors available for BNC and SMA receivers.
**Customer needs a custom designed filter for that particular frequency.
***Internal modification necessary for 70 MHz IF loop.

Appendix C

References and Resources

Satellite Aiming Guide
Triple D, Inc., 1983

Satellite Communications, Stan Prentiss.
TAB Books, Inc., 1983.

The World of Satellite Television, Mark Long & Jeffrey
   Keating.
Book Publishing Co., 1983.

The Complete Guide to Satellite TV, Martin Clifford.
TAB Books, Inc., 1984.

SPACE Dealer Certification Course, various.
SPACE, 1984.

MAGAZINES ON SATELLITE TV

Coop's Satellite Digest
P.O. Box 100858
Fort Lauderdale, FL 33310

Home Satellite Marketing
P.O. Box 504
Westport, CT 06881

Radio-Electronics
200 Park Ave. So.
New York, NY 10003
   R-E is not really a TVRO magazine, but it does have
Bob Cooper as the satellite TV editor, and it does have
monthly articles on home TVRO.

Satellite Dealer
P.O. Box 1048
Hailey, ID 83333

Satellite Orbit
P.O. Box 1048
Hailey, ID 83333

Satellite TV Opportunities Magazine
305 W. Jackson Ave.
Oxford, MS 38655

STV Magazine
501 N. Washington St.
Shelby, NC 28150

FREQUENCY COORDINATORS

Compucon, Inc.
P.O. Box 401229
Dallas, TX 75240
214/680-1000

Comsearch
11503 Sunrise Valley Dr.
Reston, VA 22091
703/620-6300

Spectrum Planning
P.O. Box 1360
Richardson, TX 75080
214/699-3536

SEMICONDUCTOR & CROSS-REFERENCE BOOKS

CMOS Cookbook, Don Lancaster
Howard W. Sams, 1977.

ECG Master Replacement Guide (Sylvania).
GE Replacement Semiconductor Guide (General Electric)
New Tone Replacement Master Guide (Technicians Component Group)

AUDIO AND VIDEO TVRO INFORMATION

Microwave Filter Co.
6743 Kinne St.
East Syracuse, NY 13057
315/437-3953
Terrestrial interference seminars, video tapes and ASTI handbook.

Conference Cassettes
4023 Lakeview Dr.
Lake Havasu City, AZ 86403
Audio cassettes of technical presentations given at most major TVRO shows. Includes material presented by Taylor Howard, Bob Luly, et al.

Shelbourne Films
Route 1
Reedsville, OH 45772
Instructional videotapes on satellite TV concepts and installations.

MAIL ORDER PARTS

The best way to get parts is to buy in quantity, but if you don't need a whole rack full of parts, then the next best route is to get them in smaller quantities from surplus dealers. These are people that buy up manufacturer's excess inventory for pennies on the dollar and then resell it to bargain hunters. The quality is usually as good as what the local OEM distributor would sell to you, but at a price that's a whole lot more reasonable. I've dealt with most of the places listed here. I find that no matter where you are, if you've got a phone, a catalog, and at least a weeks time, you can get anything you want delivered to your door.

Active Electronics
P.O. Box 9100
Westborough, MA 01518
800/343-0874
Mainly digital ICs, although they do have some analog stuff.

ACP (Advanced Computer Products)
P.O. Box 17329
Irvine, CA 92713
800/854-8230
Mainly computer parts like digital ICs, but they also carry caps, resistors, regulators, LEDs and other small parts. Very much like Active Electronics catalog.

All Electronics Corp.
P.O. Box 20406
Los Angeles, CA 90006
800/826-5432 or 800/258-6666 in CA
An interesting surplus place. Has all kinds of unique parts as well as the standard transformers, switches, relays, capacitors and connectors. You can find some good bargains but there are also some not so good ones. You have to pick and choose carefully since you're buying sight-unseen. Unless you live in LA and can drop by and see the stuff in the flesh, you can end up with some real 'interesting' paperweights if you're not careful.

Consolidated Electronics, Inc.
705 Watervliet Ave.
Dayton, OH 45420
800/543-3568 or 800/762-3412 in OH
They carry test gear, hand tools, active components, passive components, and video and audio accessories. They're sort of a one stop shopping center for audio and video service technicians.

Digi-Key
701 Brooks Ave So.
Thief River Falls, MN 56701
800/344-4539

They carry lots of ICs, diodes, capacitors, and resistor packs.

ETCO Electronics
North Country Shopping Center
Plattsburgh, NY 12901
518/561-8700

You name it, they probably got it listed somewhere. Their catalog is usually 90-some pages of entertaining reading at the very least. It's chock full of great stuff, pure junk, and fantastic bargains, all interspaced with items that they've obviously been trying to unload for years.

Fordham
260 Motor Parkway
Hauppauge, NY 11788
800/645-9518 or 800/832-1446 in NY state

Fordham is a good place to get test gear. They have good prices and fast service, and they don't do too bad on service and shipping charges either. The maximum charge is $15.00 for any order of $1,001 or more, which can't be beat. They carry many brands of oscilloscopes, DVMS, sweepers, etc.

Heathkit
Benton Harbor, MI 49022

The original DIY test equipment supplier. If you've got the time, but not the money, some of the kits are well worth building. My three year old Heath dual trace scope has more than paid for itself, as has the single trace one I built in 1965, which is now functioning as a dedicated transistor tester. They have everything from DVM kits to TV kits.

Jameco Electronics
1355 Shoreway Dr.
Belmont, CA 94002
415/595-5936

A good source for digital ICs, resistors, capacitors, switches, etc. They also make the Jim-Pak parts kits that are starting to show up at parts houses across the country.

JDR Microdevices
1224 Bascom Ave.
San Jose, CA 95128
800/538-5000 or 800/662-6279 in CA

One of my favorite parts houses, but that's probably just because they're local. They carry lots of ICs, both digital and linear, as well as capacitors, transistors and other discrete parts.

Mouser Electronics
11433 Woodside Ave.
Santee, CA 92071
619/449-2222

A general electronics parts house. They carry ICs, transistors, LEDs, coils, lamps, pots, switches, transformers, plugs, jacks and many other types of parts, over 15,000 items according to their catalog. Delivery time is usually pretty good.

New-Tone Electronics
44 Farrand St.
Bloomfield, NJ 07003

One of the more successful independent suppliers of general cross transistors, diodes and ICs. They have pretty good prices and are on par with ECG and the other general replacement parts houses.

Newark Electronics

Over 160 outlets in the U.S. and Canada, consult the telephone book for your nearest branch. Newark carries all types of new electronic parts including many TVRO type connectors, power supplies, benches, test equipment and tools, in addition to the basic transistors, resistors, capacitors and ICs. They also carry list prices, so I try to find it elsewhere first.

Radio Shack

Since Radio Shack is getting into TVRO, maybe they'll finally start carrying N connectors! On some things, the price isn't too bad, on others it is better to shop elsewhere. The best thing you can say about Radio Shack stores are they're coast to coast and have pretty much the same stock so that they can be used as a reference point. Of course, the free batteries aren't bad either. I don't think I've ever actually had to buy a flashlight battery since they started their battery card deal.

Ramsey Electronics
2575 Baird Rd.
Penfield, NY 14626
716/586-3950

And you thought John Ramsey only sold Sat-Tec receivers! He actually started out selling test gear and parts, and now has a full line of Ramsey oscilloscopes, DVMs, frequency counters and other test gear, available in kit and pre-assembled form. He used to carry surplus parts, and have a periodic flyer on what was available, but I don't know if that's still happening. I hope it is, since it was filled with lots of useful parts for TVRO equipment.

Star-Tronics
P.O. Box 683
McMinnville, OR 97128

Drop them a post card and ask to get on their mailing list. It's guaranteed to be entertaining, as they come up with some very esoteric surplus equipment. One of the surplus world's best kept secrets. If you don't keep buying stuff, they'll drop your name off their mailing list!

W.W. Grainger, Inc.
General Offices
5959 W. Howard St.
Chicago, Il 60648
312/647-8900

They're another general mail order catalog that offers over 10,300 items at 170 locations across the country. It is mainly industrial equipment like motors and air compressors, but they also carry benches, storage racks and other useful 'industrial' and installation type equipment.

The back pages of Radio-Electronics always have lots of listings for surplus places, many of which are the here today and never heard of tomorrow type, so buyer beware. It's sort of the Alice's Restaurant of electronics. You can find everything you want, from rf modulators to chassis screws, if you look hard enough.

Appendix D

Common TVRO Abbreviations

There are many abbreviations used throughout the TVRO industry. Most of which have been used somewhere in the book. The following is an alphabetical listing of the pertinent abbreviations.

ac—alternating current.
afc—automatic frequency control.
AM—amplitude modulation.
APS—antenna positioning system.
AWG—american wire gauge.
BDC—block down conversion.
ccw—counter clockwise.
CMOS—complementary metal oxide semiconductor.
C/N—carrier to noise ratio.
cw—clockwise.
dB—decibel.
DBS—direct-broadcast system.
DC—down converter.
dc—direct current.
ECL—emitter coupled logic.
EIRP—effective isotropic radiated power.
f/D—focal distance-to-diameter ratio.
FM—frequency modulation.
GHz—gigahertz.
HPF—high pass filter.
IC—integrated circuit.
i-f—intermediate frequency.
kHz—kilohertz.
LED—light emitting diode.

LHC—left hand circular.
LNA—low-noise amplifier.
LNB—low-noise blockconverter.
LNC—low-noise converter.
LNF—low-noise feed.
LO—local oscillator.
LPF—low pass filter.
LSD—least significant decimal.
MHz—megahertz.
MSD—most significant decimal.
NTSC—National Television Standards Committee.
PD—polarization device.
PLL—phase-lock loop.
rf—radio frequency.
RHC—right hand circular.
RTV—room temperature vulcanizing.
SCPC—single carrier per channel.
S/N—signal-to-noise ratio.
TI—terrestrial interference.
TTL—transistor-transistor-logic.
TVRO—television receive only.
UHF—ultra high frequency.
UUT—unit under test.
VCO—voltage-controlled oscillator.
VCR—video cassette recorder.
Vac—voltage, alternating current.
Vdc—voltage, direct current.
VHF—very high frequency.
VTO—voltage-tuned oscillator.

Index

A

ABC, 1
actuators, 180
actuators, linear, 180
actuators, other types of, 180
ADM, 6
Alcoa, 56
AM, 152
Amplica, 100, 118, 205, 237
amplifiers, 70 MHz, 97
antenna positioning system, 179
APS feedback circuits, 182
APS trouble areas, 186
APS, 179
APS, microprocessor-controlled, 186
AT&T, 1, 29, 128
audio circuits, 137
audio companding, 146
audio processing, 136
Augustin, Gene, 58
AVCOM, 7, 82, 253
AWG, 37

B

bandpass filter, 6
BDC, 9
Behar, Bob, 210
Belden, 10, 34, 35
Bendix connector, 43, 215

Bendix Corp., 43
Birdview, 27, 54, 82, 100, 103, 120, 180, 181
block conversion systems, 26
block down conversion, 9
BNC connector, 39, 45
Boman, 61, 215
BPF, 6, 137

C

C-band, 2, 28
C/N, 48
cable, checking, 38
cable, coax, 34, 219
cable, handling coax, 36
cable, splicing, 39
cable, types of coax, 35
cable/connector sealing, 38
cabling, custom, 37
Cannon connector, 43, 215
CATA, 1
CATJ, 1, 58
CATV, 38
Channel Master, 57, 114, 256, 272
Chaparral Communications, 57, 58, 61, 62, 65, 82
checking components in the field, 212
Clarke Belt, 47, 179, 211, 217, 218
CMOS, 230

Conifer, 96, 100, 114, 120, 127, 156, 256, 272
connectors, 39, 219
Cooper, Bob, 1, 58
CTI, 233

D

DBM, 6
DBS compatibility, 28
DBS, 10, 28
DC gain, 49
DC, 9
demodulation circuits, audio, 136
demodulator circuits, 115
demodulator, IC balanced, 119
demodulator, PLL, 115
detector, quadrature, 120
detector, ratio, 120
Dexcel, 18, 22, 32, 59, 78, 82, 87, 100, 103, 118, 120, 128, 141, 145, 169, 215, 216, 233, 278, 302
diodes, 223
discriminator, delay line, 120
dish gain, 48
dish materials, 51
dish mispointing, 48
dish size, 1
dish, 46

dish, matching feed to dish, 63
dishes, deep, 63
dishes, shallow, 63
dithering, 4
Dolby noise reduction, 147
double-balanced mixer, 6
down converter, 9, 77
down converter, separate, 77
downlink, 2
Draco, 194
Drake, 33, 100, 114, 120, 156, 168, 174, 256, 270, 272, 276, 342
dual conversion, 5
dual-conversion systems, 12
Duff-Norton, 180

E

EIRP, 48
emitter coupled logic, 231

F

F connector, 39
f/D ratio, 63
f/D ratio, measuring the, 65
face margin, 28
FCC, 1
FCC, 136
feedhorn, centering, 67
FETs, 226
filter, notch, 29
filters, 70 MHz bandpass, 98
FM, 3
free space loss, 46

G

G/T, 47, 63
gain, 47
gain, dish, 48
gain, LNA and DC, 49
gain, system, 46
Gamma-Feed, 82
General Transmission, 180
Gillespie, 100
ghosts, 30, 32
gold ring, 65
Gould/Dexcel, 18, 22, 32, 103, 141, 145, 169, 233, 278, 302

H

hard-line, 7
Hatfield, Andy, 253
Hero Communications, 210
high-pass filter, 137
hot spot, 63
Houston Tracker, 198
HPF, 137
hybrid components, 233
Hytek, 62

I

i-f, 6
IC pinouts, CMOS, 361-371
IC pinouts, linear, 361-371
IC regulators, 87

ICM, 6, 12, 14, 78, 100, 120
image reject mixer, 78
indicator circuits, 165
integrated circuits, 228
Intelsat, 12
interference, adjacent channel, 158
interference, atmospheric, 28
interference, combating terrestrial, 32
interference, other inband terrestrial, 32
interference, out-of-band terrestrial, 30
interference, rf, 31
interference, terrestrial out-of-band, 78
interference, terrestrial, 1, 28, 29
intermediate frequency circuits, 97
intermediate frequency, 6

J

Janeil, 56, 82
JRC, 181

K

Kaultraonics, 54
Kelvin scale, 47
KLM, 84, 87, 103, 119, 166, 180, 181, 208
Ku-band, 28

L

lamps, incandescent, 171
LED circuits, 166
LED readouts, 167
limiter circuits, 99
linear chips, 233
LNA gain, 49
LNA, 5, 71
LNA, handling and installing an, 74
LNA, troubleshooting, 76
LNB, 12, 71, 82
LNC, 12, 71, 82
LNF, 71, 82
LO leakage, 80
LO, 6
local oscillator, 6
low noise amplifier, 5, 71
low noise blockconverter, 12, 71, 82
low noise converters, 12, 71, 82
low noise feed, 71, 82
Lowrance, 28, 82
Luly, Bob, 58, 62
Luxor, 82, 100, 103, 137, 147, 159, 161, 174, 186, 188, 189, 190, 314, 334, 336, 342

M

M/A Com, 1, 61, 62, 82
Maspro, 95, 159
MCI, 29
MIC, 77

Microdyne/AFC, 1, 82
microwave communication system 29
Microwave Filter Company, 32
microwave integrated circuitry, 7
microwave, 1
Milky Way, 47
modulator, crystal controlled, 156
modulator, LC, 155
modulator, rf, 152
modulator, types of, 155
motor controllers, 181
MTI, 191, 193, 215
multiple receivers, 24
Mutual Radio, 1

N

N connector, 39
narrowband, 136
noise reduction, Dolby, 147
noise, sky, 47
NTSC, 128, 152, 154

P

PAL, 130, 154
parabolic reflectors, 46
Paraclipse, 180
PD, 9, 57
Pentec/MTI, 191, 193, 215
polarization device, 9
polarization device, attaching, 67
polarization devices, other, 61
polarization, 57
Polarotor I, 48, 57, 58, 205, 336
power supplies, 83
power supplies, antenna actuators 84
power supplies, receiver, 83
power supply failures, preventing 91
power supply filtering, 93
Prodelin, 1

R

RCA connector, 44
RCA jack, 44, 164
regulators, IC, 87
regulator protection, 91
remote control, 16, 159, 178

S

safety, 87
Saginaw Steering Gear, 180, 191
Sat-Tec, 80, 155, 339
Satcom 1, 57
satellite antenna, 46
satellite dish, 46
satellite spacing, 50
satellites, telecommunication, 1
SAW filter, 233
schematics, how to read large, 279
schematics, how to read, 279

scrambler circuits, 128
Seavy, 61, 62
SECAM, 130, 154
signal gain, 18, 46
signal loss, 18
signal-strength meter, 165
single-conversion down converters, 78
single conversion, 6
single-conversion systems, 18
site surveys, 211
skew, 57, 59
skin effect, 34
sky noise, 47
Spacenet 1, 12
sparklies, 29, 48, 217
spike supression, 95
Sprint, 29
static protection, 233
stereo broadcasting methods, 143
stereo system, Leaming, 150
stereo system, Warner-Amex, 150
stereo, discrete, 143
stereo, matrix, 145
stereo, multiplex, 146
STS, 80, 100, 188, 205, 278, 342
subcarrier specifications, audio, 136
substitution, 214
sun spots, 31
Superwinch, 180
Syntronics, 82

T

Telco, 5
terrestrial interference, 211
test bench, setting up a, 200

test gear, 211
test generator, making a dc, 205
test procedure for Dexcel LNCs, 215
test signal source, 202
transistor cross-reference books, 226
transistors, 224
transponder, 3, 30
troubleshooting a system, 210
troubleshooting Channel Master Receivers, 256
troubleshooting Conifer receivers, 256
troubleshooting demodulator circuits, 122
troubleshooting Drake Receivers, 256, 272
troubleshooting guide for antenna positioning systems, 186
troubleshooting guide for audio circuits, 150
troubleshooting guide for TVRO system, 216
troubleshooting guide, polarization device, 68
troubleshooting LNAs, 76
troubleshooting power supplies, 96
troubleshooting Sat-Tec receivers, 341
troubleshooting the i-f circuits, 102
troubleshooting the rf modulator, 161
troubleshooting tips for specific antenna positioning systems, 187
troubleshooting Winegard receivers, 256

TTL, 230
tuning circuits, 174
TVRO, history of, 1-9

U

Uniden, 61, 82
uplink, 2
USS/Maspro, 12, 15

V

varistors, 93
VCO, 12
Vertical Interval Test Signal, 131
VHF, 2
video circuits, 123
video frequency response, 130
video processing, 115
video signal level, 131
video signal synchronization, 130
video signal, 128
video signal, measuring the, 130
VITS, 131
voltage fluctuations, 93
voltage tuned oscillator, 20
VTO, 20

W

Warner Electric, 180, 191
Washburn, Clyde, 6
waveguides, 72
Wegener, 147
Wesbar, 180
wideband, 136
Winegard, 51, 82, 87, 100, 114, 156, 168, 256, 272, 342, 344, 355
wire, insulated, 37

Other Bestsellers From TAB

☐ **GETTING THE MOST OUT OF YOUR VIDEO GEAR—Quinn**

Document the celebrations of family and friends—to treasure and enjoy again and again . . . make learning fun for your children by using video . . . receive your initial investment back many times over by using your video to earn extra income. You can do all these, and countless other imaginative projects using this up-to-the-minute guide—the ULTIMATE sourcebook for every video owner who wants to take full advantage of his equipment. With it, you can be sure you're getting every dollar's worth out of your VCR, TV, video camera, and home computer! 256 pp., 88 illus. 7″ × 10″.

Paper $12.95   **Hard $19.95**
Book No. 2641

☐ **DESIGNING, BUILDING AND TESTING YOUR OWN SPEAKER SYSTEM . . . WITH PROJECTS—2nd Edition—Weems**

You can have a stereo or hi-fi speaker system that rivals the most expensive units on today's market . . . *at a fraction of the ready-made cost! Everything* you need to get started is right here in this completely revised sourcebook that includes everything you need to know about designing, building, *and* testing every aspect of a first-class speaker system. 192 pp., 152 illus.

Paper $10.95   **Book No. 1964**

☐ **BUILD A PERSONAL EARTH STATION FOR WORLDWIDE SATELLITE TV RECEPTION—2nd Edition—Traister**

You'll find a thorough explanation of how satellite TV operates . . . take a look at the latest innovations in equipment for individual home reception (including manufacturer and supplier information) . . . get detailed step-by-step instructions on selecting, assembling, and installing your earth station . . . even find a current listing of operating TV satellites (including their orbital positions and programming options). You'll also discover that the costs of satellite receiver components or kits are far less than you've imagined! Even shows how to save by using surplus components. 384 pp., 250 illus.

Paper $14.95   **Hard $21.95**
Book No. 1909

☐ **TELEVISION—FROM ANALOG TO DIGITAL —Prentiss**

Discover the leading edge of television technology in this unique sourcebook that brings together, for the first time, the information serious experimenters and professional TV engineers and technicians have been working for. Written by Stan Prentiss—one of the leading electronics writers in the U.S. and author of more than 40 highly regarded books for hobbyists and professionals—this is a thorough and comprehensive look at the most recent developments affecting TV broadcasting and receiving. 352 pp., 258 illus. 7″ × 10″.

Hard $25.00   **Book No. 1972**

☐ **AM STEREO AND TV STEREO—NEW SOUND DIMENSIONS—Prentiss**

Includes the most up-to-date AM & FM stereo information available. An in depth look at the new sound in AM radio and TV broadcasting that gives much needed advice on equipment availability and operation and provides insight into FCC regulations and equipment specifications. Treating AM stereo radio and TV multi-channel sound separately, this guide details the various systems and equipment developments with plain language descriptions of the final hardware available in both receivers and transmitters. This is truly a double-barreled bargain for all stereo enthusiasts. 192 pp., 127 illus. 7″ × 10″.

Paper $12.95   **Hard $17.95**
Book No. 1932

☐ **BUYER'S GUIDE TO COMPONENT TV—Giles**

Discover a whole new world of information and entertainment . . . new home media equipment that can literally change your lifestyle! Now, for the first time, this state-of-the-art guide provides you with the information you need to sort through all the new component TV products flooding the market . . . so that you can choose the units that suit your needs, your lifestyle, *and* your pocketbook! The authors give you complete information on brand name products—TV monitors and projection TV monitors, tuners, stereo speakers, videocassette recorders, videotapes, video cameras, satellite receivers, component furniture, and more! 224 pp., 177 illus. 7″ × 10″.

Paper $12.95   **Hard $19.95**
Book No. 1881

*Prices subject to change without notice.

Look for these and other TAB books at your local bookstore.

TAB BOOKS Inc.
P.O. Box 40
Blue Ridge Summit, PA 17214

Send for FREE TAB catalog describing over 1200 current titles in print.